Sitting posture . sitzhaltung . posture assise

Proceedings of the symposium on

sitting posture
sitzhaltung
posture assise

Edited by

E. Grandjean

Director
Department of Industrial Hygiene
and Work Physiology
Swiss Federal Institute of
Technology
Zurich, Switzerland

Published by
Taylor & Francis Ltd
10-14 Macklin Street
London WC2B 5NF

1976

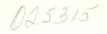

First published 1969 by Taylor & Francis Ltd, 10-14 Macklin Street, London WC2B 5NF.

Reprinted 1976.

ISBN 0 85066 029 7.

Printed and bound in Great Britain by Taylor & Francis (Printers) Ltd, Rankine Road, Basingstoke, Hampshire RG24 0PR.

Distributed in the United States of America and its territories by Halsted Press (a division of John Wiley & Sons Inc.) 605 Third Avenue. New York, N.Y. 10016.

List of contributors

Åkerblom B.
59400 *Gamleby*
Sweden

Burandt U.
Designer
Schwedlerstrasse 43
Erlangen-Bruck
Germany

Boni A.
Department of Industrial
Hygiene and Work Physiology
Swiss Federal Institute of
Technology
Zurich
Switzerland

Branton P.
Industrial Psychology
Research Unit
National Institute of
Industrial Psychology
14 *Welbeck Street*
London W.1, England

Le Carpentier E. F.
Furniture Industry Research
Association
Maxwell Road
Stevenage
Herts England

Chidsey, K. D.
Furniture Industry
Research Association
Stevenage Herts England

Dupuis H.
Max Planck-Institut für
Landarbeit und Landtechnik
Bad Kreuznach
Germany

Floyd W. F.
Department of Ergonomics and
Cybernetics
University of Technology
Loughborough
England

Grandjean E.
Director Department of
Industrial Hygiene and Work
Physiology
Swiss Federal Institute of
Technology
Zurich
Switzerland

Jones J. C.
Institute of Science and Technology
University of Manchester
Sackville Street
Manchester 1 England

Jurgens H. W.
Neue Universität
Olshausenstrasse 40-60
Kiel
Germany

De Jong J. R.
De Genestetlaan 25
Bilthoven
Holland

Kretzschmar H.
Designer
Brunnenstrasse 24
Möglingen
Germany

Leonard T.
Department of Industrial
Hygiene and Work Physiology
Swiss Federal Institute of Technology
Zurich Switzerland

Oxford H. W.
Seating Consultant
43 Gnarbo Avenue
Carss Park
New South Wales
Australia

Peters T.
Institut für Arbeitsmedizin
der Universität
Dusseldorf
Germany

Preuschen G.
Direktor
Max Planck-Institut für
Landarbeit und Landtechnik
Bad Kreuznach
Germany

Rotzler W.
Splugenstrasse 8
Zurich
Switzerland

Rieck A.
Max Planck-Institut für
Arbeitsphysiologie
Dortmund
Germany

Rizzi M.
Spezialarzt für Rheumatologie FMH
Am Schanzengraben 23
Zurich
Switzerland

Rebiffé R.
Laboratoire de Physiologie et de
Biomecanique de la Regie Nationale
des Usines Renault
Rueil-Malmaison
France

Roggeveen C.
Raadgevend Bureau
Ir. B.W. Berenschot
Amsterdam Holland

Schoberth H.
Oberarzt der Orthopädischen
Universitätsklinik Friedrichsheim
Frankfurt-am-Main-Niederrad
Germany

Serati A.
Chefarzt der SBB und der
allgemeinen Bundesverwaltung
Bern
Switzerland

Shackel B.
E.M.I. Electronics Ltd
Victoria Road
Feltham
Middlesex
England

Shipley P.
Department of Occupational
Psychology
Birkbeck College
University of London
England

Ward Joan S.
Department of Ergonomics and
Cybernetics
University of Technology
Loughborough
England

Wotzka G.
Department of Industrial Hygiene and
Work Physiology
Swiss Federal Institute of Technology
Zurich
Switzerland

Contents

Preface

An International Symposium on Sitting Posture was held from 25th to 27th September, 1968, at the Swiss Federal Institute of Technology, Zurich, under the auspices of the International Ergonomics Association. The Department of Industrial Hygiene and Work Physiology of the above Institute assumed the responsibility for organization.

On behalf of the delegates and the International Ergonomics Association we wish to express thanks to the following supporting organizations:

Giroflex Entwicklungs-A.G., Coblenz, Switzerland
Albert Stoll Giroflex A.G., Coblenz, Switzerland
Martin Stoll Giroflex, Tiengen, Germany
Mercator S.A., Brussels, Belgium
Giroflex S.A., Sao Paolo, Brazil

The purpose of the Symposium was to bring together scientists dealing with the problems of sitting posture and its anthropometric, physiological, psychological and orthopaedic aspects. The papers presented—and therefore also this publication—should give a survey of the present knowledge and convey to designers and industry biological principles in the construction of seats.

E. Grandjean
Zurich, October 1968

Vorwort

Vom 25. bis zum 27. September 1968 fand an der Eidgenössischen Technischen Hochschule in Zürich (Schweiz) ein Symposium über die Probleme der Sitzhaltung statt. Die Internationale Vereinigung für Ergonomie (International Ergonomics Association) hatte das Patronat übernommen, während das Institut für Hygiene und Arbeitsphysiologie der E.T.H. für die Organisation verantwortlich war.

Der Unterzeichnete möchte an dieser Stelle im Namen aller Teilnehmer und im Namen der International Ergonomics Association folgenden Donatoren seinen verbindlichen Dank aussprechen.

Giroflex Entwicklungs-A.G., Koblenz, Schweiz
Albert Stoll Giroflex A.G., Koblenz, Schweiz
Martin Stoll Giroflex, Tiengen, Deutschland
Mercator S.A., Brüssel, Belgien
Giroflex S.A., Sao Paolo, Brasilien

Das Ziel der Veranstaltung war die Vereinigung der Wissenschaftler, die sich mit der Anthropometrie, der Physiologie, der Psychologie und der Orthopädie der Sitzhaltung befassen. Die Vorträge—und somit auch die vorliegende Monographie—sollten einen Ueberlick über den Stand der heutigen Kenntnisse geben und der Industrie sowie Designern die biologischen Grundlagen zur Gestaltung von Sitzen vermitteln.

Prof. Dr. med. E. Grandjean
Zürich, im Oktober 1968

Zur Kulturgeschichte des Sitzes

Von W. Rotzler

Redaktor an der kulturellen Monatsschrift "du", Zurich, Schweiz

Es war der Wunsch der Veranstalter, dass diesem Gespräch unter Fachleuten eine Art kulturgeschichtlicher Prolog vorangestellt werden solle. Ein Vorspiel, das den Blick nicht nach vorwärts, sondern nach rückwärts lenkt. Solche Rückschau kann der Anthropometriker anstellen. Seine Frage an die Vergangenheit würde dann etwa lauten: Wie verhalten sich Sitzmöbel anderer Epochen zu den von uns heute erarbeiteten Daten und Forderungen ? Wie viel Wert eine derartige Rückschau für die heutige Ergonomie hätte, vermag ich nicht zu beurteilen.—Rückschau kann aber auch der Stuhlentwerfer und Stuhlhersteller halten. Die Antworten, die er aus der Historie gewinnen kann, liegen auf dem Feld der Konstruktionsprinzipien und auf dem Feld des Formalen, des Aesthetischen. Wer beispielsweise das Bedürfnis hat, einen dreibeinigen Stuhl zu entwickeln, tut gut daran, sich zu orientieren, wie andere Zeitalter dieses Problem gelöst haben.

Den Blick zurück kann aber auch der Kunsthistoriker tun, denn Sitzmöbel, wie sie sich aus verschiedenen Zeitaltern in grosser Zahl erhalten haben, sind Zeugnisse der angewandten Kunst. Ueberdies kann der Kunsthistoriker die Kunstwerke aller Zeitalter auf Stühle und auf sitzende Figuren hin untersuchen. Dabei zeigt sich, dass Darstellungen von Stühlen und von Sitzfiguren in der Kunst die Einsichten ergänzen, die an historischen Stühlen selbst gewonnen werden können. Wo sich der Kunsthistoriker nicht nur mit den Sitzmöbeln als toten Objekten, als Gebrauchsgegenständen beschäftigt, sondern auch mit den sitzenden Menschen, wird er zum Kulturhistoriker. Denn das Sitzen ist ja immer ein menschliches, ein gesellschaftliches Verhalten. Und da könnte nicht nur gefragt werden, was Sitzdarstellungen über die sitzenden Menschen verraten, sondern auch, in welche kulturellen, sozialen Muster sie sich einfügen. Ob für den Anthropometriker die Hinweise auf die Sitze und das Sitzen vergangener Zeitalter von Interesse oder gar von Nutzen ist, das zu entscheiden überlasse ich Ihnen. Was ich Ihnen hier in einem kurzen Ueberblick bieten kann, ist bloss eine summarische Faustskizze oder—bedenkt man die Fülle des Stoffes zum Thema " Geschichte des Sitzens "—eine fast frivole Schnellmalerei. Aber vielleicht versetzt Sie eine solche Ouvertüre in die rechte Stimmung für Ihre sehr viel ernstere, sorgfältigere und präzisere Arbeit.

Wir gehen in unserem zweckgerichteten Denken ganz selbstverständlich davon aus, dass der Stuhl—oder das Sitzmöbel überhaupt—ein Gebrauchsgegenstand ist, erfunden und entwickelt zum Zwecke des Sitzens. Ist das richtig? Wir besitzen aus der alt- und jungsteinzeitlichen Prähistorie unzählige Zeugnisse über eine Vielfalt von Geräten und Verrichtungen, keine aber über den Sitz und das Sitzen. Der Einwand, dass es sich da um zufällige Lücken der Ueberlieferung handle, ist kaum stichhaltig. Denn viele der noch heute lebenden Naturvölker, selbst hochstehende, kennen das Sitzen auf einer Sitzgelegenheit nicht. Sie kauern, knien oder hocken. Und einzelne Völker, die man zu den kulturell und zivilisatorisch höchst entwickelten zählt—z.B. die

Japaner—kennen den Stuhl nur als westliches Importprodukt. Noch heute kommt ein dem traditionellen Lebensstil verpflichtetes japanisches Haus ohne Sitzgelegenheiten in unserem Sinne aus.

Es besteht also Anlass zum Zweifel an dem Gedanken, der Stuhl stelle das zweckmässige Gerät zur Befriedigung eines elementaren Sitzbedürfnisses des Menschen dar.

Wie und warum aber kam es zum Stuhl, zum Sitz überhaupt? Wir haben keine Beweisstücke für seine Ursprünge. Wir können uns diese nur denken. Schon in primitiven Gemeinschaften muss es vorgekommen sein, dass einer sich aus der Gruppe löste und auf einem Natursitz, einem Stein, einem Baumstrunk Platz nahm, vielleicht gar an erhöhter Stelle. Das war nicht irgendeiner, sondern der Stammesälteste, der Häuptling. In den aus einem Baumstrunk geschnitzten "Häuptlingssitzen" afrikanischer Stämme—hockerartigen Sitzmöbeln, die mit Ahnenfiguren verziert sind—ist uns dieser ursprünglichste Stuhl überliefert. Der Stuhl ist also zunächst nicht Gebrauchsmöbel, sondern Standesabzeichen—wir würden heute sagen: Status-Symbol. Der Stuhl gehörte zu den Insignien des Mächtigen, des Herrschers; ja, er symbolisierte diesen gar selbst. Aus dem Häuptlingssitz wurde später der Herrscherthron, der Sitz des weltlichen oder kultischen Macht-und Würdenträgers.

Vor allem aus dem altnordischen Kulturbereich sind mächtige Steine als Natursitze erhalten, auf denen Stammesfürsten oder Könige Platz nahmen und, alles beherrschend, unter freiem Himmel autoritär Gericht hielten. Auch zum Symbol der richterlichen Gewalt ist der Stuhl also geworden. Im Begriff des "Vor-Sitzenden", dessen Platz an bevorzugter Stelle angeordnet ist, lebt diese uralte Bedeutung des Sitzes als Machtsymbol bis heute weiter. Und gehört nicht gerade zu heutigen Möbel-Programmen noch immer der "Direktorensessel"—also eine Repräsentationsfunktion und Machtdokumentation des Stuhls; denn anatomisch ist der Direktor ungefähr gleich gebaut wie sein Mitarbeiter oder Besucher.

Nicht nur Herrscher bestimmen die Geschicke des Menschen, sondern letztlich die Gottheiten. Da man sie, seitdem es in den frühgeschichtlichen Kulturen personifizierte Gottheiten gibt, als allmächtige Herrscher über die Menschen betrachtete, deren irdische Stellvertreter die weltlichen Herrscher waren, so wurden Gottheiten schon früh auf besonders markante, beeindruckende Sitze oder Throne gesetzt. Dieser göttliche Thron (der auch leer sein konnte und doch die Gottheit bedeutete) hat sich auch ins Christentum herüber gerettet: In der mittelalterlichen Kunst thront Gottvater im Himmel auf einem Wolkensitz oder als Dreifaltigkeit im "Gnadenstuhl"; Christus thront als Weltenrichter auf der Bank des Regenbogens oder einem fürstlichen Thronsitz; und die thronende Madonna zeigt als Himmelskönigin in einem prunkvollen Gestühl, assistiert von stehenden Heiligen, den andächtig zu ihr Emporblickenden den Christusknaben vor.

Nun ist der Stuhl selbstverständlich nicht Symbol der Gottheit, des Herrschers und des Richters geblieben, sondern früh schon, möglicherweise in Etappen, Gesellschaftsschicht um Gesellschaftsschicht von oben nach unten, zum Sitzmöbel mit wachsenden Gebrauchsfunktionen geworden. In alten Redewendungen wird davon vieles sichtbar:

Man sitzt zu Gericht, man sitzt im Rat, man nimmt Einsitz in die Zunft (und heute noch in eine Firma). Sitzen kann auch zum Merkmal des Weisen, des Gelehrten werden. Schon auf frühchristlichen Mosaiken sitzen die schreibenden Evangelisten. Und seit dem Mittelalter gibt es das Katheder des Gelehrten. Im Universitätsbetrieb sprechen wir noch heute vom " Lehrstuhl ", dem Stuhl, von dem aus gelehrt wird.—Sitzen ist vielerorts auch das Privileg des Alters. In Dörfern des Mittelmeergebietes mag man noch heute beobachten, dass lediglich die Alten auf einem Stuhl oder einer Bank sitzen, während die Jüngeren, Männer oder Frauen, sie umstehen. Selbst in unseren heutigen Höflichkeitsformen ist etwas vom Respekt rund um den Stuhl erhalten geblieben: Wenn der Höhergestellte oder eine Dame eintritt, erheben wir uns vom Sitz; erst wenn die Dame des Hauses sich gesetzt hat, dürfen auch die Gäste Platz nehmen. Und der Höfliche bietet einer Respektsperson noch immer den Stuhl an.

Sieht man von dieser Symbolfunktion des Stuhles einmal ab, so kann man das historische Dokumentenmaterial—Darstellungen sitzender Figuren in der Kunst seit den Aegyptern und erhaltene Sitzmöbel aus den verschiedenen Epochen bis heute—nach mancherlei Gesichtspunkten ordnen. Jedes Ordnungsprinzip gibt anregende Einblicke in die Geschichte des Sitzmöbels und die Geschichte des Sitzens, der Sitzgewohnheiten.

Man kann etwa vom Material ausgehen: Stein, Holz, Geflecht, Metall (Bronze schon bei den Römern, Eisen seit dem 19. Jahrhundert, Stahlrohr, Stahlstäbe und Aluminium erst in unserer Zeit) und schliesslich Kunststoffe als Material der Gegenwart.

Material heisst sofort auch Konstruktionsprinzip. Nirgends eindeutiger als beim Holz—während Jahrtausenden das Stuhl-Material par excellence— lassen sich die Konstruktionsprinzipien studieren, die für den Stuhlbau massgebend waren, es zum Teil geblieben sind. Um ein paar zu nennen: Brettkonstruktionen bei Kistenhockern und Kastensitzen (beispielsweise Thronen), Brettstuhl (Stabelle), Sprossenstuhl (mit Rahmenkonstruktion aus Rundstäben), Zargenstuhl (aus Brettstuhl und Kistensitz entstanden und aus vierkantigen Latten, den Zargen, aufgebaut). Um das Skelett, das gerade der Zargenstuhl bietet, ist in der abendländischen Entwicklung dann das immer wieder andersartige Gewand der stilbedingten Ausgestaltungen gelegt worden. Falt-, Klapp-und Scherenstuhl sind weitere alte Stuhltypen, die im Gartenliegestuhl noch fortleben. Mit dem Beginn der fabrikmässigen Massenproduktion gewinnt im 19. Jahrhundert der Bugholzstuhl an Bedeutung; seine Teile werden im Wasserdampf über Stahlformen verformt, eine Erfindung Michael Thonets, die später beim Schichtholzstuhl weiter entwickelt wurde. Eisenstühle begannen ihre Karriere als wetterfeste Gartenstühle; und Stahlrohrstühle, seit 1890 in Spitälern verwendet, erscheinen erst seit 1920 als Wohnmöbel.

Entscheidende Veränderungen im Stuhlbau brachte die industrielle Herstellung. Spezifisch schreinerisch-handwerkliche Konstruktionen mussten ersetzt werden durch Konstruktionen, die der Maschine gemäss waren. Thonets Erfindung ist da bloss ein erster Schritt. Erst im Zeitalter des Funktionalismus, in den Zwanziger Jahren, kam man dazu, konstruktiv eindeutig das Gestell vom Sitz zu trennen und für die verschiedenen Elemente auch verschiedene Materialien zu wählen. Eine Art Revolution bedeutete nach 1945 die Entwicklung der Sitzschale, die—eine organische Verbindung von Sitzfläche, Rücklehne und manchmal Armstütze—zunächst aus einem Metallstück in der Stanzpresse, später aus armiertem Kunststoff in einem Arbeitsgang gefertigt wurde. Und neuestens erscheinen auf dem Markt aus einem Stück gegossene Kunststoff-Stühle. Ihnen machen aber bereits die aufblasbaren, transparenten, aus farblosem Plastic bestehenden "Luftkissenstühle" Konkurrenz.

Man kann nun aber auch, statt vom Material und von Konstruktions-prinzipien auszugehen, unser ganzes Stuhlerbe nach Verwendungszwecken, nach Gebrauchsfunktionen ordnen. Man kommt dabei, vor allem wenn man die verschiedenen Konstruktionsformen mit einbezieht, zu einer Art Typologie des Stuhles. Ein solches Vorgehen wird aber erst anregend und aussagekräftig, wenn man Hunderte von Stühlen verschiedener Epochen zu Reihen zusammen-stellt, sei es in einer Ausstellung, sei es in einem Bilderbuch. Im Wort lässt sich eine solche Typologie nicht darstellen. Immerhin wird klar, was gemeint ist, wenn man einige durch ihre spezielle Funktion geprägte Stuhltypen nennt: Kinderstuhl, Esstischstuhl, Lehnstuhl, Gartenstuhl, Liegestuhl, Schaukelstuhl, Schreibtischstuhl, Bürostuhl, Arbeitsstuhl für Sitzberufe usw. —Aber auch etwa Klappstuhl und Stapelstuhl gehören hierher. Und nicht zuletzt der Eisenbahn-, der Auto- und der Flugzeugsitz.

Aus einer solchen Stuhl-Typologie liessen sich viele Schlüsse auf die Sitz-gewohnheiten von ganzen Zeitaltern wie von Gesellschaftsschichten ziehen. Unterstützt von dem Anschauungsmaterial, das die Kunst von sitzenden Menschen bietet, könnte man eine eigentliche "Sitz-Soziologie" entwickeln. Denn man darf wohl davon ausgehen, dass die Stühle, die in den verschiedenen Zeitaltern gebaut wurden, konkreten Sitzwünschen entsprachen.

Im Rahmen einer solchen Soziologie des Sitzens würde man etwa erkennen, dass die Geschichte des modernen Stuhls und des modernen Sitzens—charak-terisiert durch stark differenzierte Sitzwünsche und dadurch stark differenzierte Stuhlformen—ihren eigentlichen Anfang schon im 18. Jahrhundert nahm, als ein strenges Zeremoniell einem bequemen, nonchalanten und individuellen Sitzen im Rokoko Platz machte. Damals sind für die verschiedenartigsten, gelocker-ten, bequemen, manchmal frivolen Sitzbedürfnisse die verschiedenar-tigsten neuen Stuhltypen entwickelt worden.

Jede Beschäftigung mit dem Stuhl und seiner Geschichte führt zwangs-läufig auch zur Frage des Aesthetischen. Der Stuhl ist immer Bestandteil eines grösseren Ganzen, meist eines Innenraumes, einer Innenarchitektur. Seine Formensprache, vielfach unabhängig von seinem Konstruktionsprinzip, steht im Kontext der Formensprache seiner Zeit. In Stil des Stuhles spiegeln sich die Formtendenzen seines Zeitalters. Man kann Stilkunde, Stilgeschichte des Stuhles treiben. Dabei wird man erkennen, dass die Stilmerkmale umso ausgeprägter und umso rascherem Wandel ausgesetzt sind, je höher der Stuhl-besitzer auf der sozialen Rangleiter steht. Bäuerliche Stühle bewahren oft über Jahrhunderte ihre unverwechselbare Physiognomie, höfische Stühle wechseln Maske und Kostüm manchmal alle paar Jahre.

Die Stichworte "Sitzwünsche" und "Sitzfunktionen" führen zur Frage, wie dann der Stuhl den anatomischen Gegebenheiten des sitzenden Menschen angepasst sei. Selbstverständlich hat zu allen Zeiten derjenige, der einen Stuhl herstellte, daran gedacht, dass auf seinem Produkt ein Mensch sitzen wird. Er wird also, aus Instinkt und Erfahrung, gewisse Grundtatsachen des Sitzens stets berücksichtigt haben. Nun war aber "bequemes Sitzen" zum Beispiel, wie wir es fordern, keineswegs stets eine Forderung. Andere Forderun-gen waren manchmal wichtiger. Wenn im 19. Jahrhundert aus ästhetischen

Gründen die bauschigen Gewänder der sitzenden Damen sich voll entfalten müssen, so lässt man die Armlehnen zu kurzen Stummeln verkümmern, auch wenn das für die Arme nicht bequem ist. Und im frühindustriellen Zeitalter war es einem Fabrikbesitzer gleichgültig, ob seine Arbeitskräfte bequem und gesund sassen. Hauptsache war, die Sitzgelegenheiten kosteten nicht viel. Auch in diesen Dingen spiegelt sich Sitzsoziologisches!

Bewusst mit dem Sitzen beschäftigt haben sich erst die Pioniere der Jahrhundertwende. Natürlich hat es zu allen Zeiten gute, ja sehr gute Stühle gegeben. Aber erst Männer wie Henry van de Velde, Richard Riemerschmidt oder Adolf Loos machten sich Gedanken, die uns vertraut anmuten: Das Sitzmöbel wird als ein Organismus verstanden, der sich zu bestimmten Verwendungszwecken dem menschlichen Organismus anzupassen hat. Solche Ueberlegungen hatten aber noch immer mehr intuitiven als systematischen Charakter. Und von eigentlicher Sitzforschung kann da wohl keineswegs die Rede sein. Ja, selbst bei den Stuhlentwürfen der funktionalistischen Epoche, im Zeitalter des Bauhauses und der Pioniere der modernen Architektur, ist Sitzforschung noch nicht viel mehr als Rhetorik. Zwar beschäftigt man sich mit der Zweckmässigkeit, mit der Funktionstüchtigkeit eines Gegenstandes, und man trifft auch oft mit mehr Sensorium als Systematik das Richtige. Aber: letzte Instanz ist doch immer wieder das Formale. Letzte Instanz ist nicht die sitzanatomische Richtigkeit eines Stuhles, sondern ein ästhetisches Konzept. Der Barcelona-Stuhl von Mies van der Rohe ist dafür das klassische Beispiel: der schönste Stuhl unserer Zeit und zugleich der unbequemste.

Seit gut zwei Jahrzehnten leben wir im Zeitalter der Designer. Von den anatomischen und physiologischen Grundlagen des Sitzens ist nun sehr viel die Rede. Es wird auch schon an den Schulen, die Entwerfer ausbilden, allerlei Sitzforschung betrieben. Gewisse Wahrheiten über das Sitzen haben sich in der Welt herumgesprochen. Es gibt auch Stuhltypen, Büro- und Arbeitsstühle vor allem, bei denen solche Studien mit Erfolg in die Wirklickheit des Sitzmöbels übersetzt worden sind. Ueberschaut man aber die Masse der in dieser Zeit entworfenen und produzierten Stühle aller Typen, so wird man den Verdacht nicht los, dass die Sitzforschung der Designer letzten Endes immer wieder vielleicht bestechenden Formideen geopfert wurde. Manchmal auch nur einer modisch-formalistischen Idee. Stühle sind heute so etwas wie das Steckenpferd der Möbelentwerfer, es gibt darunter, neben den Eintagsfliegen, sehr viel mehr schöne als wirklich gute Stühle.

So, aus dieser etwas kritischen Perspektive gesehen, ist die wissenschaftlich betriebene anthropometrische Grundlagenforschung noch weit davon entfernt, die Praxis der Stuhlherstellung wirklich zu bestimmen. Bei reinen Zweckstühlen—Bürostühlen, Arbeitsstühlen überhaupt, vielleicht auch Schulstühlen—ist vieles, manchenorts vielleicht sogar schon ein Optimum erreicht. Bei allen anderen Kategorien von Sitzmöbeln aber steht das Aesthetische, das Visuelle nach wie vor im Vordergrund. Wie sagten wir eingangs? Stühle sind nur bedingt Gebrauchsgegenstände. Sie sind auch Symbole. Waren es zu Anfang und sind es auch heute. Für irrationale, emotionale Forderungen, die an den Stuhl eben auch gestellt werden, vermag keine Wissenschaft die genauen Masse zu liefern.

Anatomische und Physiologische Grundlagen zur Gestaltung von Sitzen

Von B. ÅKERBLOM

59400 Gamleby, Schweden

1. Einleitung

Wenn ich als Junge mit meinen Eltern Bridge spielte, wollte ich dabei immer in einem niedrigen Sessel statt auf einem gewöhnlichen Stuhl sitzen, obgleich der Tisch dann unbequem hoch war.

Und wie ermüdend war es erst in der Schule, wo man dann versuchte, sich durch die verschiedensten mehr oder weniger eigenartigen Haltungen Erleichterung, zu verschaffen.

Wenn ich nämlich auf einem gewöhnlichen Stuhl sass, schliefen oft meine Beine und Füsse ein und schmerzten ausserdem.

2. Die Bedeutung der Sitzhöhe

Zu jener Zeit kannte ich weder die Ursachen für diese Beschwerden, noch dachte ich über sie nach. Erst als ich später in den zwanziger Jahren Anatomie studierte, kristallisierte sich das Problem heraus. Im Institut bekam ich einen ungewöhnlich hohen Schreibtisch. Der dazugehörige Stuhl war auch sehr hoch, aber nicht hoch genug. Um ein Buch, das auf dem Tisch lag, bequem lesen zu können, musste ich ein paar Bücher auf den Sitz legen. Das grosse Unbehagen, dass ich dann beim Sitzen verspürte, beruhte wahrscheinlich auf der Kompression der weichen Gewebe an der unteren Seite der Schenkel, nicht zu vergessen den Nervus ischiadicus. Um diese Komplikation zu vermeiden, hätte natürlich der Sitz neidriger sein müssen, als der Unterschenkel lang war.

Ich fand heraus, dass, wenn der Stuhl meinen Beinen angepasst werden sollte, er circa 5–7 cm niedriger als gewöhnlich sein musste, d.h. ungefär 40 cm hoch.

Dass Stühle und andere Sitzgelegenheiten z.B. in Zügen und Strassenbahnen besonders bei Frauen starke Beschwerden verursachen können, wurde mir klar. Ein Blick in ein grösseres Büro, in dem viele Menschen arbeiteten, zeigte, dass gerade die Frauen, obgleich sie auf teuren, verstellbaren Bürostühlen sassen, die Lendenlehnen dieser Stühle nicht benutzten. Statt dessen sassen sie auf der Vorderkante der Sitzfläche. Diese Stellung nahmen sie unbewusst ein und vermieden dadurch den unbehaglichen Druck auf die Unterseite der Schenkel. Wahrscheinlich konnten die Konstrukteure aus Mangel an anatomischen Kenntnissen nicht verstehen, warum die Rückenlehnen dieser Stühle nicht benutzt wurden.

Zu jener Zeit herrschte die allegmeine Auffassung, dass die stützende Fläche so gross wie möglich sein sollte, d.h. Gesäss und Schenkel sollten in grössten Umfang zur Stützung des Körpers genutzt werden. Aber weiche Gewebe eignen sich nicht immer als Stütze, und besonders Muskeln und Nerven sind druckempfindlich.

Die Hauptstütze gegen die Sitzfläche sind die Tubera ischii. Die Gewebe, die sie decken, sind weniger empfindlich als die anliegenden Muskeln.

Auf Bild 1 sieht man, wie die Gewebe an der Unterseite der Schenkel dem Druck des Sitzes nachgeben. Sie geben keinerlei Stütze. Erst der Schenkelknochen kann dem Druck des Oberkörpers standhalten. Ehe dieses jedoch geschieht, werden Muskeln und Nerven stark zusammengepresst, wodurch grosses Unbehagen entsteht.

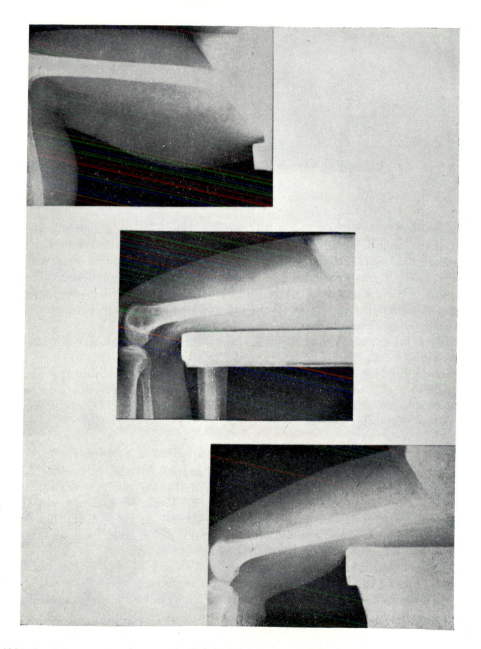

Abbt. 1.   Die Rückseite des Oberschenkels ist bei Lagerung auf der Sitzfläche starken Pressungen ausgesetzt.

Bild 2 (aus Murrell's Buch " Ergonomics ", 1963) zeigt deutlich, wie der Stuhl die Schenkel des Sitzenden eindrückt.

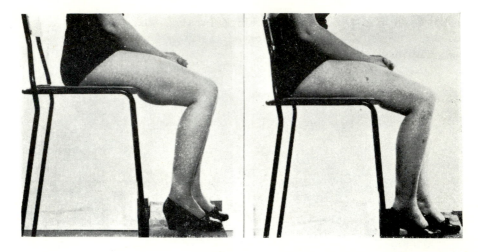

Abbt. 2.   Kompression des Schenkels auf einem zu hohen Stuhl. (Nach Murrell: *Ergonomics* 1965)

Die anthropometrischen Masse, die ich zur Hand hatte, gaben mir jedoch keine Antwort auf die Frage, wie lang der Unterschenkel des Sitzenden vom Fussboden bis zur Kniekehle ist. Messungen, vorgenommen an Menschen verschiedensten Alters und Berufes ergaben, dass die gebräuchliche Stuhlhöhe von 45–47 cm für Frauen auf 40 cm und für Männer auf 42–43 cm herabgesetzt werden sollte wenn man die Absätze einbezieht, was für die meisten Menschen eine zufriedenstellende Lösung wäre.

Später erfur ich dann, dass Bennet (1928) bereits Messungen an Schulkindern vorgenommen hatte. In Bezug auf die Stuhlhöhe kam er ungefähr zum gleichen Resultat wie 20 Jahre später ich. Nach seiner Empfehlung sollte ein Stuhl etwa 2,5–5 cm niedriger sein als der Unterschenkel des Sitzenden. Für Erwachsene schlug er eine Sitzhöhe von 40,5 cm vor. Das ist ungefähr das gleiche Mass, das ich für richtig halte.

3.   Sitzgewohnheiten bei Aussereuropäischen Völkern

Einen anderen Gesichtspunkt über die Höhe des Stuhles gibt eine Untersuchung Vendels (1959). In Ethnographischen Museen studierte er Stühle der verschiedensten aussereuropäischen Völker und fand heraus, dass die Höhe dieser Stühle (Bild 3) zwischen 15 und 30 cm variiert. Nur Häuptlinge und andere Personen von Rang benutzten Stühle mit einer Höhe von 35 cm und darüber, um dadurch ihre bevorzugte Stellung kundzutun.

Die gewöhnlichste Sitzstellung bei vielen Völkern ist das Sitzen auf dem Boden. Vendel ist der Ansicht, dass dieses eine natürliche Stellung ist, die ja auch Kinder gern einnehmen, deren Haltung dann mit Schulbeginn durch das Sitzen auf den Schulbank verschlechtert wird.

Es scheint also, als ob man Stühle ohne weiteres mehrere Zentimeter kürzer als den Unterschenkel machen könnte. Meiner persönlichen Erfahrung nach ist es für einen grossen Menschen gar nicht so unbequem auf einem niedrigen

Stuhl zu sitzen. Für einen kleinen Menschen dagegen ist es wesentlich unbequemer auf einem Stuhl zu sitzen, der zu hoch ist. Die Standardhöhe für den gewöhnlichen Stuhl sollte deshalb mit Rücksicht auf die Kurzbeinigen festgelegt werden.

Ich glaube, dass das Studium der Sitzgewohnheiten bei den verschiedenen Völkern der Welt für die hier erörterten Probleme sehr wertvoll sein könnte.

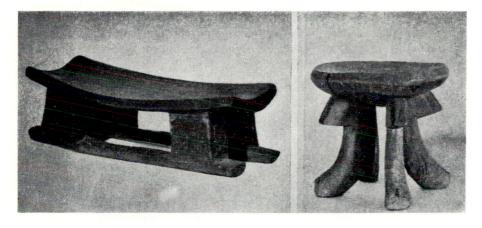

Abbt. 3. Ein amerikanischer und ein afrikanischer Stuhl. (Nach Vendel: *Att sitta* 1959)

4. Eine neue Auffassung der Muskelfunktion

Die sitzende Stellung wird in Uebereinstimmung mit dem Stehen bis zu einem gewissen Grad von den Muskeln ausbalanciert. Für das Verständnis dieses Mechanismus sind daher einige grundlegende Kenntnisse der Muskelfunktion erforderlich.

Früher war man vorwiegend der Auffassung, dass die Bewegung eines Gelenkes nicht nur von den Muskeln, welche die Hauptarbeit zu leisten haben, somdern auch durch eine gewisse Gegenaktivität der Antagonisten der anderen Gelenkseite ausbalanciert wird. So, glaubte man, kam die feine Kombination der Kräfte zustande, welche für die Präzision einer Bewegung ausschlaggebend war. Man stellt sich die Frage: Warum eine solche Energierveschwendung?

Eigentümlicherweise konnte man auch der Ansicht begegnen, das es die Muskeln sind, die die extreme Beuge-und Strecklage eines Gelenkes begrenzen und dadurch den Bewegungsumfang bestimmen. Diese Auffassung über die Funktion der Muskeln beruhte wahrscheinlich auf einer falschen Auslegung dessen, was Sherrington *postural tonus* nannte (Basmajian 1962). Dies stimmt jedoch nur bei decerebrierten Tieren, dagegen nicht wenn das Gehirn intakt ist. Vom intakten Gehirn gehen inhibierende Impulse aus, die die Antagonisten daran hindern, unnötige Arbeit zu leisten.

Meine elektromyographischen Untersuchungen (Åkerblom 1948) zeigten, dass die Antagonisten der für eine Bewegung hauptsächlich verantwortlichen Muskeln in der Regel keine Aktivität aufweisen. Ungefähr zur gleichen Zeit machte auch Allen (1948) dieselbe Beobachtung, die bald von anderen Foschern bestätigt wurde.

Ausserdem zeigt sich keinerlei Aktivität in den Muskeln, wenn ein Gelenk in seiner extremen Stellung durch andere Kräfte gehalten wird (Åkerblom 1947,

1948). Wahrscheinlich nehmen die Bänder oder auch die Knochen selbst die Belastung auf sich. Erst wenn die Bänder einer besonders heftigen oder andauernden Belastung ausgesetzt werden, treten die Muskeln in Funktion. Selbstverständlich ist die normale Funktion der Muskeln notwendig, damit die Bänder ihre Stärke behalten und ihre Aufgabe erfüllen können.

5. Die Vorwärts Zusammengesunkene Sitzhaltung

Hieraus lässt sich folgern, dass die vornübergebeugte Haltung, die der Sitzende so oft einnimmt, eine ausgesprochene Ruhestellung sein muss. Wird die Wirbelsäule so weit wie möglich vorgebeugt, sind die Rückenmuskeln ganz entspannt (Bild 4 (e)). Die Stützgewebe, Knochen, Zwischenwirbelscheiben und Bänder tragen die Belastung.

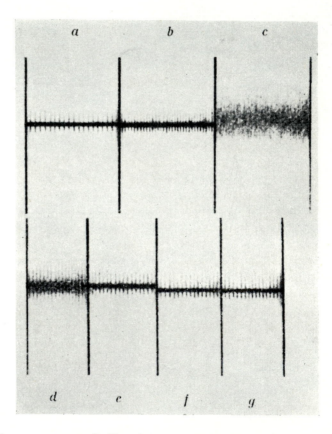

Abbt. 4. Electromyogramme mit Hautelektroden von den Rückenmuskeln (die Herzschläge treten in den meisten EMG deutlich hervor). (a) Liegend, (b) Ruhiges Stehen, (c) Vorgeneigt stehend, (d) Geradesitzend ohne Lehne, (e) Vorgeneigt und zusammengesunken sitzend (ohne Lehne), (f) Mit Lenden-Brustrückenlehne sitzend, (g) Nur mit Lendenlehne sitzend, (Nach Akerblom: *Den goda arbetsstolens problem*, 1947).

Hat man keine Rückenlehne, kann man diese Stellung leichter beibehalten, wenn der Stuhl niedrig ist. Dafür sprechen z.B. die niedrigen Stühle ohne Rückenlehnen, die von den nichteuropäischen Völkern benutzt werden (Bild 3). Ein bequemer Hocker sollte deshalb niedriger als ein gewöhnlicher Stuhl sein. Ich würde vorschlagen, nicht höher als 38 cm.

6. Die Notwendigkeit des Stellungwechsels

Die maximal vorgebeugte Haltung beim Sitzen mit oder ohne Benutzung eine geraden Rückenlehne ist die einzige, die auf einem solchen Stuhl die völlige Entspannung der Rückenmuskulatur ermöglicht. Wie günstig diese Ruhestellung auch ist, kann man sie dennoch nicht allzulange beibehalten. Die Stützgewebe des Rückens ermüden allmählich. Um das zu vermeiden, muss die Haltung der Wirbelsäule ab and zu verändert werden.

Dieses stimmt mit der Tatsache überein, auf die Vernon (1924) hingewiesen hat, dass man nämlich seine Stellung hin und wieder ändern muss, um eine schnelle Ermüdung zu vermeiden. Entsprechend dieser Erkenntnis müssten Stühle so konstruiert sein, dass sie verschiedene entspannte Haltungen ermöglichen.

7. Die Rückenlehne

Vom Standpunkt der Bequemlichkeit aus gesehen, hat man der Rückenlehne die grösste Bedeutung zukommen lassen. In der medizinischen Literatur wurde oft kritisiert, dass gewöhnliche Stühle meistens eine gerade Rückenlehne haben. Man hat vorgeschlagen, die Rückenlehne entsprechend der S-form der Wirbelsäule zu konstruieren. Hierbei ist die Bedeutung eine feste Stütze im Lendenbereich besonders betont worden. Dieser Forderung ist man bei der Konstruktion moderner Bürostühle nachgekommen.

Es sind auch gebogene Lehnen konstruiert worden, die einen grösseren Teil des Rückens abstützen. Ich habe eine Lehne vorgeschlagen, die entweder nur die Lendengegend oder aber auch den Rücken einschliesslich des Brustteils stützt. Sie vereint also zwei verschiedene Stützmöglichkeiten. Diese Konstruktion scheint auch gelungen zu sein. Bei diesen beiden Haltungen sind die Rückenmuskeln gut entspannt (Bild 4 (*f*) und (*g*)).

Der untere Teil der Rückenlehne—gegenüber dem Sacrum—sollte ausgebuchtet sein oder ganz ausgespart werden, damit das Gesäss genügend Platz hat.

Wenn der Rumpf gut gestützt werden soll, sollte die Neigung der Lehne gegen die Horizontale circa 115° betragen. Bei Sesseln könnte dieser Winkel um 5° grösser sein.

Bei dieser kombinierten Rückenlehne soll die Lendenstütze verschiedene Stellungen ermöglichen. Sie muss daher schwach gebogen und möglichst gepolstert sein, um den verschiedenen Haltungen der Wirbelsäule gerecht zu werden.

Tabelle 1. Der Abstand vom Stuhlsitz bis zu den zwei untersten Zwischenwirbelscheiben

		16	17	18	19	20	21	22	23	24 cm
L^5-S^1	Frauen	5	10	6	4	3	1			
	Männer	4	9	11	4	4				
L^4-L^5	Frauen				6	10	7	3	2	1
	Männer				6	8	8	7	2	1

Die Stützung des Lendenrückens sollte in seinem unteren Teil erfolgen, dort nämlich, wo die Belastung der beweglichen Wirbel am grössten ist. Natürlich ist die Höhe des Lendenabschnittes—gemessen von der Sitzfläche—individuell

verschieden (Tabelle 1).   Da es sich jedoch nur um einen relativ kleinen Teil
der gesamten Körperlänge handelt, kann man von einer individuellen Anpassung dieses Teiles absehen.   Die Toleranz ist auch hier relativ gross.

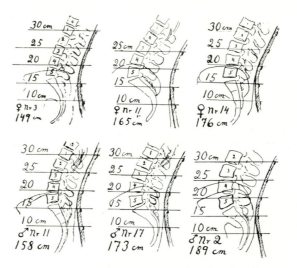

Abbt. 5.   Der Abstand vom Stuhlsitz bis zu den Lendenwirbeln.

Bild 5 zeigt die Abmessungen des Lendenabschnittes bei einige Individuen
beider Geschlechter.   Man sieht, dass der Punkt der stärksten Vorwölbung der
Lendenlehne etwa 17–21 cm oberhalb der Sitzfläche liegen muss.

Viele meiner Patienten mit Rückenschmerzen haben mir erzählt, dass sie
beim Sitzen auf einem Stuhl mit gerader Rückenlehne instinktiv ein kleines
Kissen oder den Vorderarm im Lendenbereich plazierten.   Sie berichteten mir
von der wohltuenden Wirkung, die sie dabei verspürten.

8.   Die Neigung der Sitzfläche

Der Tendenz zum Nachvorngleiten bei Benutzung der Rückenlehne muss
man durch eine entsprechende Rückwärtsneigung der Sitzfläche entgegenwirken.   Eine Neigung von 5–7° sollte ausreichend sein.   Ist der Reibungs-
widerstand grosss, z.B. bei Stoffbezügen, kann die Neigung auch geringer sein.

9.   Drei Verschiedene, Bequeme Sitzhaltungen auf Demselben Stuhl

Auf dem vorhergehend beschriebenen Stuhl kann man sitzen (Bild 6):
1. in der nach vorn zusammengesunkenen Haltung mit oder ohne Benutzung
   der Rückenlehne,
2. mit Abstützung nur im Lendenbereich.
3. mit Abstützung sowohl des Lenden- als auch des Brustrückens.

Da diese drei Möglichkeiten als gute Ruhestellungen für den Rücken angesehen werden können, kannmansie als die Grundruhestellungen beim Sitzen
bezeichnen.

Hierbei sollte betont werden, dass keine der überhaupt denkbaren Sitzhaltungen schädlich ist, wenn man nicht zu lange darin verharrt.   Das Bedürfnis,
die Haltung zu ändern, ist individuall verschieden.   Der eine ermüdet bei

einer bestimmten Haltung früher als der andere. Genau so gut kann man annehmen, dass die Druckempfindlichkeit der weichen Gewebe variieren kann.

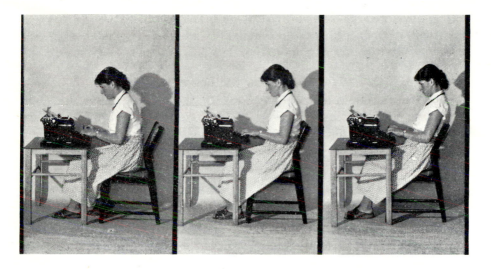

Abbt. 6.   Drei verschiedene, bequeme Sitzhaltungen auf demselben Stuhl.

10.   Der Plats für die Beine

Die Bequemlichkeit beim Sitzen hängt auch davon ab, wieviel Platz die Beine haben. Man muss sie bequem austrecken, nach beiden Seiten hin bewegen und unter dem Stuhl halten können. Eine einfache Fussbank ist also kein genügender Ausgleich, wenn der Stuhl zu hoch ist.

Ausserdem sollte man die Möglichkeit haben, die Schenkel hin und wieder auf dem Sitz ruhen lassen zu können. Dieser sollte deshalb eine gewisse Tiefe besitzen. Unglücklicherweise haben besonders Sessel und Sofas oft so tiefe Sitze, dass kurzbeinige Menschen die Rückenlehne nicht erreichen können. Deshalb sollt ein Sitz nicht tiefer als 45 cm sein. Etwa 40 cm sollten ausreichend sein.

11.   Polsterung

Sehr wichtig für die Bequemlichkeit ist fernerhin die Polsterung der Sitzmöbel. Diese darf nicht zu weich sein. Besonders die Lendenregion braucht eine feste Stütze, und die Sitzhöcker dürfen nicht so tief in den Sitz einsinken, dass die sie umgebenden Msukeln einem zu starken Druck ausgesetzt werden.

12.   Einstellbare Stühle

So viel über den gewöhnlichen Stuhl. Für besondere Zwecke können Stühle anders konstruiert werden. Ein Beispiel dafür ist der Bürostuhl mit Lendenlehne und der Möglichkeit, die Sitzhöhe zu verändern. Man darf jedoch nicht vergessen, dass die meisten Menschen entweder nicht wissen, wie man diesen Stuhl einstellt, oder aber auch sich nicht darum kümmern.

13.   Das Verhältnis Stuhl–Tisch

In diesem Zusammenhang sollte betont werden, dass sich die Höhe eines Tisches nach der Höhe eines Stuhles zu richten hat, und nicht umgekehrt, wie man es leider oft erlebt. Bei der von mir vorgeschlagenen Stuhlhöhe könnte die Höhe des dazugehörigen Tisches 67–69 cm und die lichte Höhe vom Boden aus mindestens 53 cm betragen, damit die Knie gegnügend Platz haben.

Für gewisse Zwecke können Tische natürlich niedriger sein, Schreibmaschinentische z.B. 60 cm hoch (die Verstellbarkeit der Tischhöhe kann sogar wichtiger sein als die der Stuhlhöhe).

14. Die Macht der Gewohnheit

Die persönliche Gewöhnung beeinflusst in hohem Grade die Einstellung eines Individuums zu einem Stuhl. Die meisten Menschen sind konservativ. An einen gewissen Stuhltyp von Kindheit an gewöhnt, sind sie Neuheiten gegenüber misstrauisch. Man kann sich vorstellen, dass sich besonders die Langbeinigen gegen Stühle wehren, die niedriger sind als die, die sie früher benutzten.

15. Schulbänke

Schon in der Schule werden die Sitzgewohnheiten gebildet. Die Erwachsenen würden sich dem alten Stuhltyp gegenüber sehr ablehnend verhalten, wenn sie schon in ihrer Kindheit und Schulzeit körperangepasste Stühle der vorgeschlagenen Art benutzt hätten.

Tabelle 2. Die Länge der Unterschenkel im Verhältnis zu der Körperlänge bei 770 Kindern von 6–16 Jahren

Körper-Länge, cm:	Unterschenkellänge, cm:										
	28	29 30	31 32	33 34	35 36	37 38	39 40	41 42	43 44	45 46	47 48
115–120	2	11									
121–130	1	17	66	17							
131–140			17	71	46	6					
141–150				6	59	60	18				
151–160					1	25	101	43			
161–170							33	71	31	3	
171–180								3	39	16	
181–185										5	3

Tabelle 3. Der Abstand vom Stuhlsitz bis zu der letzten Zwischenwirbelscheibe bei 89 Kindern

cm :	11	12	13	14	15	16	17	18	19
7 J.	2	4	15	4	3				
8 J.	2	4	6	4	6				
9 J.			10	2	2				
10 J.				2	2				
11 J.				5	8	2	3	2	1

Tabelle 2 zeigt die Unterschenkellänge bei Schulkindern und Tabelle 3 den Abstand vom Stuhlsitz bis zu der letzten Zwischenwirbelscheibe.

Der unterschiedlichen Grösse der Kinder könnte man durch einstellbare Stühle gerecht werden. Wie jedoch soll man sicher sein, dass diejenigen, die die Stühle verstellen sollen, dieser Aufgabe auch wirklich nachkommen? Ausserdem können solche Stühle Lärm verursachen und sind teuer.

Wie Bennet (1928) schlage ich darum Schulstühle ohne bewegliche Teile vor. Das Problem, die richtige Grösse zu wählen, ist damit natürlich nicht gelöst.

Es muss genau beobachtet werden. wann das wachsende Kind einen grösseren Stuhl braucht. Diese Beobachtung sollte möglichst erleichtert werden. Vielleicht würden drei verschiedenen Grössen ausreichen (Bild 7). Stuhl und Tisch könnten dann so lange benutzt werden, wie der Sitzende die Knie bequem unter dem Tisch halten kann. Wenn letzteres nicht mehr möglich ist, ist es Zeit zum Tauschen.

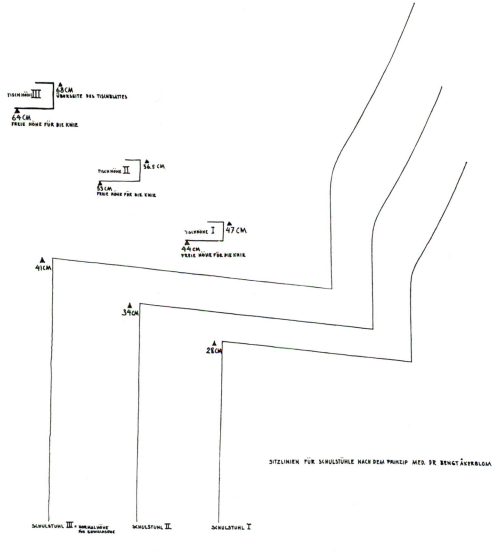

Abbt. 7. Prinzipskizze für Schulbänke.

16. Theorie und Praxis in der Forschung

Sowohl Hersteller als auch Konsumenten wünschen ein schnelles Resultat unserer Forschungen. Alle Ergebnisse von Versuchen im Laboratorium sind jedoch nicht so ausschlaggebend wie die praktische Erfahrung und sie verlangt viel Zeit. Wenn man einen Stuhltyp in der Praxis ausprobieren will, sollten sich die Versuchspersonen dazu verpflichten, diesen eine relativ lange Zeit zu

benutzen, etwa ein halbes Jahr oder länger.   Wenn sie nach dieser Zeit ihre
alte Stühle wieder prüfen, zeigt der Vergleich besser, welcher Stuhltyp zu
bevorzugen ist.

Da aber die Sitzgewohnheiten schon in der Kindheit gebildet werden und
sich später nicht gern ändern, sollte man eigentlich mit den Versuchen schon
bei kleinen Kindern beginnen.   Es würde dann einige Jahrzehnte dauern, bis
man ein sicheres Resultat erzielt.   Deshalb sollte man nicht schon jetzt
bestimmte Masse für Stühle und Tische festlegen.

Nach allgemeiner Erfahrung ist das Sitzen vielfach mit Beschwerden verbunden.   Die Stühle
sind oft so hoch, dass der Sitz Muskeln und Nerven der Schenkel von unten her zusammenpresst,
was Schmerzen und Gefühllosigkeit in den Beinen zur Folge hat.   Um dieses zu vermeiden
schlägt der Autor eine Sitzhöhe von circa 40 cm vor.

Die Lehne ist selten so geformt worden, dass sie dem Rücken gute Ruhestellungen und die
Möglichkeit zwischen diesen zu wescheln erlaubt.   Der Autor schlägt daher vor, dass sie entsprechend der Rückenkrümmung geformt wird.   Dadurch erhält man nicht nur die Möglichkeit einer
guten Stütze für den Lenden- und Brustrüken unter Beibehaltung der natürlichen Schwingung
der Wirbelsäule, sondern kann auch die Lehne nur als Lendenstütze wie bei gewöhnlichen
Bürostühlen benutzen.

Elektromyographische Untersuchungen zeigen, dass man bei diesen beiden Haltungen gute
Entspannung der Rückenmuskulatur erzielt.   Dieses ist auch bei der zusammengesunkenen
Stellung mit vorgeneigter Wirbelsäule der Fall.   Diese drei Haltungen können daher als Grundru-
hestellungen beim Sitzen bezeichnet werden.

Es werden Versuche mit nach diesen Prinzipen konstruierten Schulbänken vorgeschlagen.
Die Schulkinder sollten von Anfang an gute Stühle haben.

Der Autor betont die Wichtigkeit einer Kombination der Theorie und des Laborversuches
mit praktischen Versuchen von genügend langer Dauer, bevor ein bestimmter Stuhltyp als
Standard akzeptiert werden kann.

It is a common experience that sitting on chairs often causes trouble.   Chairs have for instance
been so high that the seat has compressed the muscles and nerves of the thigh, causing pains and
numbness in the legs.   To avoid this the author has suggested a seat height of about 40 cm.

Back-rests have seldom been designed in a way to give a good support to the back and the
possibility of change between different resting positions.   The author suggests that it should be
formed with a bend corresponding to that of the vertebral column.   This gives the possibility
of sitting not only with a good support for the lumbar and dorsal regions, but also of sitting with a
simple lumbar support, just as on ordinary office chairs.

Electromyographic investigations show us that we get good relaxation of the back muscles
in both these positions, as well as when we sink down with a ventriflexed spine.   Thus these
three positions can be called the basic resting positions when sitting.

It is suggested that school furniture designed on these principles should be tested.   School
children should have correctly designed chairs from the beginning.

The author emphasizes the importance of combining theory and laboratory trials with practical
tests of sufficient duration, before a certain type of chair is accepted as standard.

Il est bien connu que la position assise sur des chaises est souvent à l'origine de troubles.   Par
exemple la hauteur des sièges est telle que les muscles et nerfs de la cuisse sont comprimés, ce qui
conduit à des douleurs et des engourdissements des jambes.   Afin de pallier cet inconvénient,
l'auteur recommande une hauteur de 40 cm.

Le dossier est rarement construit de manière à fournir un bon soutien au dos ou à permettre de
changer d'inclinaison pour présenter plusieurs positions de repos.   L'auteur recommande des
dossiers présentant une courbure identique à celle de la colonne vertébrale.   Ceci permet de
asseoir non seulement avec un bon soutien des régions lombaire et dorsale, mais aussi avec un
simple soutien lombaire comme sur les sièges de bureau ordinaires.   Des investigations électro-
myographiques ont montré un bon relâchement des muscles dorsaux dans les deux positions, comme
dans une posture avec la colonne penchée vers l'avant.   Ainsi ces positions peuvent être considérées
comme les positions de repos idéales pour un sujet assis.

L'auteur suggère de tester du matériel scolaire d'après ces principes.   Il est important que les
enfants d'âge scolaire disposent dès leur jeune âge de sièges bien adaptés.

Enfin l'auteur met l'accent sur l'importance qu'il y a à combiner des essais théoriques et de
laboratoire avec une expérimentation pratique d'assez longue durée avant de considérer un modèle
de chaise comme modèle standard.

Literatur

ÅKERBLOM, B., 1947, Den goda arbetsstolens problem. *Affärsekonomi*, **20**, 8–10, 34–38.

ÅKERBLOM, B., 1948, *Standing and Sitting Posture* (Stockholm: A. B. NORDISKA BOKHANDELN).

ALLEN, C. E. L., 1948, Muscle action potentials used in the study of dynamic anatomy. *Brit. J. phys. Med.*, **11**, 66–73.

BASMAJIAN, J. V., 1962, *Muscles Alive* (Baltimore: THE WILLIAMS & WILKINS COMPANY).

BENNET, H. E., 1928, *School Posture and Seating* (Boston: GINN and COMPANY).

MURRELL, K. F. H., 1965, *Ergonomics: Man in His Working Environment* (London: CHAPMAN & HALL).

VENDEL, S., 1959, Att sitta. *Nytt och Nyttigt*, 4, 7–17.

VERNON, H. M., 1924, The influence of rest-pauses and changes of posture on the capacity for muscular work. *Medical Research Council Rep.* **29**, 28–55.

Anthropometric and Physiological Considerations in School, Office and Factory Seating

By W. F. FLOYD and JOAN S. WARD

Department of Ergonomics and Cybernetics, Loughborough University of Technology, England

British Standards Institution recommendations regarding the dimensions of school, office and factory chairs and seats (B.S. 3030: Part 3: 1959; B.S. 3044: 1958; B.S. 3079: 1959; B.S. 3893: 1965; B.S. 4141: 1967) are still the main bases upon which seating suited to the anthropometric and physiological needs of the user is determined in Britain.

We have earlier published corroborative evidence in the school, office and factory situations (Floyd and Ward 1964, 1967) that seating based upon the principles embodied in the British Standards recommendations does confer postural benefits upon the user. It is not intended in this paper either to dwell upon the recommendations referred to or to repeat the findings of the publications mentioned but to draw attention to the need for further work in this area and to give an account of some preliminary studies undertaken in Loughborough in the past year or so.

We have stated, particularly in connection with school furniture (Floyd and Ward 1964), that ' . . . the benefits of anthropometrically-based designs (of school furniture) cannot be fully exploited unless there is strict discipline over their choice and use . . . ' and also (Floyd and Ward 1967) that ' . . .the school-children of today are the next generation of industrial employees. What they require for good posture in the schoolroom is, when translated into the adult situation, a sound guide to what they will require in the factory . . . '

Habits, as we all know, once acquired are tenaciously held. This is undoubtedly as true of sitting habits as it is, say, of postural habits in sleep or skilled movements. It is also a fact that the longer a particular habit has endured the more difficult it is to change or abandon it. This being so, it would appear to be of the greatest importance to instil, and maintain, good sitting habits as early in the individual's life as possible. Our contention is that the schoolroom is a most suitable place for this habituation to occur to the individual's maximum advantage.

In Britain, where a range of sizes of school desks and chairs to suit the dimensions of different age groups of both girls and boys has been recommended, it should be possible to provide the majority of children with furniture best suited to their anthropometric and physiological requirements and to ensure that they make the fullest postural use of it. This appears to be true also in Australia which has already adopted school furniture standards (Oxford 1966) and may also apply shortly in India.

Sitting habits thus acquired in the schoolroom are likely to be carried over into later commercial or industrial employment. The demand for better fitting equipment, to be effective, must come from the user—it cannot be imposed solely by authority, however well-intentioned. If an individual's early training has taught him good postural habits at correctly dimensioned and designed furniture, he is likely to require that his subsequent commercial or industrial work place be such that he can maintain the same postural habits. We conclude, therefore, that it is of great importance to place emphasis upon the further study and analysis of postural behaviour in the schoolroom.

The initial study by Floyd and Ward (1964) of the behavioural responses of schoolchildren to two different designs of furniture (the main results of which are summarized in Figure 1) prompted us to look particularly at the use of the chair backrest. From Figure 1 it can be seen that the backrest was used only about 50 per cent of the time with the ' old ' or traditional furniture and only up to about 60 per cent of the time with the ' new ' furniture. We also noted that the posture most frequently associated with writing on the horizontal surface of the desk was one in which the trunk was slumped and tilted forward, with the arms supporting the body upon the desk, the backrest unused. It could not be inferred that it was principally the writing posture that precluded greater use of the backrest because the data from the study were not in a form to answer this question. In order, therefore, to determine more precisely the amount of time spent in different classroom activities, including writing, and in different postures, an investigation was undertaken by Cowen (as yet unpublished).

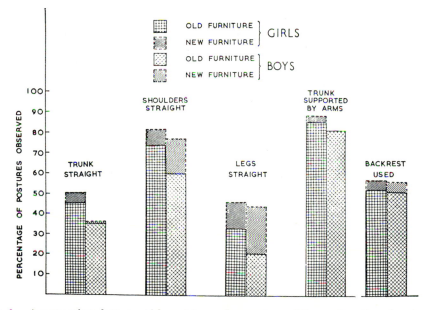

Figure 1. A comparison between girls and boys of percentage of times when the advantageous postures of the body segments were found in the 'old' and 'new' furniture.

Forty-two girls and 42 boys, of mean age 17·2 years, in the upper classes of two secondary schools were studied. The mean stature of the boys was 174·5 cm., that of the girls 163·1 cm., i.e. slightly above the averages quoted by Barkla (1961) for the British population aged between 18 and 40 years of age. The mean shod lower leg length (the dimension upon which the recommended height of the chair seat above the floor is based) was 45·2 cm for boys and 41·7 cm for the girls. The chairs upon which they were seated in their classrooms were British Standard Size E having a seat height of 44·5 cm.

These 84 pupils were observed during periods when they were undergoing instruction in French, English, Geography, History and Mathematics. These lessons were selected as not requiring special accommodation or facilities (such as are required by Chemistry or Art) and thus the children sat at their usual desks and chairs. After a preliminary period of observation the activities displayed by these pupils were observed to fall into the following well-defined categories: listening; following text; reading; looking up (concentrating attention on teacher); writing; speaking; preparation for lesson (assembling books etc.); reaching for contents of satchel, coat or desk; standing up and sitting down; and absent temporarily from classroom.

The postural behaviour of these same pupils observed in parallel with the above activities was classified under the following headings, in which combinations of the postures of trunk and arms, and use or otherwise of the backrest, could be recorded: standing; trunk forward; trunk upright; trunk backwards;

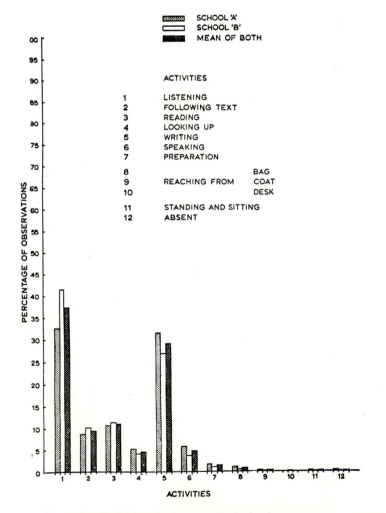

Figure 2. Activities of school pupils during lessons.

trunk leaning towards floor; backrest unused; backrest used; arms—one supported on desk; arms—both supported on desk; and arms—neither supported on desk.

From analyses of the records made (over a total of 20 lesson periods in one school and 24 lesson periods in the second school), of the activities of the 84 pupils, listening (Activity 1) was observed to occupy the greatest proportion of time, viz., between 35 and 40 per cent of the total time. Listening was followed closely by writing (Activity 5) (nearly 30 per cent of the total time). As can be seen from Figure 2, these two activities occupied the major proportion of the total time. Following a text (Activity 2), or reading (Activity 3), took up about 10 to 15 per cent of the time, while the other activities (speaking, preparation for lessons etc.) occupied the remainder.

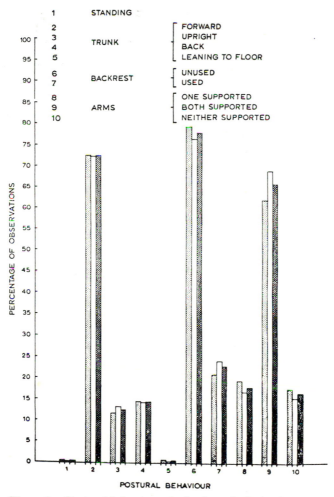

Figure 3.   Postural behaviour of school pupils during lessons.

Figure 3 shows the proportions of time that postural behaviour was observed to occur in parallel with these activities.   From the Figure it can be seen that 3 aspects of postural behaviour are most frequently observed.   These are (a) sitting without support from the backrest (Posture 6), (b) the trunk inclined forwards (Posture 2) and (c) both arms supported (Posture 9) by leaning them on the desk. This combined posture appears to be imposed by the necessity to write on a horizontal surface, as appeared a likely inference from the earlier schoolroom study. Writing, however, was observed to be an activity occupying only about 30 per cent of the total time, whereas this desk-supported posture was observed to occur between 65 per cent and nearly 80 per cent of the total time. The conclusion must be that it is a posture adopted even when the contraints of the task do not impose it. From the analysis it is also evident that use of the backrest was most often observed at a time when only one arm was resting on the desk or when the arms were not in contact with the desk at all.

Subsequent to this study of activity and postural behaviour a pilot electro-myographic investigation was made on one schoolboy to obtain recordings from some of the muscles considered to be active in the seated working position— muscles in addition to the erectores spinae which have been often investigated in connection with the erect posture. The schoolboy was of the same age and within the same stature range as the 42 schoolboys previously studied.

The myographic apparatus available permitted only 8-channels of recording. It was considered essential to record muscular activity simultaneously from both sides of the trunk, so that unilateral activities or movements (such as writing or leaning an arm upon the desk) could be observed. Four pairs of muscles on the left and right sides of the trunk were therefore studied. Figure 4 illustrates the siting of the electrodes over the muscles selected. They were placed in two positions over the trapezius muscle (on the cervical portion and on the mid-clavicular portion), on the post-axillary fold of the latissimus dorsi and on the lumbar region of the erectores spinae muscles.

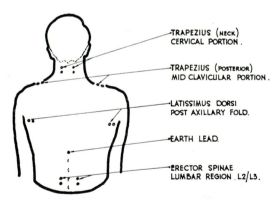

Figure 4.   Siting of surface electrodes.

Figures 5 and 6 show the major myographic findings. In Figure 5 it will be noted that the muscular activity is symmetrical, i.e. that the same posture is maintained by both sides of the body. Sitting very erect in the chair, not using the backrest, with the legs bent at a right angle at the knee and the arms relaxed in the lap, shows an increase in activity in the erectores spinae muscles. The central illustration, where the subject is sitting erect comfortably, supported by the backrest and with the trunk slightly rounded, but in other respects in a posture similar to the first illustration, shows almost minimum activity in all 4 pairs of muscles.   This ' sitting comfortably erect ' position appears to confirm Åkerblom's (1948) statement that ' . . . the muscles of the lumbar region can be maximally relaxed in three sitting positions which can be regarded as resting positions from an anatomical and mechanical point of view, viz . . . . sitting with a lumbar support . . . '

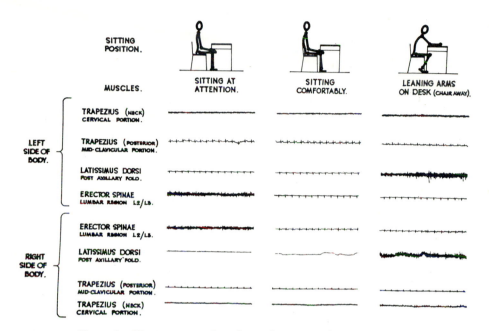

Figure 5.   Electromyography of seated postures, showing symmetry.

The third posture shown in Figure 5, with the body supported by the arms leaning upon the desk and the trunk in a forward relaxed position, again confirms Åkerblom's view that this position of the trunk confers maximal relaxation on the muscles of the lumbar region.   At the same time, however, the left and right latissimus dorsi muscles and, to a less extent, the trapezius muscles in the cervical region, are brought into play—the latissimus dorsi to an extent even greater than that of the lumbar muscles when the subject sits erect without support for the back.

Figure 6 demonstrates the effects of asymmetrical seated postures upon the muscles studied.   The first and second illustrations show the subject seated in an erect comfortable position but with, firstly, the right arm only, and secondly, the left arm only, supported on the desk.   In these two myographic recordings the switch in activity from the right latissimus dorsi and erector spinae muscles particularly (when the right arm is on the desk) to the left corresponding muscles when the left arm is on the desk, can be seen clearly. In both these positions there is an increase in activity in the trapezius muscles in the shoulder region.

The third illustration in Figure 6 shows the myographic trace when the subject adopted his habitual position for writing at the desk.   In this left-handed subject the only muscle from which no increase in activity was recorded was the left erector spinae.   All other muscles showed a marked increase in activity, as might be expected from an activity requiring movement of a limb and of the trunk. The increased activity of the latissimus dorsi muscle on the side of the trunk opposite to the moving limb indicates tension in muscles acting against gravity to support the seated trunk.

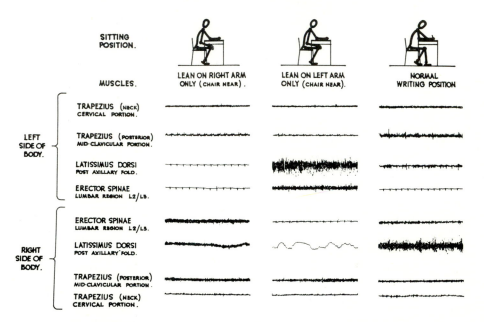

Figure 6.   Electromyography of seated postures, showing asymmetry.

It will be recalled from the histograms (Figures 2 and 3) of activities and postural behaviour observed directly on 84 school pupils that up to 30 per cent of their time in the classroom was spent in writing. As much as 80 per cent of the time was spent by some of them in postures where the trunk was slumped forward and the arms were both resting on the desk surface whether they were writing or not. This position, although it does not provoke much, if any, activity in the muscles of the lumbar region, certainly does so in the latissimus dorsi muscles particularly and also in the trapezius muscle in the neck region. Can this posture therefore really be considered to be one of rest?

The results of our preliminary findings lead us to believe that further studies in myography may well be most valuable in determining the postural behaviour, in addition to basic anthropometric requirements, that should be considered in the dimensions and design of seats and chairs.

The authors wish to thank Mr. M. Cowen and Mr. B. Saville for their help in this study.

Anthropometric data and analyses of the behaviour of schoolchildren with the multimoment technique during lessons are the basis for the considerations of the problems of school, office and factory seats. Characteristic postures of schoolchildren are: trunk straight, shoulders straight, legs straight, trunk supported by arms, and backrest used. The backrest was used about 50 per cent of the time in traditional furniture, and about 60 per cent of the time with new furniture. The most frequent sitting behaviour was a desk supported posture.

With one school child myograms were recorded in the characteristic sitting posture. The lowest electrical activity occurred when the backrest was used.

Des données anthropométriques et des analyses effectuées grâce à la technique multimoment sur le comportement d'enfants à l'école sont à la base de ces considérations sur les problèmes posés par les sièges d'école, de bureau ou d'usine. Certaines postures caractéristiques des enfants à l'école ont été déterminées. Ce sont: le tronc redressé, les épaules droites, les jambes allongées, le tronc soutenu par les bras et l'usage du dossier. Le dossier est utilisé pendant environ 50 p. 100 du temps avec le matériel usuel et 60 p. 100 du temps avec le nouveau matériel. Le comportement le plus fréquent est une position avec soutien sur le pupitre.

Sur un enfant on a enregistré des myogrammes en position assise caractéristique; l'activité électrique est minimale quand le dossier est utilisé.

Anthropometrische Daten and Analysen des Verhaltens von Schulkindern während des Unterrichts mit der Multimoment-Methode dienten als Grundlage für Überlegungen zur Problematik von Schul-, Büro- und Fabrik-Sitzen. Charakteristische Haltungen von Schulkirdern sind: Dumpf gerade, Schultern gerade, Beine gerade, Rumpf von Armen gestützt, und Rückenlehne benutzt Die Rückenlehne wurdi bei den herkömmlichen Möbeln ungefähr 50% der Zeit, bei neuen Möbeln etwa 60% benutzt. Die häufigste Sitzhaltung war eine pult—unterstützte Haltung.

Bei einem Schulkind wurden in der charkteristischen Haltung Myogramme aufgenommen. Die geringste elektrische Aktivität trat bei Benutzung der Rückenlehne auf.

References

ÅKERBLOM, B., 1948, *Standing and Sitting Posture* (Stockholm: A-B NORDISKA BOKHANDELN).

BARKLA, D., 1961, The estimation of body measurements of British population in relation to seat design. *Ergonomics*, **4**, 123–132.

BRITISH STANDARDS INSTITUTION, 1959, B.S. 3030: Part 3, *Pupils' Classroom Chairs and Tables*.

BRITISH STANDARDS INSTITUTION, 1958, B.S. 3044, *Anatomical, Physiological and Anthropometric Principles in the Design of Office Chairs and Tables*.

BRITISH STANDARDS INSTITUTION, 1959, B.S. 3079, *Anthropometric Recommendations for Dimensions of non-Adjustable Office Chairs, Desks and Tables*.

BRITISH STANDARDS INSTITUTION, 1965, B.S. 3893, *Office Desks, Tables and Seating*.

BRITISH STANDARDS INSTITUTION, 1967, B.S. 4141, *Industrial Seating*.

FLOYD, W. F., and WARD, J. S., 1964, Posture of schoolchildren and office workers. *Ergonomics, Proceedings of 2nd I.E.A. Congress, Dortmund*, p. 351.

FLOYD, W. F., and WARD, J. S., 1967, Posture in industry. *Int. J. Prod. Res.*, **5**, 213–224.

OXFORD, H. W., 1966, The problem of misfit furniture. *Furniture Review Committee, Department of Education, N.S.W. Australia*.

Anthropometric Data for Educational Chairs

By H. W. Oxford

Seating Consultant, Gnarbo Ave., Sydney, New South Wales, Australia

1. Introduction

I sit, but not too well,
My teacher knows not how.

Since 1950 the N.S.W. Department of Education has been supplying assortments of chairs and tables, based on anthropometric data, to every infant, primary and secondary grade in order to accommodate pupils of different size in each class.

This paper is a report of what has been done.

Part One deals with the size and growth of Australians after the age of three.

Part Two provides details of present-day chairs based on anthropometric, educational, and postural requirements, as well as first-hand observations of practices overseas. It also contains the results of questionnaires submitted to 1760 primary and secondary pupils about their seat-height preferences and sitting habits, and the chair preferences of 1346 infant children.

New South Wales, which comprises an area of 309,432 sq. miles, is approximately one-tenth the size of Australia, or twenty times the area of Switzerland. 504,000 primary school pupils and 240,000 secondary school students are taught by 32,000 teachers in 2,650 public schools under the control of the Department of Education and its regional offices. The production, distribution and maintenance of every item of school furniture is controlled from Sydney where the Department owns one of the largest wood-furniture establishments in Australia.

Prior to 1903, all school children with the exception of those in kindergartens, sat on movable forms at long fixed desks. In that year, four sizes of dual desks were introduced for screwing to floors. Only one size was supplied to a classroom, and the same size to more than one grade.

After World War II, it was decided to cease the supply of fixed desks and to introduce movable chairs and tables.

1.1. *1949 Anthropometric Survey*

In preparation for this, five measurements were taken of 13,000 school children to determine the sizes of chairs and tables required for all classes. The measuring was done by doctors and nurses of The School Medical Service and the random samples were arranged by the Research Division of the Department.

Three facts emerged from the survey. First, the single-height desks in classrooms were not closely related to the mean measurements of the pupils using them. Secondly, it seemed necessary to use eight heights of chairs ranging from 10 in. (254 mm) to 18 in. (457 mm), sizes ' A ' to ' H '. Thirdly, differences in the lower-leg measurement of pupils in *each* class called for the supply of a particular assortment of chairs and tables to each grade. Unfortunately, the 1949 data were never processed for publication.

In 1954, it was decided to reduce the eight sizes to six because 1 in. (25 mm) gradations were considered more theoretical than practical. Since then, chairs have ranged from 10 in. (254 mm) to 17·5 in. (445 mm) with 1·5 in. (38 mm) rises—sizes 'A' to 'F' respectively. Anything in excess of 38 mm rises in seat height would make the formation of good sitting habits more difficult. A child's learning ability could also be impaired.

1.2. *1965 Anthropometric Survey*

All the anthropometric data supplied in this paper were derived from this survey. The earlier one did not include tertiary students and adults, and the five measurements that were taken were inadequate for specialist furniture.

Prior to the second survey, special sets of silk-screen measuring equipment, square, long ruler, short ruler and plywood for sitting data were made at the Department's Teaching Aid establishment. The equipment was sent to 64 primary schools and 60 secondary schools throughout the State, and to three teachers' colleges.

One hundred and twenty student-teachers of physical education and college lecturers were instructed and given demonstrations in the use of the equipment and the recording of results. Considerable importance was attached to raising both arms, checking heels, knees, scapula, and so on.

Altogether, over thirteen thousand subjects in infant, primary, secondary, tertiary and adult groups were included, and eleven measurements were recorded of each subject in terms of sex, class, age, and locality. Females were measured by women teachers. All subjects removed shoes, coats and bulk from pockets.

The measurements were:

standing erect—height, bent elbows to floor, scapula to finger tips with arms outstretched;

sitting erect— sitting eye height, seat to scapula, seat to bent elbow(s), lower leg from bare heel to popliteal angle, length of upper thigh, buttock to popliteal, thickness of thigh, and sitting width of body.

Each of the 1,400 record sheets was carefully scrutinized and then collated in terms of age and locality, grade and locality.

2. Part One: Anthropometric Data

Figure 1 shows that five-year-old children in 1965 were 6·3 cm taller than in 1908 and that nearly all the increase in total stature between those years occurred before children began school. Between the ages of 5 and 17, the difference increased to 6·5 cm in the case of boys, but regressed slightly to 4·8 cm in the case of girls. It is possible that migrant children accounted for this regression.

Boys of 17 in 1908 were 10·3 cm taller than girls. In 1965, they were 12 cm taller than girls of the same age—an indication that the gap between the sexes is widening. This gap presents problems in educational seating, especially when small girls and tall boys are accommodated in the same room.

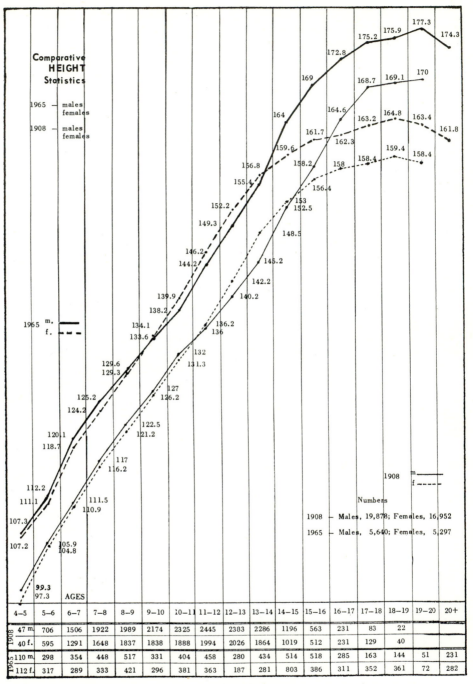

Figure 1.   Comparative heights 1908–1965 in cm.

Both surveys show that maximum stature for males and females was attained at the age of 19 and 18 respectively.   There was also a definite indication that the growth spurt in girls occurred earlier.   In 1908, girls exceeded the mean height of boys between the ages of 11 and 14.   In 1965, it occurred between the ages of 9 and 13, that is, in the late primary and early secondary stages.

The proportional relationship between the stature of girls and boys and their anticipated maximum stature at the age of 18 and 19 respectively, is

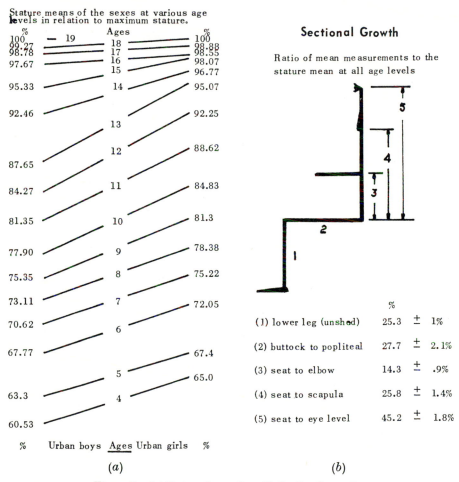

Rates of Growth

Stature means of the sexes at various age levels in relation to maximum stature.

Sectional Growth

Ratio of mean measurements to the stature mean at all age levels

	%
(1) lower leg (unshod)	25.3 ± 1%
(2) buttock to popliteal	27.7 ± 2.1%
(3) seat to elbow	14.3 ± .9%
(4) seat to scapula	25.8 ± 1.4%
(5) seat to eye level	45.2 ± 1.8%

%   Urban boys Ages Urban girls   %

(a)   (b)

Figure 2.   (a) Rates of growth ; (b) Sectional growth.

shown in Figure 2a.   Girls have reached 65 per cent of their total stature at the age of four, and 95 per cent at the age of thirteen.   After that, the average girl can expect to grow an additional 8 cm in height.

Boys, on the other hand, attain 60 per cent of their total stature at the age of four, and 95 per cent at the age of fifteen—two years later than girls.   After that, they too, on the average, can expect to grow a further 8 cm in height (Figure 1).

Figure 2b, which provides a reasonably accurate method of determining sectional measurements of the body when sitting in relation to maximum stature, was prepared after a close analysis of mean sectional measurements and mean stature of both sexes at all age levels.   No worthwhile divergence was noted at any stage.

2.1. *Mean-Dimensional Puppets*

These are based on five mean measurements and drawn to scale.

They are a useful starting point in designing educational chairs; but, as the title implies, they are not intended to represent an ' average ' individual, because ' average ' people do not exist.

In the 1965 survey, for example, out of 578 kindergarten children, one child had three measurements out of eleven equal to the means; and out of 775 college students, only one student had two measurements equal to the means.

Ten per cent of kindergarten children had a lower-leg measurement equal to the mean, and ten per cent had a seat-to-elbow measurement equal to the mean, but not one child had both means—the two measurements that are closely related to chair and table heights.

2.2. *Seat Heights*

Having regard to the importance of providing suitable educational chairs to all pupils in each class and not merely the central group, it is important to appreciate the range of differences which exist in the length of lower-leg at every age and every class level.

The difference between the shortest and longest lower-leg measurement at the age of five was equivalent to 40 per cent of the shortest measurement; at twelve it was 42 per cent; and at eighteen it was 44 per cent.   In kindergarten, sixth class, and through secondary grades, the differences were 44 per cent, 52 per cent and 52 per cent respectively.   In stature, however, the percentages were 20 per cent for kindergarten, 40 per cent for sixth class, and 50 per cent for secondary grades (Figure 3).

Physical differences persist through life.   Out of 1,020 tertiary students and adults in the 1965 survey

2% had an unshod lower-leg measurement between	33 cm and 36·5 cm;
27% ,, ,, ,, ,, ,, ,, ,,	37 cm and 40·5 cm;
53% ,, ,, ,, ,, ,, ,, ,,	41 cm and 45 cm;
18% ,, ,, ,, ,, ,, ,, ,,	over 45 cm.

2.3. *Table Heights*

As table height is frequently related to the position of the elbow when a person is sitting, Table 1 shows the actual number that had a particular combined measurement (shod-lower-leg and seat-to-elbow) out of 1,232 persons between the ages of 17 and 65.   The adults were teachers.

Some seat-to-elbow measurements varied considerably.   For example, one woman 160 cm tall measured 32 cm, which was 2 cm more than a man 197 cm tall.   Another woman 155 cm in height had a seat-to-elbow measurement of 15 cm, which was identical with that of a man 190·5 cm in height.

MEAN – DIMENSION PUPPETS

(based on the smallest and largest
height and width measurements of
all pupils in each group)

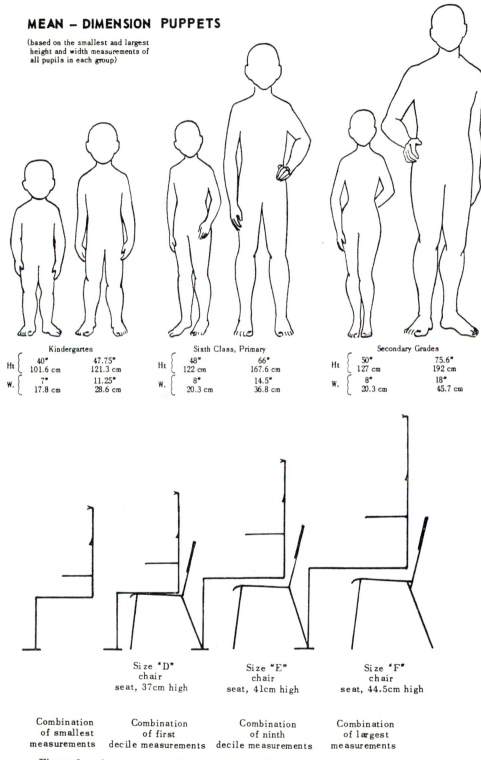

	Kindergarten	
Ht	40" 101.6 cm	47.75" 121.3 cm
W.	7" 17.8 cm	11.25" 28.6 cm

	Sixth Class, Primary	
Ht	48" 122 cm	66" 167.6 cm
W.	8" 20.3 cm	14.5" 36.8 cm

	Secondary Grades	
Ht	50" 127 cm	75.6" 192 cm
W.	8" 20.3 cm	18" 45.7 cm

Size "D"
chair
seat, 37cm high

Size "E"
chair
seat, 41cm high

Size "F"
chair
seat, 44.5cm high

Combination
of smallest
measurements

Combination
of first
decile measurements

Combination
of ninth
decile measurements

Combination
of largest
measurements

Figure 3.   above—comparative sizes in standing position:
below—diagrammatic representation of secondary pupils when sitting.

D

Table 1. The numbers out of 1,232 subjects who had the same combined measurement for shod lower leg and seat to elbow.

Combined measurement		Males	Females	Combined measurement		Males	Females
cm	in.			cm	in.		
56	22	–	1	71	28	145	85
58·5	23	2	15	73·5	29	122	31
61	24	2	58	76	30	47	2
63·5	25	5	148	78·5	31	23	1
66	26	49	223	81	32	4	–
68·5	27	79	189	86·5	34	1	–

As a result of the survey, the height of tables supplied to teachers in New South Wales was reduced from 30 in. (76 cm) to 28 in. (71 cm); and chairs from 18 in. (45·7 cm) to 17 in. (43·2 cm).

3.   Part Two: Educational Chairs

It is a comparatively simple objective process to obtain and collate anthropometric data of pupils.   It is far from simple, however, to apply these measurements scientifically so that they meet modern educational and postural needs, facilitate production and distribution, satisfy the economist, and, above all, ensure that chairs and tables fulfil their rightful role as *teaching aids* instead of being mere pieces of equipment.

Educational chairs and tables can only be regarded as satisfactory if they play their maximum role in facilitating learning.   It is not sufficient that they are located in a school.   In order to merit the term 'educational' they must encourage good sitting posture and aid the user in the acquisition of knowledge.

'Misfit' chairs and tables (desks) exist in every country for a number of reasons—inadequate sizes for all pupils, lack of knowledge, traditional thinking, poor design and construction, unsuitability (e.g. fixed desks) to meet the needs of modern education, replacement costs, etc.

It is not easy to overcome some of these problems, especially when it comes to replacing desks in existing schools.   Traditional thinking cannot be excused, however, if it means opposition to corrective measures and indifference to sitting posture for the sake of expediency.

3.1. *Chair and Table Design*

Educational chairs must be available in several heights.

Having decided the height rises (e.g. 38 mm), the other chair measurements must then be related to the various mean data applicable to that particular length of lower leg, and carefully checked.

The depth of the seat (buttock to popliteal) is just as important as the height, because deep chairs, combined with excessive seat slope, force a child to 'perch'. The 1965 survey revealed a need to decrease the seat depth in order to encourage pupils to use the backrest.   The 1968 questionnaires, however, showed that pupils were extremely critical of backrests.   These need to be compound curved in the interests of good posture.

Blind children require chairs that have higher backrests—and splayed back legs—in order to feel secure.

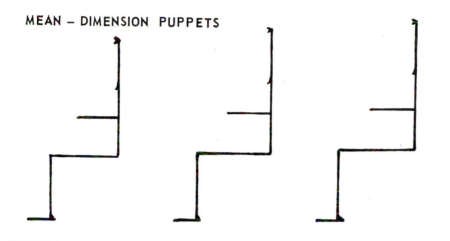

MEAN – DIMENSION PUPPETS

AGE – 4 to 5	5 to 6	6 to 7
seat to eye	seat to eye	seat to eye
m. 50.4 cm s.d. 2.6 cm f. 49.2 cm s.d. 3.8 cm	m. 52.2 cm s.d. 3.0 cm f. 51.3 cm s.d. 2.5 cm	m. 54.8 cm s.d. 3.0 cm f. 53.1 cm s.d. 2.8 cm
seat to scapula	seat to scapula	seat to scapula
m. 28.7 cm s.d. 2.2 cm f. 28.1 cm s.d. 2.1 cm	m. 29.5 cm s.d. 2.1 cm f. 28.8 cm s.d. 2.1 cm	m. 30.8 cm s.d. 2.5 cm f. 30.6 cm s.d. 2.5 cm
seat to elbow	seat to elbow	seat to elbow
m. 16.2 cm s.d. 1.5 cm f. 15.5 cm s.d. 1.5 cm	m. 16.2 cm s.d. 1.7 cm f. 16.1 cm s.d. 1.8 cm	m. 16.9 cm s.d. 2.0 cm f. 16.7 cm s.d. 1.8 cm
popliteal to buttock	popliteal to buttock	popliteal to buttock
m. 27.5 cm s.d. 2.4 cm f. 28.9 cm s.d. 2.3 cm	m. 29.8 cm s.d. 2.6 cm f. 31.3 cm s.d. 2.4 cm	m. 32.2 cm s.d. 2.2 cm f. 33.2 cm s.d. 2.2 cm
heel to popliteal	heel to popliteal	heel to popliteal
m. 26.5 cm s.d. 1.3 cm f. 26.2 cm s.d. 1.3 cm	m. 28.1 cm s.d. 1.9 cm f. 27.6 cm s.d. 1.5 cm	m. 30.0 cm s.d. 1.8 cm f. 29.2 cm s.d. 1.8 cm
Kindergarten	Kindergarten, 1st Class	K'n, 1st, 2nd

Figure 4.   Combination of mean anthropometric measurements.

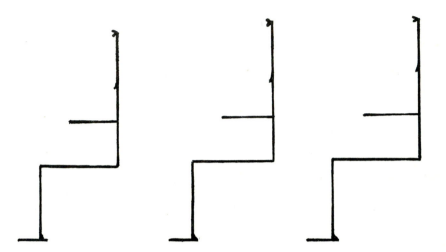

7 to 8	8 to 9	9 to 10
seat to eye	seat to eye	seat to eye
m. 55.9 cm s.d. 3.1 cm f. 55.5 cm s.d. 3.3 cm	m. 58.3 cm s.d. 2.9 cm f. 57.7 cm s.d. 2.7 cm	m. 59.8 cm s.d. 2.7 cm f. 59.1 cm s.d. 3.1 cm
seat to scapula	seat to scapula	seat to scapula
m. 32.0 cm s.d. 2.3 cm f. 32.0 cm s.d. 2.6 cm	m. 32.8 cm s.d. 2.6 cm f. 32.8 cm s.d. 2.3 cm	m. 33.7 cm s.d. 2.1 cm f. 34.2 cm s.d. 2.4 cm
seat to elbow	seat to elbow	seat to elbow
m. 17.9 cm s.d. 1.9 cm f. 17.2 cm s.d. 2.1 cm	m. 18.4 cm s.d. 1.9 cm f. 18.1 cm s.d. 1.9 cm	m. 18.8 cm s.d. 2.2 cm f. 18.4 cm s.d. 2.4 cm
popliteal to buttock	popliteal to buttock	popliteal to buttock
m. 34.2 cm s.d. 2.3 cm f. 34.7 cm s.d. 2.3 cm	m. 35.3 cm s.d. 2.3 cm f. 35.7 cm s.d. 2.4 cm	m. 35.9 cm s.d. 2.7 cm f. 37.4 cm s.d. 3.0 cm
heel to popliteal	heel to popliteal	heel to popliteal
m. 31.2 cm s.d. 2.3 cm f. 31.3 cm s.d. 2.4 cm	m. 32.6 cm s.d. 2.7 cm f. 32.7 cm s.d. 2.2 cm	m. 33.5 cm s.d. 2.4 cm f. 33.7 cm s.d. 2.3 cm
1st, 2nd, 3rd	1st, 2nd, 3rd, 4th	2nd, 3rd, 4th, 5th

Figure 4. (*continued*)

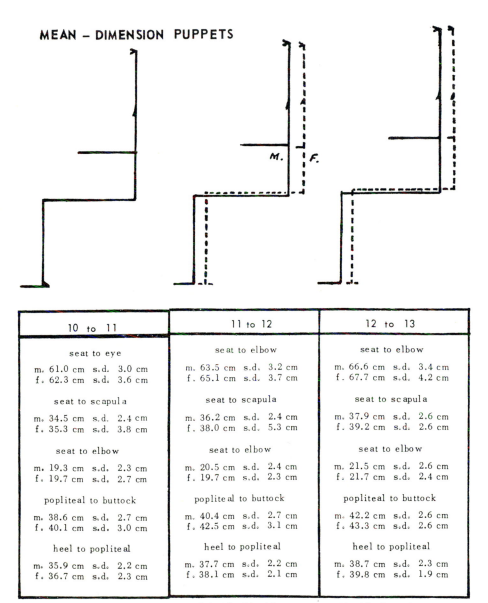

MEAN – DIMENSION PUPPETS

10 to 11	11 to 12	12 to 13
seat to eye	seat to elbow	seat to elbow
m. 61.0 cm s.d. 3.0 cm f. 62.3 cm s.d. 3.6 cm	m. 63.5 cm s.d. 3.2 cm f. 65.1 cm s.d. 3.7 cm	m. 66.6 cm s.d. 3.4 cm f. 67.7 cm s.d. 4.2 cm
seat to scapula	seat to scapula	seat to scapula
m. 34.5 cm s.d. 2.4 cm f. 35.3 cm s.d. 3.8 cm	m. 36.2 cm s.d. 2.4 cm f. 38.0 cm s.d. 5.3 cm	m. 37.9 cm s.d. 2.6 cm f. 39.2 cm s.d. 2.6 cm
seat to elbow	seat to elbow	seat to elbow
m. 19.3 cm s.d. 2.3 cm f. 19.7 cm s.d. 2.7 cm	m. 20.5 cm s.d. 2.4 cm f. 19.7 cm s.d. 2.3 cm	m. 21.5 cm s.d. 2.6 cm f. 21.7 cm s.d. 2.4 cm
popliteal to buttock	popliteal to buttock	popliteal to buttock
m. 38.6 cm s.d. 2.7 cm f. 40.1 cm s.d. 3.0 cm	m. 40.4 cm s.d. 2.7 cm f. 42.5 cm s.d. 3.1 cm	m. 42.2 cm s.d. 2.6 cm f. 43.3 cm s.d. 2.6 cm
heel to popliteal	heel to popliteal	heel to popliteal
m. 35.9 cm s.d. 2.2 cm f. 36.7 cm s.d. 2.3 cm	m. 37.7 cm s.d. 2.2 cm f. 38.1 cm s.d. 2.1 cm	m. 38.7 cm s.d. 2.3 cm f. 39.8 cm s.d. 1.9 cm
2nd, 3rd, 4th, 5th, 6th	3rd, 4th, 5th, 6th	5th, 6th, Secondary

Figure 4. (*continued*)

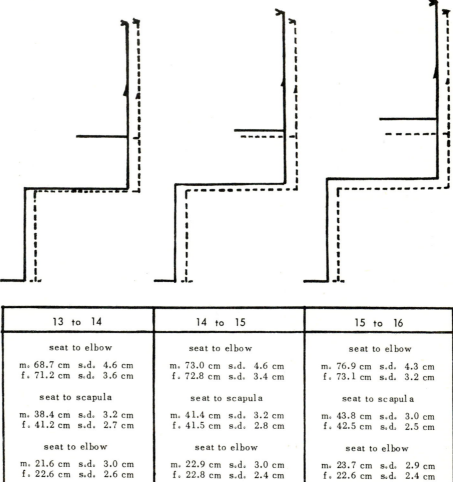

13 to 14	14 to 15	15 to 16
seat to elbow	seat to elbow	seat to elbow
m. 68.7 cm s.d. 4.6 cm	m. 73.0 cm s.d. 4.6 cm	m. 76.9 cm s.d. 4.3 cm
f. 71.2 cm s.d. 3.6 cm	f. 72.8 cm s.d. 3.4 cm	f. 73.1 cm s.d. 3.2 cm
seat to scapula	seat to scapula	seat to scapula
m. 38.4 cm s.d. 3.2 cm	m. 41.4 cm s.d. 3.2 cm	m. 43.8 cm s.d. 3.0 cm
f. 41.2 cm s.d. 2.7 cm	f. 41.5 cm s.d. 2.8 cm	f. 42.5 cm s.d. 2.5 cm
seat to elbow	seat to elbow	seat to elbow
m. 21.6 cm s.d. 3.0 cm	m. 22.9 cm s.d. 3.0 cm	m. 23.7 cm s.d. 2.9 cm
f. 22.6 cm s.d. 2.6 cm	f. 22.8 cm s.d. 2.4 cm	f. 22.6 cm s.d. 2.4 cm
popliteal to buttock	popliteal to buttock	popliteal to buttock
m. 43.7 cm s.d. 3.0 cm	m. 46.7 cm s.d. 3.0 cm	m. 47.9 cm s.d. 3.2 cm
f. 44.9 cm s.d. 2.6 cm	f. 46.4 cm s.d. 2.7 cm	f. 47.0 cm s.d. 2.7 cm
heel to popliteal	heel to popliteal	heel to popliteal
m. 40.6 cm s.d. 2.4 cm	m. 41.9 cm s.d. 2.4 cm	m. 43.7 cm s.d. 2.4 cm
f. 39.4 cm s.d. 1.8 cm	f. 39.5 cm s.d. 2.1 cm	f. 39.6 cm s.d. 2.2 cm
5th, 6th, Secondary	6th, Secondary	Secondary

Figure 4. (continued)

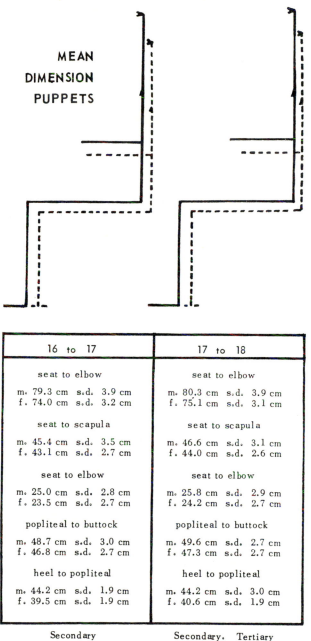

MEAN
DIMENSION
PUPPETS

16 to 17	17 to 18
seat to elbow	seat to elbow
m. 79.3 cm s.d. 3.9 cm f. 74.0 cm s.d. 3.2 cm	m. 80.3 cm s.d. 3.9 cm f. 75.1 cm s.d. 3.1 cm
seat to scapula	seat to scapula
m. 45.4 cm s.d. 3.5 cm f. 43.1 cm s.d. 2.7 cm	m. 46.6 cm s.d. 3.1 cm f. 44.0 cm s.d. 2.6 cm
seat to elbow	seat to elbow
m. 25.0 cm s.d. 2.8 cm f. 23.5 cm s.d. 2.7 cm	m. 25.8 cm s.d. 2.9 cm f. 24.2 cm s.d. 2.7 cm
popliteal to buttock	popliteal to buttock
m. 48.7 cm s.d. 3.0 cm f. 46.8 cm s.d. 2.7 cm	m. 49.6 cm s.d. 2.7 cm f. 47.3 cm s.d. 2.7 cm
heel to popliteal	heel to popliteal
m. 44.2 cm s.d. 1.9 cm f. 39.5 cm s.d. 1.9 cm	m. 44.2 cm s.d. 3.0 cm f. 40.6 cm s.d. 1.9 cm
Secondary	Secondary. Tertiary

Figure 4. (*continued*)

Seats should be curved slightly at the front, but this should not be overdone. Dished seats and saddle seats restrict freedom of movement. All chairs must be strong, durable, preferably stackable, and easy to maintain.

Tables should be related to the height-range of chairs and the top should be as large as possible. Four-legged tables should *not* be used in a classroom because ingress and egress are not possible without moving either the table or the chair. On the other hand, classroom tables must be strong, durable, and easy to maintain. Sloping tops improve posture.

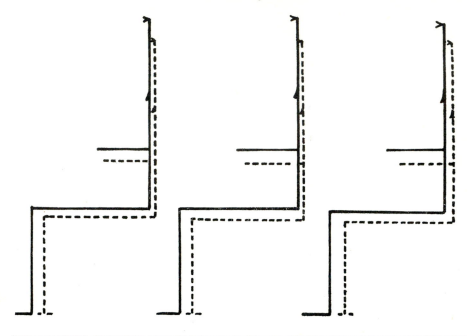

18 to 19	19 to 20	Over 20
seat to eye	seat to eye	seat to eye
m. 80.1 cm s.d. 3.6 cm	m. 80.8 cm s.d. 3.5 cm	m. 79.5 cm s.d. 3.7 cm
f. 74.7 cm s.d. 3.0 cm	f. 75.5 cm s.d. 3.2 cm	f. 75.6 cm s.d. 3.5 cm
seat to scapula	seat to scapula	seat to scapula
m. 46.1 cm s.d. 2.9 cm	m. 46.3 cm s.d. 2.7 cm	m. 45.8 cm s.d. 2.9 cm
f. 43.8 cm s.d. 2.7 cm	f. 44.5 cm s.d. 2.6 cm	f. 43.2 cm s.d. 3.0 cm
seat to elbow	seat to elbow	seat to elbow
m. 25.3 cm s.d. 3.4 cm	m. 24.9 cm s.d. 3.4 cm	m. 26.2 cm s.d. 2.7 cm
f. 23.6 cm s.d. 2.7 cm	f. 23.8 cm s.d. 2.7 cm	f. 23.8 cm s.d. 2.6 cm
popliteal to buttock	popliteal to buttock	popliteal to buttock
m. 50.3 cm s.d. 3.5 cm	m. 50.5 cm s.d. 3.5 cm	m. 49.1 cm s.d. 3.2 cm
f. 46.8 cm s.d. 3.3 cm	f. 46.3 cm s.d. 3.2 cm	f. 46.1 cm s.d. 3.1 cm
heel to popliteal	heel to popliteal	heel to popliteal
m. 44.3 cm s.d. 2.2 cm	m. 44.4 cm s.d. 2.3 cm	m. 43.2 cm s.d. 2.1 cm
f. 41.0 cm s.d. 2.2 cm	f. 40.1 cm s.d. 2.2 cm	f. 40.2 cm s.d. 1.9 cm
Secondary, Tertiary	Tertiary	Tertiary. Adults

Figure 4.　*(continued)*

Unnecessary noise must be avoided in a classroom because of the bad overall effect it has on the educative process. Metal tube should be filled with foam to combat resonance; and chair and table tips should be of rubber or soft polythene—never nylon.

Education authorities which insist on a deep book container above the child's thighs—often a mere repository of 'dead' material and rubbish—are guilty of hastening fatigue, establishing bad postural habits, and thus interfering with the child's ability to learn. Some private schools in N.S.W. insist on book bowls 8 in. (20 mm) deep.

Children wilt on very hot days; they also wilt if they are forced to sit in cramped positions for long periods.

Book containers should be as shallow as possible for posture reasons, or eliminated altogether. Lift-up tops are unsatisfactory.

Seventy per cent of secondary pupils in the 1968 survey stated that a book recess was not necessary. Most secondary teachers in N.S.W. are strongly opposed to a receptacle, especially in view of the frequent movement of classes. Secondary tables in N.S.W. were reduced in height in 1965 from $29\frac{1}{2}$ in. (75 cm) to $27\frac{1}{2}$ in. (70 cm) when the book recess was not included.

Table 2. Structural details of wood and metal tables and chairs manufactured and supplied to schools by the Department of Education, N.S.W., Australia (based on 1965 Anthopometric Survey)

CHAIRS	Size	"A"	"B"	"C"	"D"	"E"	"F"
Height of seat	inches	10	$11\frac{1}{2}$	13	$14\frac{1}{2}$	16	$17\frac{1}{2}$
	mm.	255	290	330	368	406	445
Width of seat	inches	12	12	12	14	14	14
	mm.	305	305	305	355	355	355
Depth of seat	inches	9	$10\frac{1}{2}$	12	$13\frac{1}{2}$	$14\frac{1}{2}$	$14\frac{1}{2}$
	mm.	228	267	305	343	368	368
Space between seat and back support	inches	$5\frac{1}{2}$	6	7	8	8	8
	mm.	140	152	178	203	203	203
Depth of back support	inches	$9\frac{1}{2}$	11	$12\frac{1}{2}$	14	15	$15\frac{1}{2}$
	mm.	240	280	318	355	380	393
Space from floor to front rung *	inches	$5\frac{3}{4}$	$6\frac{3}{4}$	$7\frac{3}{4}$	$9\frac{1}{2}$	$10\frac{1}{2}$	$11\frac{1}{2}$
	mm.	145	172	197	240	267	292
TABLES	Heights						
Infant — wood (top to bottom of book recess)	inches	18	$20\frac{1}{2}$	23	-	-	-
	mm.	457	520	585			
Primary — wood (top to bottom of book recess)	inches	-	$20\frac{1}{2}$	23	$25\frac{1}{4}$	$27\frac{3}{4}$	-
	mm.	-	520	585	640	705	
Secondary — metal (no book recess)	inches	-	-	-	-	$25\frac{3}{4}$	$27\frac{1}{2}$
	mm.	-	-	-	-	655	700

Size of table top: Infant 36" x 18" Primary 42" x 18" Secondary 24" x 18"
* Infant and primary chairs have rungs.
 All secondary school tables and chairs have metal frames. Only Science chairs have rungs.

Infant and primary sizes are identified by means of embossed stamping and coloured discs; secondary sizes have different frame colours.

3.2. *Need for Assorted Sizes*

A chair satisfies postural requirements when the pupil is able to sit against the backrest and the height of the seat is the same as the shod-lower-leg. It satisfies comfort and educational requirements *when the sitter can sit for long periods without becoming aware of the chair.*

When pupils sit on chairs that are too low, there is a marked tendency for them to slouch, to push the feet forward and become a nuisance to others, and, generally, to adopt bad postural habits.

When chairs are too high, pupils are forced to 'perch' on the front of the seat, to sacrifice the comfort of the backrest, to lean on the table, to fidget, and, if the table is too high, to sit with arms and elbows in a state of tension and the head too close to the work surface.

Pupils are never grouped according to size, so marked physical differences exist in stature, height of knee, and width of body in every classroom.

Figure 3 illustrates the smallest and largest pupils in kindergarten, sixth class, and secondary grades. Table 4 shows the percentage of lower-leg measurements in each grade in relation to chair sizes.

The full range of lower-leg measurements recorded for each grade in 1965 is shown in Table 3.

Table 3. Lower-leg measurement

Grade	Mean		Full range		10th–90th percentile range		Difference in full range	
	in.	mm	mm	mm	in.	mm	in.	mm
K'n	10·64	270	228	330	1·25	32	4·0	102
1.	11·37	290	247	343	1·5	38	3·75	96
2.	11·85	302	247	367	2·0	51	4·75	120
3.	13·00	330	273	393	1·75	45	4·75	120
4.	13·23	336	273	407	2·25	57	5·25	134
5.	14·25	362	292	412	1·75	45	4·75	120
6.	14·75	375	292	445	2·25	57	6·0	153
Secondary grades			337	508	3·5	89	6·75	171

Helpful to education

Easy ingress and egress
Seat equal to shod-lower leg
Slight slope in top
Simple metal construction
Encourages good sitting posture
Economical to purchase and maintain
Suction pads attached to table base
   to counteract noise and restrict move-
   ment of the table

Harmful to education

Four-legged-table restricts ingress
   and egress
Chair too high
Eyes too close
Book container results in poor posture
Disciplinary problems increased
Fatigue accelerated

Figure 5. Good and bad classroom furniture.

It will be seen from Table 3 that there was little difference between some of the class means in Col 2, e.g. between 3rd and 4th grades; and not a great difference, except in the secondary group, in Col 4. There were, however, considerable differences in Col 3 and 5. These differences between short and tall pupils necessitate chairs and tables of more than one size in every classroom, including secondary rooms. Otherwise, the educational and postural needs of these pupils are being ignored.

To-day, N.S.W. is the only state in Australia where more than one size is supplied to each secondary grade. This was begun when the 1949 survey showed that 6 per cent of students had a shod-lower-leg measurement less than 14 in. (355 mm), and 86 per cent measured less than the domestic-chair-height 18 in. (457 mm).

At the primary level, N.S.W. and South Australia supply three sizes to each class, Western Australia supplies two, in Tasmania it is left to the discretion of the headmaster, in Victoria and Queensland only one size is supplied. Victoria, however, has begun to experiment with the supply of three sizes to all grades, including secondary.

Experience in N.S.W. over the past eighteen years has shown that *there are no administrative problems* involved in providing appropriate assortments of chairs to every classroom. What is more, both teachers and pupils now appreciate the need for them. At first, many teachers did not take kindly to the elimination of traditional and regimental rows of fixed desks. The use of soft polythene suction pads on tables has helped to maintain a sense of orderliness and improve the aesthetic appearance of a room.

4. 1968 Survey. Chairs Preferred by Pupils

Early in 1968, 1,346 children in Infant classes were asked by teachers to sit on four chairs ' A ', ' B ', ' C ', ' D ' (against the backrest) and say which size they preferred to use every day. Teachers were asked not to influence pupils, but to note whether each child could rest both feet flat on the floor.

Seven hundred primary pupils in 21 classes in a number of schools, and 1,060 secondary students of all grades in co-educational High Schools were asked to answer separate questionnaires about seat-height preferences, sitting posture, chairs for homework, whether table tops should be horizontal or sloping, and whether he or she in secondary school thought it necessary to have a book recess beneath the top of the table.

Unfortunately, it is not possible in the space of this article to do more than summarize some of the results.

Infants. Normally, teachers instruct children which size chair to use. When they were free to make their own selection it became obvious at once that psychological factors dominated their preferences. For example, 20 per cent of kindergarten children said they preferred the largest size ' D ', which is never supplied, although 95 per cent dangled their feet.

Teachers noted that 31 per cent of kindergarten, 22 per cent of First Class, and 20 per cent of Second Class could not rest both feet flat on the floor. The percentages shown in Table 4 were obtained after these had been downgraded one size.

However, notwithstanding these corrections, the orderly manner in which high chairs were preferred from grade to grade made it obvious that infant children (possibly from home influences) like to feel pressure on the lower thigh when sitting. Second Class pupils showed a greater preference for ' D ' than Third Class pupils.

Table 4.   Relationship between seat-heights based on anthropometric (shod lower-leg) data (1965) and seat-heights preferred by pupils (1968)

Primary Grades

Class	Age Range (Yrs Mths)	Samples 1965 – 1968	"A" 10" to — 260 mm. to %	"B" 11½" to — 295 mm. to %	"C" 13" to — 330 mm. to %	"D" 14½" to — 368 mm. to %	"E" 16" to — 406 mm. to %	"F" 17½" plus 445 mm. to %
K'n	4 . 6 to 6 . 1	590	34	55	11	—		
		390	22	37	40	1		
1st	4 . 11 to 7 . 9	558	18	48	34	—		
		497	9	29	53	9		
2nd	5 . 10 to 10 . 1	702	9	37	42	12		
		459	1	7	62	30		
3rd	7 . 0 to 10 . 6	927	2	17	61	20		
		165		10	70	20		
4th	8 . 3 to 10 . 10	720	—	13	37	43	7	—
		161			65	30	5	
5th	9 . 3 to 12 . 6	857	—	—	21	48	31	
		203	—	—	15	55	30	
6th	10 . 2 to 13 . 11	895	—	—	12	40	41	7
		175	—	—	5	30	65	not used in tests

Secondary and Tertiary Grades

Form		Samples				D	E	F
1	12	604				26	47	23
		192			4		41 *	59
2		620	1965			21	49	30
		227	Averages for Secondary Students (shod heel to popliteal angle)				41 **	59
3		914	"D" 16%; "E" 40%; "F" 44%			16	41	43
		240					45 ***	55
4	to	709	1968			13	37	50
		264	"Preference" Averages ◈				28	72
5		657	"E" 40% "F" 60%			11	34	55
		60					37	63
6	17 +	666	Boys Girls			7	32	61
		78	33% 67% 48% 52%				37	63

Adults

			1966 Survey			D	E	F
Teachers	20 to 60	382	74% of men and 63% of women preferred the seat to be lower than the shod leg.			14	42	44

* 23%;   ** 16%;   *** 9%, stated that "E" chairs were too high.

◈ "E" 40% should read "D" 4%   "E" 36% if allowance is made for *, ** and ***.

The marked differences between the 'preferred' sizes and 'anthropometric' percentages in early secondary classes could result from psychological factors. By contrast, the two surveys are closely related in later maximum-stature groups.

4.1. *The Questionnaires*

Three per cent of primary pupils were unaware that different size chairs were in the classroom; and 30 per cent of those who were aware of this had never tried to find out which size was the most comfortable.

The following facts clearly emerged from the two questionnaires.

1. Pupils prefer to sit so that they are conscious of some degree of pressure from the front of the seat. Most Sixth Class boys, and most girls and boys in the early Secondary grades preferred chairs that were higher than the length of the shod-lower-leg, a few as much as 5 cm higher. Senior Secondary students, however, in common with teachers, showed a preference for seat heights that were approximately the same as the length of the shod-lower-leg, sometimes less in height.

2. Notwithstanding the above, every Primary and Secondary group showed an orderly progression from size to size in terms of the 1965 Survey, and a *need for an assortment* of three sizes in all Primary classrooms, and at least two sizes in secondary rooms.

Forty-seven per cent of Secondary girls (including 43 per cent in senior grades) and 30 per cent of Secondary boys (including 30 per cent in senior grades) expressed a definite preference for the second-size chair, size ' E ', 16 in. (406 mm) high.

3. The percentage answering: ' When sitting, do you usually sit at the front of the seat, or against the backrest?' was as follows:

	Prim.	Sec.		Prim.	Sec.
Generally, at the front	55	35	Against the back	45	65
When reading, at the front	48	35	Against the back	52	65
When singing, at the front	26	17	Against the back	74	83

Four per cent of Secondary students preferred to sit always at the front of the seat, and 22 per cent preferred to sit always against the backrest. There was no obvious relationship between scholastic achievement and either of these groups, but a high percentage of prefects and team captains were in the second group.

4. Fifty per cent of Primary pupils and 64 per cent of Secondary students expressed a preference for a table top that was sloping, not horizontal.

5. The percentage answering: ' When writing, how close is your chin to the paper ?' (a ball-point pen is 6 in. (15 cm)) was as follows:

	Primary	Secondary
15 cm or less	34	9
15 cm to 30 cm	60	84
30 cm and over	6	7

Observation showed that sustained concentration causes most pupils to lower the head.

6. Sixty-five per cent of Primary pupils and 70 per cent of Secondary students showed a preference for home tables and chairs to school-height tables and chairs when doing homework. As this conflicted with the desire for lower chairs at school, the writer was supplied with each pupil's reasons in writing. Many gave three or four[3] reasons. Education authorities should never underestimate the ability of pupils to evaluate comfort factors.

Fifty per cent of pupils referred to the advantage of soft seats and soft backrests as an aid to concentration;

Thirty-four per cent of pupils referred to the disadvantage of austere chairs;

Twenty-seven per cent of the reasons reflected a liking for large tables at home in order to spread books, etc.

Many referred to other factors—better concentration at home, relaxation, the existence of chair rungs for supporting the feet, more space for knees, less need to bend so far, and so on.

Only 8 per cent of the reasons given by primary pupils, and 11 per cent of those by secondary students referred specifically to the need for higher chairs and tables.

At one Primary school where pupils preferred reasonable chair heights at school, but home sizes for homework, it was found that 50 per cent had been provided with special tables by their parents.

There is probably a real educational need for pupils to use adjustable foot-stools when doing homework.

The writer believes that much more attention should be given to the comfort of school chairs, and to the elimination of restrictive features in tables and desks. ' Hard ' seats never help a child to pay attention to the teacher. Affluent authorities should supply students who sit for long periods with chairs that have firmly-upholstered seats and softly-upholstered backs.

5. Assortments Required

Next in importance to determining the structural details of each size of chair and table (Table 2), comes the variety of sizes to be supplied to each grade to meet the needs of pupils of different stature.

It is not wise to leave this matter to the discretion of headmasters, although they should be encouraged to attend to the needs of extremely short and tall pupils.

Assortments are strictly controlled in New South Wales. They are sent according to the grade stated by the headmaster, and checked at regular intervals by Furniture Maintenance Officers.

As most education authorities and manufacturers carry stocks of chairs and tables, it is a simple matter to supply assortments. No increased expenditure is involved. Chairs or desks in old schools should be interchanged in the interests of education.

Table 5.   Assorted sizes of chairs and tables required by each grade

Grade	A	B	C	D	E	F
Primary			Percentages			
Kindergarten	25	50	25			
First		50	40	10		
Second		30	50	20		
Third		15	60	25		
Fourth			50	40	10	
Fifth			20	50	30	
Sixth				40	50	10
Secondary						
First Form				10	50	40
Second	E 55				60	40
Third	F 45	E 50			50	50
Fourth		F 50			40	60
Fifth	E 40				40	60
Sixth	F 60				35	65

Table 5, based in the first instance on the schedule at present in use in N.S.W., corresponds very closely to the sizes of pupils in each grade, and to the percentage preference shown by pupils in the 1968 survey to sit on chairs equal in height to the shod-lower-leg (Table 4).

6. Conclusion

School chairs are *not* entitled to be ' educational ' in character unless they are suited to the size of pupils using them, encourage good sitting posture, and aid the process of learning.   Mere existence in a school is not enough.

Any chair which is badly proportioned, too large or too small for a particular child, and which interferes with a pupil's concentration, like a ruler which is chipped and buckled, is no aid to education.

It is truly unfortunate for school children everywhere that the importance of school chairs and tables as teaching aids is imperfectly understood by teachers, educators, administrators and manufacturers.   There is no place in modern education for ' misfit ' furniture.

When pupils of marked difference in stature are required to use chairs and desks of uniform height for long periods, it is impossible to provide each pupil with an equal opportunity of forming good postural habits, of concentrating, and of learning to the best of his or her ability.

It is *not* a matter of providing different sizes in terms of progression from year to year, but of providing different heights of suitable chairs and tables to the various sizes of pupils in *every* grade.

The writer gratefully acknowledges the permission of the Department of Education, New South Wales, to submit the questionnaires and to publish the results of the various surveys he directed when he was an officer of that Department.   The views expressed in this paper, especially following the 1968 questionnaires, do not necessarily reflect the views of the Department.

In 1949, five anthropometric measurements were taken of thirteen thousand primary and secondary pupils in public schools in New South Wales, Australia, for the purpose of designing school chairs.   The survey showed that desks of uniform height in classrooms were unsatisfactory for a large percentage of pupils because of their differences in stature.

Since then, every grade in N.S.W. has been supplied with more than one height of seating. Eight sizes of chairs were introduced in 1950 and distributed in various assortments in terms of grade.   In 1954, the number of sizes was reduced from eight to six in number, ranging in seat height from 10 in. (260 mm) to $17\frac{1}{2}$ in. (445 mm) at $1\frac{1}{2}$ in. (38 mm) rises, but the number of sizes issued to each class remained the same.

In 1965, a more extensive anthropometric survey was conducted to obtain eleven measurements of 12,000 pupils of all grades.   Four hundred teachers were included.   This paper is largely based on the data obtained from that survey.

It was found that nearly all the increase in maximum stature between 1908 and 1965 occurred before school age.   After the age of five, the rate of growth in 1965 was only slightly more than in 1908.   Boys and girls reach 95 per cent of maximum stature at the age of fifteen and thirteen respectively.

The most important fact that emerged from both surveys, so far as seating was concerned, apart from chair sizes, was that pupils differ in size to such an extent that a chair and table satisfactory for one pupil could be quite unsatisfactory for another in the same class.   In 1965, for example, there was a range of 10 cm, 15 cm, and 16·5 cm in the lower-leg measurement of kindergarten, sixth grade and secondary students.

Three thousand pupils of all ages in N.S.W. made it clear in 1968, that they did not wish to use chairs of equal height in any classroom.   They preferred to sit on chairs equal to the length of the shod-lower-leg, or about 5 mm to 8 mm in excess of it.   Their preferences could not be satisfied unless three sizes of chairs were supplied in suitable assortments (see previous section) throughout Infants and Primary schools, and at least two sizes were supplied to Secondary schools. Eleven per cent of First Form students, or 4 per cent of all secondary students in 1965, were not as tall as the biggest pupils in the First Class of the Infants school.

En 1949, cinq mensurations anthropométriques portant sur 13.000 élèves de l'enseignement primaire et secondaire ont été entreprises dans la Nouvelle Galles du Sud, en Australie, afin de réaliser des chaises pur écoliers. L'étude a montré que dans les salles de classe, les pupitres à hauteur uniforme ne convenaient pas à un pourcentage élevé d'écoliers, à cause des différences entre les tailles.

Depuis, dans chaque degré d'enseignement en N.G.S., on a mis en place des sièges de différentes hauteurs. Huit dimensions différentes de chaises ont été introduites en 1950 et distribuées en divers lots selon le degré de l'enseignement. En 1954, le nombre de dimensions différentes a été ramené à six, allant d'une hauteur du siège de 260 mm à une hauteur de 445 mm au pas de 38 mm, mais le nombre de dimensions mises en place dans chaque salle de classe était resté le même.

En 1965, il a été entrepris une étude plus approfonide comportant le relevé de II mensurations chez 12.000 écoliers de tous les degrés. Quatre cents instituteurs y ont été inclus. Dans cet article, on examine les données issues de cette étude.

Il s'avère que presque tout l'accroissement de la stature maximale observé entre 1908 et 1965 se produit avant l'âge scolaire. Apres l'âge de cinq ans, la vitesse de croissance en 1965 n'est que légèrement supérieure à celle de 1908. Les garcons et les filles atteignent 95 p. 100 de leur stature maximale, respectivement à l'âge de 15 et de 13 ans.

Le fait le plus important qui ressort de deux études, du moins en ce qui concerne la posture assise et indépendamment de la hauteur des sièges, est que les écoliers présentent des différences de tailles telles qu'un chaise et une table de hauteur donnée qui conveinnent de manière satisfaisante à un enfant, peuvent être, complètement inadaptées à un autre enfant de la même classe. En 1965 par exemple, il y avait des rangs de 10 cm, 15 cm, et 16,5 cm dans les mensurations des jambes chez les enfants du Jardin d'Enfants, chez ceux du sixième degré et chez les élèves du Secondaire.

Les chaises pour écloiers ne présentent aucun caractère " éducatif ", si elles ne sont pas adaptées à la dimension corporelle de l'enfant, si elles n'entraînent pas une bonne posture assise et ne facilitent pas l'apprentissage. Leur simple présence dans une école n'est pas suffisante.

Toute chaise qui est mal proportionnée, qui est trop grande, ou trop petite pour un enfant et qui perturbe les capacités de concentration de l'élève est, comme une règle entaillée et gauchie, d' aucune aide pour l'éducation.

Il est vraiment malheureux pour les enfants des écoles que l'importance des chaises et des tables d'école en tant qu'instruments d'education, soit sous-estimée par les maîtres, les éducateurs, les administrateurs et les constructurs. Il n'y a pas de place, dans un enseignement moderne, pour des fournitures " lassiées pour compte ".

Lorsque de élèves, présentant des différences staturales importantes, doivent se servir, pendant de longues périodes, de chaises et de pupitres de hauteur égale, il est impossible de faire acquérir, à chaque élève. de bonnes habitudes posturales, de les amener à se concentrer et de les instruire en fonction de leurs capacités.

Trois mille élèves de tout âge, en N.G.S., se sont prononces, en 1968, contre les chaises de hauteur iunforme dans les salles de classe. Ils préféraient être assis sur des chaises dont la hauteur est égale à la longueur de la jambe chaussée ou même 5 à 8 mm au-dessus. Leurs préférences ne pouvaient être satisfaites qu'après l'introduction de trois nouvelles dimensions de chaises en proportion convenable (voir chapitre précédent), dans les écoles maternelles et dans les écoles primaires et d'au moins deux dimensions nouvelles dans les écoles secondaires. Onze pour cent des élèves de première année du Secondaire ou quatre per cent de tous les élèves du Secondaire n'étainet, en 1965, pas aussi grands que les enfants les plus grands des écoles Enfantines.

Il ne s'agit pas d'introduire des dimensions croissantes de chaises à mesure que l'on franchit les années scolaires, mais de fournier à chaqu classe un assortiment de hauteurs de chaises et de tables.

1949 wurden an 13 000 Primär—und Sekundär—Schülern in New South Wales (Australien) in öffentlichen Schulen 5 anthropometrische Masse aufgenommen, um Schulsitze zu entwerfen. Diese Enquete zeigte, dáss Pulte einheitlicher Höhe in Schulklassen für einen grossen Prozentsatz von Schülern wegen verschiedener Grössen ungeeignet sind.

Seither wurde jede Schulklasse in New South Wales mit mehr als einer Sitzhöhe versorgt. 1950 wurden wurden 8 Stuhlgrössen eingeführt und in verschiedenen Zusammenstellungen auf die verschiedenen Klassen verteilt. 1954 wurde die Zahle der Grössen von 8 auf 6 reduziert, deren Sitzhöhen zwischen 260 mm und 445 mmm mit Stufen von 38 mm Unterschiede lagen. Die Zahl der Grössen, die jeder Klasse zugeteilt wurde, blieb dieselbe. 1965 erfolgte eine ausgedehntere anthropometrische Studie, um 11 Maße von 12 000 Schülern einschliesslich 400 Lehrern zu erhalten. Die Veröffentlichung stützt sich hauptsächlich auf diese Enquete.

Es wurde gefunden, dass nahezu der gante Zuwachs der Maximalgrösse zwischen 1908 und 1965 vor dem Schulalter erfolgte. Nach dem Alter von 5 Jahren war die Wachstumsgeschwindigkeit 1965 nur wenig grösser als 1908. Knaben und Mädchen erreichten 95% der Maximalgrösse im Alter von 15, bzw. 13 Jahren. Die wichtigste Tatsache, die sich hinsichtlich der Sitze aus diesen Enqueten ergab, war, von der Sitzhöhe angesehen, folgendes: Schüler unterscheiden sich in der Grösse so stark, dass eine Stuhl oder Tisch, der einen Schüler befriedigt, für einen anderen Schüler derselben Klasse ganz unbefriedigend sein kann. 1965 zum Beispiel bestand ein Bereich der Unterschenkelmaße von 10 cm, 15 cm und 16·5 cm im Kindergarten, in der 6. Klasse und bei fortgeschrittenen Studenten.

3000 Schüler aller Altersstufen machten es 1968 in New South Wales klar, dass sie in keinem Klassenraum Stühle gleicher Höhe benutzen wollten. Sie zogen es vor, auf Stühlen zu sitzen, die gleichhoch wie die Länge des beschuhten Unterbeins, oder 5 mm höher waren. Ihre Wünsche konnten nicht befriedigte werden, wenn nicht 3 Stuhlgrössen in passender Zusammenstellung (siehe oben) in allen Kinder- und Prärschulen, und wenigstens 2 Stuhlgrössen in Sekundärschulen vorgesehen wurden. 11% der Studenten in den ersten Semestern oder 4% aller Studenten in den späteren Semestern waren 1965 nicht so gross wie die grössten Schüler in der 1, Klasse der Kinderschule.

E

Anthropometrische und Physiologische Grundlagen zur Gestaltung von Büroarbeitssitzen

Von T. Peters

Institut für Arbeitsmedizin der Universität, Dusseldorf, Deutschland

1. Einleitung

In der Bundesrepulik Deutschland ist der gewerbeärztliche Dienst kraft Gesetzes als Schußgutachter in das Berufskrankheitenverfahren eingeschaltet. Diese Aufgabe verpflichtet ihn, nicht nur auf medizinisch diagnostischem Gebiet aktiv tätig zu sein, sondern vor allem auch die Arbeitsplatzbedingungen zu studieren, unter denen es zu den angezeigten Berufskrankheiten gekommen ist. Ohne die Kenntnis der Arbeitsplatzbedingungen bzw. der davon auf den arbeitenden Menschen ausgehenden Einflüsse kann die Frage nach dem Causalzusammenhang zwischen einer Erkrankung und Arbeitseinflüssen praktisch nicht beantwortet werden. Dies um so weniger, als die Berufskrankheitsbilder zunehmend uncharakteristischer werden, also aus dem medizinischen Befund oder der Symptomatologie allein kaum noch Rückschlüsse auf bestimmte Arbeitsbelastungen gezogen werden können. Außerdem finden wir gleichartige Belastungsmomente nicht nur bei der Arbeit, sondern auch im privaten Bereich, weshalb eine Wertung der einzelnen Komponenten zwingend notwendig ist. Da sich die privaten Einflüsse weitgehend unserer Kenntnis entziehen bzw. deren Wertigkeit für ein pathologisches Geschehen nur indirekt, z.B. durch Analogieschlüsse, beurteilt werden kann, ist das Studium der Arbeitseinflüsse von besonderer Wichtigkeit, um wenigstens für diesen Bereich sagen zu können: der Faktor Arbeitseinfluß ist so bedeutend, daß er als eigentliche oder wesentliche Ursache eines Krankheitsgeschehens angesehen werden muß oder daß er als Ursachenmoment in diesem Sinne nicht in Frage kommt.

Im Jahre 1952 wurden mit der 5. Berufskrankheitenverordnung erstmalig Sehnenscheiden- und Sehnenansatzerkrankungen unter bestimmten Voraussetzungen zu Berufskrankheiten erklärt. Dabei handelt es sich um ein Krankheitsbild, das bei Büroarbeit relativ häufig vorkommt bzw. als solche Erkrankung gemeldet wird. Die im Rahmen der bereits genannten Tätigkeit als gewerbeärztlicher Schlußgutachter gesammelten Erfahrungen haben gelehrt, daß in der Mehrzahl der Fälle ein derartiges Krankheitsbild—zumindest im engeren Sinne—überhaupt nicht besteht und daß entgegen der üblichen Auffassung über die Pathogenese derartiger Krankheitsbilder eine mechanische Ursache kaum oder überhaupt nicht als eigentliches oder wesentlich teilursächlicher Moment in Fage kommt.

Ich will mich im folgenden auf die Bekanntgabe der Erfahrungen und Untersuchungsergebnisse beschränken, die ich beim Studium der Arbeitsbedingungen in Büros sammeln konnte, unter denen es zu pathologischen Erscheinungen im Schulterarmbereich (einschließlich der Sehnenscheiden oder Sehnenansätze) kam. Die Kenntnis dieser Arbeitsbedingungen waren ja auch der Anlaß meiner Untersuchungen an 1 166 Büroangestellten in Hamburg, Nordrhein-Westfalen und München in den Jahren 1965 bis 1968.

2. Problemstellung

Wenn man in den Büros die Damen an ihren Arbeitsplätzen sitzen sieht, stellt der aufmerksame Beobacher fest, daß die de facto Situation weit von dem Soll, also dem arbeitsphysiologischen Optimum, entfernt ist. Die Ursachen hierfür sind vielfältig. Ein Ursachenmoment hebt sich aber deutlich von anderen ab: der Zwang der ' äußeren Verhältnisse '. Die Mehrzahl der Berufskrankheiten bei Büroarbeit sind nicht Folge einer mechanischen Überbeansprunchung, sondern Folge schlechter Arbeitshaltungen, für die wiederum die Arbeitsplatzsituation, insbesondere die Arbeitsmittel, von entscheidendem Einfluß sind. (Burkhard 1966, Dörling 1966; Edholm 1967; McFarland 1963; Gärtner 1966; Hartwell *et al.* 1964; Junghanns 1964; Kirn und Hahn 1966; Kroemer 1965; Mittelmeier 1967; Nottbohm 1967; Peters 1958, 1967, 1968; Schoberth 1962; Schöllner 1967; Schröter 1961; Stier 1966; Timm 1965; Weichardt 1966). Die Arbeitsplatzsituation wird von Bauherren, Architekten, Innenarchitekten, Organisatoren oder Einkäufern geschaffen, meist nach anderen als nach arbeitsphysiologischen Gesichtspunkten. Und auch viele Hersteller von Arbeitsmitteln sehen vordergründig meist nur die Elemente Konstruktion, Design und Herstellungskosten, während arbeitsphysiologische Aspekte gar nicht oder nur am Rande eine Rolle spielen. So werden Fakten geschaffen und am Schluß erst der Mensch an ' seinen ' Arbeitsplatz gesetzt. Und wenn er nicht zufällig in die Gegebenheiten paßt, wird er solange ' gequetscht ', bis die Situation für ihn halbwegs erträglich ist. (Peters 1964). Was dabei herauskommt, ist eben jene arbeitsphysiologisch unbefriedigende Situation, der man augenfällig überall begegnen kann.

Um zu einem arbeitsphysiologisch optimalen Arbeitsplatz zu kommen und dadurch den Zwang der ' äußeren Verhältnisse ' in eine Richtung zu lenken, die nicht zu Krankheitserscheinungen Anlaß gibt, gilt es, den umgekehrten Weg des bisher Üblichen zu gehen, nämlich nicht den Menschen an technische Gegebenheiten zu adaptieren, sondern letztere an die Gegebenheiten des Menschen. Voraussetzung solchen Vorgehens ist natürlich die genaue Kenntnis der biologischen Gegebenheiten des Menschen. (Kroemer 1964; Lehmann und Stier 1961; Schnelle 1955).

3. Untersuchungen

Bei den untersuchten 1 166 Büroangestellten handelte es sich außschließlich um weibliche Personen. Das untersuchte Kollektiv wurde nicht nach der speziellen Art der Büroarbeit (z.B. Arbeit an Schreibmaschinen, Rechenmaschinen, Lochkartenmaschinen, Vorzimmertätigkeit etc.) aufgeteilt. Die Messungen erfolgten im Stehen und Sitzen ohne Schuhe. Das Maß Scheitel-Fußsohle wurde mit fixiertem Meßband ermittelt, die übrigen Maße mittels an definierten Meßpunkten angelegtem Zollstock, Zeichenwinkel (rechtschenkliges Dreieck) und Reißschiene. Als Meßstuhl diente ein hydraulisch stufenlos einstellbarer Stuhl ohne Rollen der Fa. Drabert & Söhne, Minden/Westfalen mit planer, vorne abgerundeter Sitzfläche und dünner Polsterauflage, deren Eindrückung mittels eines Holzbrettchens, das bei jeder Messung zwischen Gesäß und Stuhlfläche lag, ausgeglichen wurde. Die Oberfläche dieses Brettchens war auch der fixe Meßpunkt für alle Parameter, die als Bezugspunkt die Sitzfläche haben. Da auch bei niedrigster Einstellung der Stuhlsitzfläche in der Mehrzahl der Fälle eine reproduzierbare eindeutige Meßhaltung nicht erreicht werden konnte, wurden bei jeder Messung die Füße auf eine Fußstütze Type FB 2 der Fa. Christoph Stoll, Waldshut/Baden gestellt, deren Stellfläche horizontal auf 10 cm Höhe über Fußboden eingestellt war. Diese Differenz der Beinaufstellfläche gegenüber dem Fußboden ist bei der nachstehenden Wiedergabe aller Meßwerte berücksichtigt, so daß sich alle Maßangaben auf den Füßboden beziehen.

Die Meßhaltung im Stehen (Abb. 1) war einmal eine bewußt aufgerichtete Haltung, vergleichbar mit dem 'Stillgestanden' bei Soldaten und eine entspannte, sog. natürliche Haltung. Bei beiden Messungen durften sich die Untersuchten nicht anlehnen, sondern nur die Wand, an der gemessen wurde, berühren.

Die Meßhaltung im Sitzen (Abb. 1a) war ebenfalls einmal aufgerichtet das andere mal natürlich. In beiden Haltungen wurde nicht angelehnt, die Lehne nur berührt. Die Unterarme wurden bei senkrecht herabhängenden Oberarmen und rechtwinklig gebeugten Ellenbogengelenken waagerecht gehalten. Die Handflächen standen bei gestreckten adduzierten Fingern senkrecht.

Die Oberschenkel verliefen bei rechtwinklig gebeugtem Kniegelenk waagerecht, die Unterschenkel standen senkrecht auf der Stellfläche bzw. dem Füßboden.

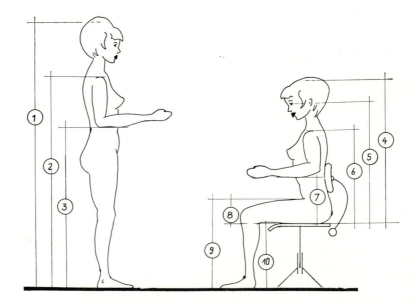

Abb. 1 und 1 a. Messhaltung im Stehen und Sitzen. Die Zahlen bezeichnen die in Tabelle A wiedergegebenen Parameter.

Die unter diesen Voraussetzungen an 1166 weiblichen Büroangestellten ermittelten Meßwerte sind in Tab. 1 wiedergegeben. Die einzelnen Parameter sind mit Zahlen gekennzeichnet. Meßpunkt für die Schulterhöhe war das Acromion scapulae im Bereich des Acromio—clavicular—Gelenkes. Meßpunkt für die Augen war eine durch die Nasenwurzel gedachte Linie.

Die Ellenbogenhöhe über Fußsohle und Sitzfläche wurde in Unterarmmitte über der facies medialis der Ulna gemessen. Meßpunkt für den Parameter Oberschenkel—Sitzfläche war die Oberschenkelstreckseite distal der Leistenbeuge, für den Parameter Oberschenkel—Fußsohle die Oberschenkelstreckseite proximal (4 Querfinger) des Kniegelenkes. Der Abstand Kniekehle—Fußsohle wurde von den beiden tendines des musculus bicipitis femoris auf der Kniegelenkbeugeseite gegen den Füßboden gemessen.

4. Ergebnis

Tabelle 1. Aufstellung der an 1166 weiblichen Büroangestellten ermittelten Meßwerte im Stehen (Parameter 1–3) und Sitzen (Parameter 4–10)*

	Parameter			Arithmetischer Mittelwert (mm)	Standardabweichung (mm)
1	Scheitel	Fußsohle	aufrecht	1634	69
	Scheitel	Fußsohle	natürlich	1632	46
2	Acromion	Fußsohle	aufrecht rechts	1321	50
	Acromion	Fußsohle	aufrecht links	1324	39
	Acromion	Fußsohle	natürlich rechts	1313	51
	Acromion	Fußsohle	natürlich links	1316	45
3	Ellenbogen	Fußsohle	aufrecht rechts	1001	46
	Ellenbogen	Fußsohle	aufrecht links	999	51
	Ellenbogen	Fußsohle	natürlich rechts	994	34
	Ellenbogen	Fußsohle	natürlich links	992	42
4	Scheitel	Sitzfläche	aufrecht	837	26
	Scheitel	Sitzfläche	natürlich	829	22
5	Augen	Sitzfläche	aufrecht	742	37
	Augen	Sitzfläche	natürlich	731	33
6	Acromion	Sitzfläche	aufrecht rechts	527	27
	Acromion	Sitzfläche	aufrecht links	529	14
	Acromion	Sitzfläche	natürlich rechts	517	23
	Acromion	Sitzfläche	natürlich links	518	28
7	Ellenbogen	Sitzfläche	aufrecht rechts	203	23
	Ellenbogen	Sitzfläche	aufrecht links	206	22
	Ellenbogen	Sitzfläche	natürlich rechts	193	25
	Ellenbogen	Sitzfläche	natürlich links	196	24
8	Oberschenkel	Sitzfläche	rechts	141	14
	Oberschenkel	Sitzfläche	links	140	13
9	Oberschenkel	Fußsohle	rechts	507	34
	Oberschenkel	Fußsohle	links	513	31
10	Kniekehle	Fußsohle	rechts	393	23
	Kniekehle	Fußsohle	links	396	23

5. Folgerungen

Bezüglich der Gestaltung von Büroarbeitssitzen ergeben sich aus den anthropometrischen Untersuchungen folgende Konsequenzen:

(1) Die Sitzfläche soll einen Höhenverstellbereich haben, der den Abstand Kniekehle—Fußsohle (Parameter 10) von ≈ 40 cm mit der Standardabweichung von 2,3 cm berücksichtigt. Da noch eine Absatzhöhe für Schuhe einkalkuliert werden muß, die mit 4 cm im Mittel anzusetzen ist, sollte der *Verstellbereich für die Sitzflächenhöhe von 40 cm bis 51 cm gehen.* Die 40 cm rechtfertigen sich aus der Erfahrung, daß kleine Damen Schuhe mit höherem Absatz bevorzugen. Sofern die Konstruktion des Stuhles (z.B. der

---

* Für die freundliche Beratung und Unterstützung bei der Vorbereitung der Messungen und deren Auswertung sei der Programmiergruppe des Rechenzentrums beim Versorgungsamt Düsseldorf besonders gedankt.

Verstellmechanismus in der Stuhlsäule) einen derartigen Verstellbereich nicht zuläßt, kann der obere Wert herabgesetzt werden, da sehr große Damen erfahrungsgemäß niedrigere Absatzhöhen bevorzugen. Dabei sollte aber eine obere Sitzflächenhöhe von 48 cm in jedem Falle erreichbar sein.

(2) Die Sitzfläche soll horizontal verlaufen, von vorne gesehen Leicht Konkav geformt sein, eine stark abgebogene Vorder Kante haben und im hinteren Sitzbereich Leicht abwärts geneigt verlaufen (Abb 2).

Derartige Sitzflächen sind bei richtiger Höheneinstellung (vgl. Ziffer 1) nicht nur die bequemsten, sondern auch die unproblematischsten. Sie verhindern Druck auf die Oberschenkelbeugeseite und erlauben die größte Beweglichkeit beim Sitzen, d. h. ein arbeitsphysiologisch günstiges wechselndes Sitzverhalten.

(3) Die Rückenlehne soll höhenverstellbar und mit einem Pendelgelenk am Lehnenträger befestigt sein. Der Lehnenträger soll um eine Querachse direkt unter der Sitzfläche drehbar (wie auf Abb 1a und 2 dargestellt) bzw. verschiebbar sein (wie auf Abb 3). Bei gefederten Lehnenträgern (Blattfedern, Haarnadelfedern) darf die Federung nicht zu weich sein.

(4) Stühle mit Rollen sind unproblematisch, wenn durch richtige Einstellung der Sitzflächenhöhe ganzflächig aufgesessen wird bzw. werden kann. Bei weichen Fußböden sollen harte Rollen (Stahl, Kunststoff), bei harten Boden weiche Rollen (Gummi) verwendet werden.

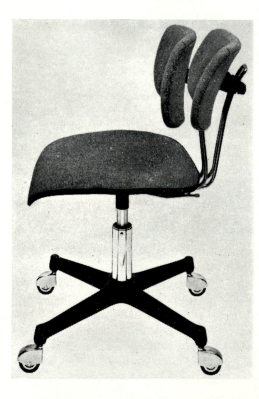

Abb. 2. Stufenlos höhenverstellbarer Stuhl der Fa. Drabert Söhne, Minden/Westfalen mit nach vorne abgerundeter Sitzfläche und leichter Abwärtsneigung des hinteren Sitzflächenteiles. Lehnenträger stufenlos adjustierbar.

Wenn auch nach dem gestellten Thema nur auf die Gestaltung von Büroarbeitssitzen eingegangen werden soll, so sei abschließend doch auch noch etwas über die Arbeitstischhöhen gesagt, weil eine isolierte Betrachtung von Stuhl oder Tisch kaum möglich ist, bzw. das arbeitsphysiologische Optimum ohne Betrachtung des Arbeitsplatzes als Ganzes—als ' unit '—kaum erreicht werden kann.

Nach den Messungen beträgt der Abstand Ellenbogen—Sitzfläche (Parameter 7) $\approx$ 20 cm mit der Standardabweichung von $\approx$ 2,4 cm. Die Addition der Meßwerte für den Abstand Kniekehle—Fußsohle (Parameter 10) und den Abstand Ellenbogen-Sitzfläche (Parameter 7) ergäbe einen arithmetischen Mittelwert (AM-Wert) für die Höhe der Arbeitsfläche über Fußboden von ca 60 cm, unter Berücksichtigung der Standardabweichung für sehr große Damen (AM − Wert + 3 mal Standardabweichung) einen Wert von ca 73 cm. Zieht man die durchschnittliche Tastaturhöhe einer Schreibmaschine von 11 cm von letztgenanntem Wert ab, so ergäbe sich eine arbeitsphysiologisch richtige Schreibmaschinentischhöhe von 62 cm, für Arbeiten am Schreibtisch ohne Maschinen von 73 cm. Da aber wegen der Schuhe und evtl. benützten Diktiergeräteschaltern noch ein Zuschlag erforderlich ist, ist eine Schreibmaschinentischhöhe von 65 cm und eine Schreibtischhöhe von 75 cm als optimal anzusehen. (Burkhard *op. cit.*; Grandjean und Burandt 1966; Grandjean 1967; Kaminsky und Pilz 1963; Kellermann *et al.* 1964; Kroemer 1963(a) und (b), Lehmann 1961).

Da diese Arbeitstischhöhen den Erfordernissen der größten Büroangestellten entsprechen (AM − Wert + 3 mal Standardabweichung), muß man für die

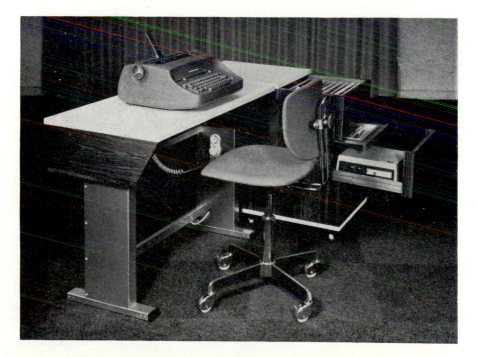

Abb. 3. Schreibmaschinentisch mit Beistellschrank und Büroarbeitsstuhl der Fa. Rosendahl und Schmid, Wuppertal/Barmen. Der Maschinentisch ist mittels eines kleinen Hebels unter der Tischplatte stufenlos höhenverstellbar von 55 cm bis 79 cm.

Mehrzahl der Büroangestellten auch bei 65 cm hohen Schreibmaschinentischen und 75 cm hohen Schreibtischen noch einen Ausgleich durch Fußstützen schaffen, um ein optimales Sitzverhalten zu erreichen.

Der Zwang, Fußstützen einsetzen zu müssen, zeigt, daß alle starren Lösungen nur für einen Teil der Büroangestellten halbwegs befriedigende Arbeitsplatzverhältnisse schaffen können. Wie die anthropometrischen Messungen zeigen, erstreckt sich die individuelle Streuung auf einen relativ breiten Bereich, weshalb flexible Lösungen aus arbeitsmedizinischer Sicht besser sind. Während verstellbare Büroarbeitssitze heute schon fast eine Selbstverständlichkeit sind, ist dies für Arbeitstische leider noch nicht der Fall. Die von arbeitsmedizinischer Seite immer wieder erhobene Forderung nach höhenverstellbaren Tischen fand bisher wenig Gegenliebe.

Inzwischen konnten aber von 2 Firmen in der Bundesrepublik stufenlos höhenverstellbare Tische entwickelt und produziert werden, die eine vollständige Anpassung der Arbeitstischhöhen an die individuellen Gegebenheiten erlauben. Sie sind auf Abb. 3 v. 4 gezeigt.

Die Erfahrungen mit diesen höhenverstellbaren Schreibmaschinentischn, die in unserem Institut seit etwa 1 Jahr praktisch eingesetzt sind und nach allen Richtungen erprobt wurden, sind so positiv, daß ich mir abschließend die Feststellung erlauben darf: Ohne höhenverstellbare Schreibmaschinentische kann die Arbeitshaltung nur in den wenigsten Fällen optimiert werden.

Abb. 4. Stufenlos höhenverstellbarer Schreibmaschinentisch der Fa. Gutmann/Unterkirnach. Verstellbereich von 60 cm bis 78 cm. Die Schreibmaschine ist in die Tischplatte eingelassen, um den Arbeitsunterbau so niedrig wie möglich zu halten.

Arbeitsphysiologisch richtig gestaltete Büroarbeitssitze können nur dann zu voller Geltung im weitesten Sinne des Wortes kommen, wenn auch bei der Gestaltung von Arbeitstischen anthropometrische und arbeitsphysiologische Grundlagen Beachtung finden.

Es wird über die Ergebnisse anthropometrischer Untersuchungen an 1166 weiblichen Büroangestellten in Hamburg, Nordrhein-Westfalen und München berichtet. Die Messungen wurden durchgeführt, um die derzeitigen biometrischen Gegebenheiten bei der Gestaltung von Büroarbeitssitzen berücksichtigen zu können. Die ganzheitliche Betrachtung des Büroarbeitsplatzes zwingt dazu, auch die Arbeitstischhöhen den biometrischen Gegebenheiten anzupassen.

The results of anthropometric measurements on 1166 female office employees in Hamburg, North Rhine-Westphalia and Munich are reported. The measurements were taken, to bring the knowledge of the biometric situation up to date and to make it possible for these biometric data to be incorporated in the construction and design of office chairs.

The aspect of an office working place as a ' unit ' requires the height of working table plates to be adjusted to the biometric situation too.

On rend compte, dans cet article, des résultats relatifs aux mensurations anthropométriques effectuées sur 1166 employées de bureau, à Hambourg, en Rhénanie de Nord-Westphalie et à Munich.

Ces mensurations ont été effectuées afin d'accroître nos connaissances du matériel biométrique et pour pouvoir appliquer ces données à la conception et la construction des chaises de bureau.

Le poste de travail de bureau doit être considéré comme formant un tout; il importe alors également d'ajuster la hauteur de la table de travail à la réalité biométrique.

Literatur

Burkhard, F., 1966, Physiologische und psychologische Gesichtspunkte bei der Gestaltung von Büroarbeitsplätzen. *Arbeitswissenschaft*, **3**, 67–71.

Dörling, E., 1966, Kleine Ursachen-Grosse Wirkung. *DAG Zeitschrift ' Der Angestellte '*, **9**, 29–31.

Edholm, O. G., 1967, *Probleme der Arbeitswissenschaft* (München : Kindler G.M.B.H.).

Gärtner, H., 1966, Arbeitshygiene im Büro. *Arbeitsmed. Sozialmed. Arbeitshyg.*, **10**, 341–346.

Grandjean, E., 1967, *Physiologische Arbeitsgestaltung. Leitfaden der Ergonomie.* 2. Auflage (Thun und München : Ott Verlag).

Grandjean, E., und Burandt, U., 1966, Gesundes Sitzen im Büro. *Bürotechnik und Organisation*, **5**, 478–482.

Hartwell, S. W., Lassen, R. D., und Posch, J. L., 1964, Tenosynovitis on woman in industry. *Cleveland Clinic Quart.*, **31**, 115–118.

Junghanns, H., 1964, Wirbelsäule und Arbeit. *Arbeitsmedizin*, **3**, 60–62.

Kaminsky, G., und Pilz, E., 1963, *Gestaltung von Arbeitsplatz und Arbeitsmittel. RKW Reihe Arbeitsphysiologie-Arbeitspsychologie.* (Berlin : Beuth-Vertrieb G.M.B.H.).

Kellermann, F. Th., van Wely, P. A., und Willems, P. J., 1964, Mensch und Arbeit in der Industrie. Philips Technische Bibliothek. *N.V. Philips' Gloeilampenfabrieken Eindhoven (Niederlande)*.

Kirn, A., und Hahn, J., 1966, Haltuug aud Fehlhaltung. *Werkstattechnik*, Heft 9 und 10.

Kroemer, K. H. E., 1963 a, Welche Forderungen werden an Arbeitsstühle gestellt. *Arbeitswissenschaft*, **6**, 197–199.

Kroemer, K. H. E., 1963 b, Über die Höhe von Schreibtischen. *Arbeitswissenschaft*, **4**, 132–140.

Kroemer, K. H. E., 1964, Heute zutreffende Körpermaße. *Arbeitswissenschaft*, **2**, 42–45.

Kroemer, K. H. E., 1965, Zur Verbesserung der Schreibmaschinentastatur. *Arbeitswissenschaft*, **1**, 11–16.

Lehmann, G., und Stier, F., 1961, Mensch und Gerät. In *Handbuch der gesamten Arbeitsmedizin* (von Baader), **1**, 718–788 (Berlin : Urban & Schwarzenberg).

Lehmann, G., 1961, Die Körperstellung bei Ruhe und Arbeit. In *Handbuch der gesamten Arbeitsmedizin* (von Baader), **1**, 825–856 (Berlin : Urban & Schwarzenberg).

McFarland, R. A., 1963, Ergonomics—the study of man at work ; with special reference to the psychological factors in the practice of industrial hygiene. *Am. ind. Hyg. Assn. J.*, **24**, 209–221.

MITTELMEIER, H., 1967, Forschungsauftrag zur Frage von Arbeitsschäden am Haltungs- und Bewegungsapparat durch Büromaschinenarbeit. *Auszugsweise veröffentlicht in Zeitschrift ' Die Angestelltenversicherung '*, **8**, 219–233.

NOTTBOHM, L., 1967, Arbeitsplatzgestaltung als Mittel zur Gesunderhaltung. *Arbeitswissenschaft*, **6**, 69–72.

PETERS, T., 1958, Ist die Schreibmaschine die Ursache von Armerkrankungen ? *Bürotechnik und Organisation*, **1**, 16–19.

PETERS, T., 1964, Aktuelle Fragen des Arbeitsschutzes von Angestellten. *Verhandlungen der Deutschen Gesellschaft für Arbeitsschutz*, **8**, 42–45.

PETERS, T., 1967, Arbeitsmedizinische Erkenntnisse und Erfahrungen beim Studium sogenannter Berufskrankheiten im Büro. Das Büro in arbeitsmedizinischer Sicht. *IBM Org. Brief Nr. 2*

PETERS, T., 1968, Wodurch werden Angestellte krank ? *Bürotechnik und Organisation*, **2**, 132–136.

SCHNELLE, H. H., 1955, *Längen- Umfangs- und Bewegungsmaße des menschlichen Körpers* (Leipzig : J. A. BARTH).

SCHOBERTH, H., 1962, *Sitzhaltung, Sitzschaden, Sitzmöbel* (Berlin : SPRINGER).

SCHÖLLNER, D., 1967, Aufbrauchschäden an Wirbelsäule und Gelenken. *Arbeit und Leistung*, **4/5**, 69–73.

SCHRÖTER, G., 1961, Aufbrauch und Abnutzungskrankheiten. In *Handbuch der gesamten Arbeitsmedizin* (von BAADER), B II/2 (Berlin : URBAN & SCHWARZENBERG).

STIER, F., 1966, Optimale Gestaltung von Arbeitsmitteln. *Arbeitsmed. Sozialmed. Arbeitshyg.*, **9**, 303–306.

TIMM, H., 1965, Die Bedeutung der Abduktion für die Entstehung der Epicondylitis. *Zbl. f. Arbeitsmed. u. Arbeitsschutz*, **Z**, 30–33.

WEICHARDT, H., 1966, Die Anpassung des Arbeitsplatzes an den Menschen aus arbeitshygienischer Sicht. *Arbeitsmed. Sozialmed. Arbeitshyg*, **11**, 393–397.

Methods and Results of Seating Research

By J. C. JONES

Design Research Laboratory, University of Manchester Institute of Science & Technology, England

1. Introduction

The objectives of the research reviewed here are twofold: firstly, to generate data to be used in the designing of car seats, industrial seats, lecture theatre seats and the like, and, secondly, to develop research methods that are quick enough and realistic enough to be used by engineers and designers. Some of this work has appeared only in internal reports and conference presentations; other parts of it have appeared in design journals and in an M.Sc. dissertation (Dutch 1965).

2. Fitting Trials

2.1. *Introduction*

The method of fitting trials is a development of the ' method of experimental trials ' originally proposed by Morant (1954) as a means of determining the dimension of pilots' cockpits. In Morant's method all dimensions of the workspace are adjusted together to fit each member of a sample of the user population. In the method of fitting trials each dimension is adjusted independently while all the other dimensions remain at an initial setting. The purpose of this change of procedure is to provide fixed reference points from which the size of the designer's room for manoeuvre can be plotted. Morant's method generates a set of workspace dimensions to suit each subject: these sets of dimensions are difficult to combine as each incorporates unrecorded logical dependancies between dimensions. The output from the method of fitting trials is a set of reach and comfort limits from which the design constraints can easily be inferred and compared.

The objectives of both Morant's method, and this modification of it, are, firstly, to determine acceptable standards of reach and comfort for seat and workspace and, secondly, to apply these standards to all but the exceptionally large or small members of the user population.

2.2. *Method*

The procedure, but not the results, is fully described in a previous paper (Jones 1963). A brief description is as follows.

The user population is identified in terms of the distributions of both height and weight (Table 1). Subjects representing extreme and average body sizes are selected. Ideally there should be both naive subjects and experienced users for each of the selected height/weight combinations.

The designers of the seat or workspace have first to identify what they believe to be the critical dimensions and then to design a simulator in which each dimension can be adjusted independently over a wide range. In the

Table 1. The extreme height/weight combinations of a user population abstracted from Kemsley (1952). The figures in the last two columns refer to the percentiles by weight of people of the heights given in the first column

British population 18–40 yrs	Stature (in.)	Weight (lbs)	
		1st percentile	99th percentile
Males	Short	Short-Light	Short-Heavy
1st percentile	61	86	167
	Tall	Tall-Light	Tall-Heavy
99th percentile	73	126	230
Females	Short	Short-Light	Short-Heavy
1st percentile	58	78	185
	Tall	Tall-Light	Tall-Heavy
99th percentile	70	111	223

simulator shown in Figures 1 and 2 the angles and positions of seat squab, backrest, steering wheel and footrest can be adjusted independently of each other. The initial settings of the apparatus are found by adjusting each dimension until a user of average size can carry out each of the task actions with no more difficulty or discomfort than the designers believe to be acceptable for the class of equipment that is being designed. It might, for instance, be decided to eliminate even slight discomforts from a long-range passenger aircraft seat but to tolerate all but extreme discomforts in the dimensioning of seats for short-distance public transport. It is essential that each critical dimension of the simulator can be adjusted quickly (i.e. within a few seconds) so that the subject can experience a new setting before his memory of the last one has decayed. Each subject is then asked to carry out a critical task action (such as turning the steering wheel through 180°) for each of a range of settings of each adjustable dimension. The settings are adjusted in equal steps of about one inch, or of a few degrees, over a range that extends into zones that are definitely out of reach, or are intolerably uncomfortable to the subject. At each setting the subject gives yes/no answers to the questions listed in Table 2 while the range of adjustment is traversed in both the ascending and descending directions.

Table 2. Questions and answers for fitting trials

Adjusted dimension (in.)	Ascending adjustment		Descending adjustment	
	Questions	Responses	Questions	Responses
15	Is this tolerable?	No	Is this tolerable?	No
14	,, ,, ,,	No	,, ,, ,,	No
13	,, ,, ,,	No	,, ,, ,,	Yes (max)
12	,, ,, ,,	No	Is this better?	Yes
11	,, ,, ,,	Yes (max)	,, ,, ,,	Yes
10	,, ,, ,,	Yes	,, ,, ,,	Yes
9	,, ,, ,,	Yes	,, ,, ,,	Yes (opt)
8	Is this better?	No	,, ,, ,,	No
7	,, ,, ,,	Yes (opt)	Is this tolerable?	Yes
6	,, ,, ,,	Yes	,, ,, ,,	Yes
5	,, ,, ,,	Yes	,, ,, ,,	Yes (min)
4	Is this tolerable?	Yes (min)	,, ,, ,,	No
3	,, ,, ,,	No	,, ,, ,,	No
2	,, ,, ,,	No	,, ,, ,,	No
1	,, ,, ,,	No	,, ,, ,,	No

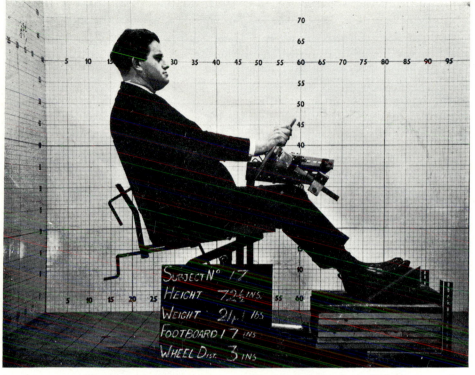

Figure 1.   A tall heavy man seated on a simulator of a car interior.

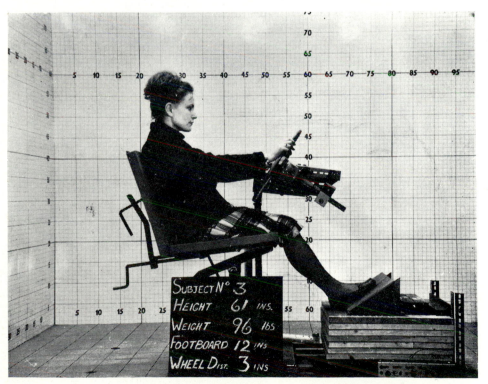

Figure 2.   A short light woman seated on a simulator of a car interior.

The dimensions marked 'max', 'opt' and 'min' in Table 2 are the upper limits, the most comfortable positions and the lower limits. As in the psychophysical *method of limits* there is a tendency to get lower readings when ascending and higher readings when descending.

The next step is to select final dimensions that fit all or most of the subjects tested. Usually a single dimension can be found that falls between the distributions of maxima and minima for nearly all subjects, as in Table 3. Occasionally these distributions overlap by a large amount and a range of adjustment has to be retained in the final design, as in Tables 4 and 5. If there is a gap between distributions of maxima and minima the distribution of optima can be taken into account.

The last step is to set the simulator to the final dimensions. The subjects are then asked to sit in the simulator and to re-enact the critical actions of reaching and seeing that they performed during the fitting trials. If any difficulties remain they are dealt with by making further small adjustments, keeping as far as possible within the distributions of maxima and minima for each dimension.

The purpose of the second fitting is to separate user sensitivity to relationships between dimensions from user sensitivity to adjustments made one-at-a-time. In Morant's method these two stages are combined with the result that the designer's room for manoeuvre is very much more difficult to discern as data such as that in Tables 3, 4 and 5 cannot be generated.

Table 3. Distribution of fitting trial dimensions for car seat height. Final dimension 12 inches

Seat height (in.)	6·5	7·5	8·5	9·5	10·5	11·5	12·5	13·5	14·5	15·5	16·5	17·5	18·5
Number of maxima recorded					1	1	1	7	8	15	4	3	2
Number of optima recorded			1	2	3	7	18	5	5	1			
Number of minima recorded	1	5	9	12	10	4	1						

Table 4. Distribution of fitting trial dimensions for steering wheel distance. Final range of adjustment 0 inches to −6 inches

Steering wheel distance (in.)	+4	+3	+2	+1	0	−1	−2	−3	−4	−5	−6	−7	−8	−9	−10
Number of maxima recorded						3	2	2	2	8	14	5	3	2	1
Number of optima recorded					5	5	5	11	8	2	1	3	2		
Number of minima recorded	1	2	7	9	7	3	5	4	2	1	1				

Table 5. Distribution of fitting trial dimensions for footrest distance. Final range of adjustment 10 inches to 20 inches

Footrest distance (in.)	7	8	9	10	11	12	13	14	15	16	17	18	19	20	21	22	23	24
Number of maxima recorded				1		3		2		4	9	2	2		6	8	3	2
Number of optima recorded		1	1	3	1	1		4	9	3		2	6	2	7	2		
Number of minima recorded	5	1	1	1	4	5	4	5			5	6	2	3				

2.3. Results

The final results of a fitting trial appear in Table 6 which shows the dimensions of a car interior designed to accommodate 98 per cent of British males and females (of the extreme height-weight combinations of Table 1) with little or no discomfort (Dutch *op. cit.*). It was decided to make the controls adjustable and to keep the front seats fixed thus avoiding the reduction of rear passenger space when the front seats are pushed back. This decision may not be reasonable for present day cars but is surely worth considering for future vehicles in which all the controls may be transmitted through flexible electrical or hydraulic connections.

The results of other fitting trials appear in Jones & Moffatt (1958) and Jones (1961).

Table 6. Comparison of fitting trials dimensions for car interior compared with those of Domey and McFarland (1963). Heights and lengths are in inches

Descriptions	Fitting trials	Domey and McFarland
Seat height		
Floor to front edge of compressed seat	12	10–14
Seat length		
Junction of compressed squab and backrest to front edge of squab	17	18
Seat angle		
Inclination of rear of squab below front edge	7°	7°
Backrest angle		
Angle between compressed back and vertical	18°	17°–27°
Interior height		
Floor to roof lining above head (2 in. head clearance)	50·5	50
Steering wheel distance		
Distance of lower rim of wheel forward of front edge of seat	0 to − 6	− 8
Steering wheel height		
Height of lower rim of wheel above front edge of compressed seat	8·5	7
Steering wheel angle		
Angle of wheel to vertical	30°	—
Footrest distance		
Horizontal distance from front edge of seat to intersection of footrest and floor	10 to 20	—
Footrest angle		
Angle of footrest to horizontal	37·5°	—
Rear passenger foot space		
Horizontal distance from rear seat edge to toe	21·25	—

2.4. *Discussion*

As can be seen from Table 6 the results of fitting trials agree well with those obtained by Domey & McFarland (1963). The latter's results were obtained by comparing American car dimensions with anthropometric data. The British car dimensions appearing in Table 10 are, however, considerably smaller than those obtained by fitting trials. It appears that the small cars common in Europe are definitely ' too small for comfort '.

3. Seat Discomfort Index

3.1. *Introduction*

This is an account of a method of measuring changes in discomfort when sitting is continued for long periods. The method was developed in order to compare the comfort of car seats but it can be adapted to any situation in which body weight is supported by pressure-resisting pads or surfaces, e.g. domestic chairs, hospital beds or bicycle saddles. The principle is that of training seat testers to discriminate reasonably consistently between different sensations in parts of the body that are subject to discomfort. These areas include both the body parts that are subjected to pressure and nearby areas that also become uncomfortable.

3.2. *Method*

A small number of testers are trained to recognise the following sensations:

0	No sensation
1	Conscious (of contact with seat)
2	Numbness
3	Ache
4	Pain

in different parts of the body. They are trained to do this by sitting on a hard seat that is mounted low enough to induce all these sensations, up to No. 4 pain,

Table 7. Discomfort sensations recorded at quarter-hourly intervals during a road test of a typical small car seat

Body part	Start	1st hour	2nd hour	3rd hour	4th hour	5th hour
Shoulder blades	0	0 0 0 1	1 1 2 1	2 2 2 2	2 3 2 2	2
Ribs	0	0 0 1 1	1 1 2 2	2 2 2 2	3 3 3 3	3
Lumbar region	0	0 0 0 0	1 1 1 2	2 2 2 3	3 3 3 3	3
Sacral region	0	0 0 0 1	1 1 2 2	2 3 3 3	3 3 3 4	4
Ischial tuberosities	0	0 0 1 1	2 2 2 3	3 3 3 4	4 4 4 4	4
Buttocks	0	0 0 1 1	2 2 2 3	3 3 3 3	3 4 4 4	4
Rear thigh	0	0 0 0 1	1 2 2 2	3 3 3 3	3 3 4 4	4
Mid thigh	0	0 0 0 0	1 1 1 2	2 3 2 2	3 3 3 3	3
Front thigh	0	0 0 0 0	1 1 1 1	2 2 2 2	2 3 3 3	3
Highest level recorded at each interval	0	0 0 1 1	2 2 2 3	3 3 3 4	4 4 4 4	4

Seat discomfort index (time of onset of discomfort level 3) = 2 hours within an hour.   The training seat was in this case composed of the plywood moulding of a Finmar/Jacobsen chair mounted on top of the front seat of a small car.   Any hard seat mounted a few inches from the floor with restricted leg room would be likely to have the same effect.   Testers are taught a standard procedure of recording the sensation number at each of the body parts listed in Table 7.   They make these recordings at five minute intervals when sitting on the training seat and at fifteen minute intervals when testing normal seats. Precautions are taken to conceal previous readings when each new set of sensations is recorded.   Testers are not permitted to discuss body sensations during tests but are free to converse about anything else, to read, or to watch the traffic outside the vehicle.   Training was in this case carried out in a stationary car parked in sight of town traffic and the subsequent road trials were carried out in cars travelling non-stop for five hours on a motorway.

3.3. *Results*

A typical set of results for one tester appears in Table 7.   It can be seen that discomfort levels rise fairly regularly throughout the journey.   It is found that only experienced testers can detect sensation 2 and below but that sensation 3 is obvious to anyone who is not preoccupied by something else.   Sensation 4 is hard for anyone to ignore.   The index of discomfort is chosen to indicate only those discomforts that the majority of users could detect.   It is therefore set at the time of onset of sensation 3 in any body part.   (A time, rather than a sensation number, is used for the index so that the data for statistical calculations is on an interval scale).   The first of two successive 3's or 4's are needed before sensation 3 is deemed to have been reached.   The discomfort index for the seat and tester of Table 6 is therefore 2 hours (when sensation 3 is reached in both the buttocks and in the ischial tuberosities.).

3.4. *Discussion*

The most interesting results obtained so far are compared in Table 8.   The first row shows the result of a single test, the onset of sensation 3 in the third hour when sitting on a hard upright seat of conventional height.   When this same seat is placed (with legs removed) on a small car seat sensation 3 began in the first hour for 23 out of 25 testers.

Table 8.   Frequency of onset of sensation 3 for various seats.   Each figure indicates the number of journeys

Kind of seat	Number of testers recording onset of sensation 3 in each hour					
	1st hour	2nd hour	3rd hour	4th hour	5th hour	6th hour
Hard seat at 17 inches height at desk	0	0	1	0	0	0
Training seat (hard surface on small car seat)	23	2	0	0	0	0
Typical small car seats e.g. BMC Mini and A.40	0	3	5	0	0	0
Typical large car seats e.g. BMC A110 and Ford Zephyr	0	1	5	5	6	4

When the seat is mounted in a car it is at less than half its normal height and sitters are prevented by the shape of the car interior from stretching their legs or adopting a variety of body postures.   The dramatic increase in **discomfort**

F

when the height of a hard seat is reduced casts much doubt on the previously accepted advice that seat pressure should as far as possible be confined to the ischial tuberosities. This appears to be sensible advice only if the seat is relatively high.

The comparison between rows two and three of Table 8 is equally interesting. The only difference between the seats is the addition of a hard surface in row two.   The probability of the difference between the frequency distributions 23, 2, 0, 0, 0, 0, and 0, 3, 5, 0, 0, 0, occurring by chance is less than one in ten thousand (Mann-Whitney U Test).   This is further evidence in favour of soft surfaces for low seats.

The last two rows show that the discomfort index discriminates reliably between car seats of different classes of vehicle.   The probability of the difference between the two distributions 0, 3, 5, 0, 0, 0, and 0, 1, 5, 5, 6, 4 occurring by chance is less than one in a thousand (Mann-Whitney U Test).   It is suggested that this difference occurs because the greater seat height and leg room in large cars permits larger angles of the joints at the hip and the knee and does not impose the pressures that occur within the body at these joints when the sitter is in a small car.

The times of onset of sensation 3 in seats of any one class of car do not as yet show statistically significant differences.   This result suggests that the discomfort index is not sensitive to the fine differences about which seat designers are often concerned.   The data available so far suggests that the sensitivity of the index could be improved if it were releated to the body sizes of the testers as well as to the times of occurrence of sensation 3.   Detailed design information can often be extracted by treating each body area separately, as in Table 7.

The repeatability of this method of measuring discomfort is better than might be expected.   The discomfort indices for three testers who have tested the same seats on two successive occasions appear in Table 9.   The high gross cost per tester-hour (£5 or more) suggests that it is cheaper to improve sensitivity by selecting and training testers than by increasing the number of tests.

Table 9.   Differences in onset of sensation 3 on retesting

Tester	Vehicle	Time of onset of sensation 3 (quarter hour intervals)		Difference
		First test	Second test	
A	1959 Ford Zephyr	14	10	− 4
	Renault R8	18	17	− 1
B	1959 Ford Zephyr	7	18	+ 11
	BMC 1100	12	13	+ 1
C	1959 Ford Zephyr	11	11	0
	BMC 1100	12	14	+ 2

4. Sitting Posture Envelopes

Table 10 is an attempt to combine the results of fitting trials and other data into a simple dimensional index that will permit easy comparison of one workspace with another.   The index used is the sum of the height and the length of a rectangle enclosing a 95th percentile man sitting in the workspace (allowing 2 inches clearance above his head in cases where no roof exists and measuring to the car roof lining when there is one).   As can be seen from the table the three small cars that have been measured in this way differ more in interior dimensions than they do in the sitting envelope index A + B.   This suggests that a given envelope determines a given level of discomfort even though the height and length of the enclosing rectangle varies.   The insensitivity of the discomfort index described earlier to differences between car seats in the same class of vehicle suggests that the relationship between the sitting envelope index and the discomfort index is worth further investigation.

Table 10.   Comparison of dimensions A and B for fitting trials, for existing cars and for lecture theatre seating

Source of data	Dimension A Heels to roof (in.)	Dimension B Toes to back (in.)	Sitting envelope index A + B
Fitting trials for car seats (assuming 2 in. headroom) (Dutch 1965)			
(a) driver	50·5	49	99·5
(b) front passenger	50·5	52	102·5
(c) rear passenger	50·5	44	94·5
Typical small cars (Dutch 1965) BMC Mini minor			
(a) driver	45·5	45·25	90·75
(b) front passenger	45·5	51	96·5
(c) rear passenger	43	41·5	84·5
Ford Cortina			
(a) driver	44·25	47·25	91·5
(b) front passenger	44·25	52·25	96·5
(c) rear passenger	44·5	46·25	90·75
Rootes Imp			
(a) driver	44	47·5	91·5
(b) front passenger	44	54	98
(c) rear passenger	46	40	86
Fitting trials for lecture theatre seating (adding 2 in. above top of head of 95th percentile man) (Jones 1961)	54·5	44	98·5
Typical lecture theatre seating (Jones 1961)	54·5	36	90·5

It can be seen that the sitting envelope indices obtained by fitting trials are all greater than those of existing small cars and of existing lecture threatres. The mean differences are 8·1 inches for car drivers, 5·5 inches for front passengers, 7·4 inches for rear passengers and 8 inches for lecture theatre seats. These differences suggest that the simplest way of making cars and lecture theatres comfortable for a large proportion of the population is to increase the sitting envelope index by about 8 inches.   If, for other reasons, the workspace cannot be enlarged, the discomforts of sitting within a small sitting envelope for long periods will have to be relieved in some other way.   It could be inferred from Table 7 and Table 8 that one way of doing this is to prevent the accumulation of pressure discomfort on lower back and buttocks.   This might entail the use of a spring-assisted or power-operated seat in which the pressure pattern on the body could be altered during the journey by changing posture and by changing the positions of body supports.

Acknowledgements are made to A.E.I. Ltd., and Cox of Watford Ltd., who sponsored much of the work described here and to W. A. Moffatt, A. J. Ward, R. C. Gray, N. Gough, W. G. Dutch and G. Moss with whose collaboration it was carried out.

Two methods of assessing sitting comfort are described. The first, the method of fitting trials, is a modification of Morant's method of determining cockpit dimensions. Morant's method is modified so that the sensitivity of users to changes in each work-space dimension can be recorded before the relationships between dimensions are investigated. An application of the method to car interior dimensions indicates that small cars are considerably below the size needed to *accommodate* 98% *of the population*. The fitting trial results agree with Domey and McFarland's recommendations for car interior dimensions.

The second method provides a measure of seat discomfort. Trained testers record sensations in parts of the body at intervals during a journey. The resulting discomfort index discriminates reliably between seats in different classes of vehicle but not between the seats in different makes of vehicle of the same class.

The sum of the length and the height of the rectangle enclosing the sitter is proposed as another index of sitting comfort. This index is nearly constant for the front seats of three different makes of small car. The index for workspaces determined by fitting trials is consistently larger than the index for small cars and for lecture theatre seating.

Les auteurs décrivent les méthodes qui permettent d'évaluer la sensation de confort en position assise.

La première méthode, celle des essais d'ajustement dérive de la méthode de Morant pour la détermination des dimensions d'un cockpit. La modification respose sur le fait que la sensibilité des usagers aux changements de chaque dimension de leur espace de travail peut être testée avant d'étudier les relations entre ces dimensions. Une application de cette méthode aux dimensions intérieures des voitures montre que les petites voitures sont sensiblement en dessous des normes qui permettraient leur adaptation à 98 p. 100 de la population. Cette méthode des essias d'ajustement recoupe les recommandations de Domey et McFarland sur les dimensions intérieures.

La seconde méthode utilise la mesure de la sensation d'inconfort en position assise. Des sujets entraînés notent leurs sensations aux différents endroits du corps à divers moments de la journée. L'indice d'inconfort qui a résulté de cette étude permet une différenciation valable entre les sièges de différents types de véhicules, mais non entre les sièges de différents véhicules du même ordre.

La somme de la longueur et de la hauteur du rectangle où s'inscrit le sujet assis peut être proposée comme un autre indice de confort en position assise. Cet indice est à peu près constant pour les sièges avant de trois sortes de petites voitures. L'indice pour un espace de travail déterminé par les essais d'ajustement est plus grand que celui utilisé pour une voiture ou un fauteuil de théâtre.

Zwei Methoden zur Ermittlung des Sitzkomforts werden beschrieben. Die erste, die Methodr von Anpassungsversuchen, ist eine Modifikation von Morants's Methode zur Bestimmung dee Flufzeugkanzel—Dimensionen. Morants's Methode wurde so modifiziert, dass die Empfindlichkeit der Benutzer gegen gegen Änderungen jeder Arbeitsraum—Dimension aufgezeichnet werden kann, bevor die Beziehungen zwischen den Dimensionen untersucht werden. Eine Anwendung dieser Methode auf Innen-Dimensionen eines Kraftfahrzeugszeigt, dass Kleinwagen erheblich unter der Grösse liegen, die notwendig ist, um 98% per Bevölkerung bequem in ihnen unters zubringen. Die Resultate der Anpassungsversuche stimmen mit Domey und McFarland-Emphehlungen für die Wagen-Innen-Dimensionen überein.

Die zweite Methode sieht eine Messung des Sitzkomforts vor. Geübte Prüfer registrieren die Empfindungen in verschiedenen Teilen des Körpers in Zwischenzeiten während einer Fahrt. Der resultierende Diskomfort-Index unterscheidet zuverlässig zwischen Sitzen verschiedener Fahrzeugmerken, aber nicht zwischen Sitzen verschiedener Fahrzeuge derselben Marke.

Die Summe der Länge und Höhe des Rechtecks, das den Sitzenden einschliesst, wird als ein weiterer Index des Sitzkomforts vorgeschlagen. Dieser Sitzindex ist für die Vordersitze dreier verschiedener Marken eines Kleinwagens nahezu konstant. Der Arbeitsraum—Index, der mit Anpassungsversuchen bestimmt wurde, ist deutlich grösser als als der Index für Kleinwagen und für Hörsaalsitze.

References

DOMEY, R. G., and McFARLAND, R. A., 1963, The operator and vehicle design. In *Human Factors in Technology* (Edited by E. BENNET *et al.*) (New York: McGRAW HILL), 247–267.

DUTCH, W. G., 1965, Interior dimensions of small cars determined by the method of fitting trials. *M.Sc. Dissertation, University of Manchester Institute of Science & Technology Library.*

JONES, J. C., 1960, Fitting for action. *Design*, **135**, 38–42; **137**, 49–52.

JONES, J. C., 1961, Seating in the lecture theatre: theoretical considerations and practical problems. In *Modern Lecture Theatres* (Edited by C. J. DUNCAN) (Newcastle-upon-Tyne: ORIEL PRESS), 110–116.

JONES, J. C., 1963, Fitting trials: A method of fitting equipment dimensions to variations in the activities, comfort requirements and body sizes of users. *Architects' Journal*, **137**, 321–325.

JONES, J. C., 1964, Methods of assessing seat comfort. *Proceedings of the Industrial Section of the Ergonomics Research Society*.

JONES, J. C., 1967, Layout of workspaces. *Ergonomics for Industry No.* 11 (London: MINISTRY OF TECHNOLOGY).

JONES, J. C., and MOFFATT, W. A., 1958, The results of fitting trials for a new winder control unit. *Engineering Report No. ID5, Industrial Design Office, A.E.I. Ltd., Manchester.*

KEMSLEY, W. F. F., 1952, Body weights at different ages and heights. *Annals of Eugenics*, **16**, 4.

MORANT, G. M., 1964, Body size and work spaces. In *Symposium on Human Factors in Equipment Design* (Edited by W. F. FLOYD and A. T. WELFORD) (London: H. K. LEWIS).

Investigations for the Development of an Auditorium Seat

By G. Wotzka, E. Grandjean, U. Burandt, H. Kretzschmar and
T. Leonhard

Department of Hygiene and Applied Physiology,
Swiss Federal Institute of Technology, Zurich, Switzerland

1. Problem and General Experimental Plan

In view of the present-day growth in requirements for new lecture theatres in universities, it seemed useful to embark upon a new study of the question of a suitable auditorium seat.

The auditorium seats available in the market have been developed empirically; economic, design, formal and anthropometric considerations being decisive. A number of authors have, in addition, stipulated orthopaedically based seat forms designed especially to combat undesirable sitting postures at school. Mention may be made, from among them, of Åkerblom (1948), Schoberth (1962), Schneider and Lippert (1961), Keegan (1962) and Oxford (1966). These authors' recommendations are based on theoretical aspects; nonetheless the seat profiles they have derived reveal substantial differences.

In the development of an auditorium seat, we proceeded from the following premises. An orthopaedically proper seat is designed to avoid unnatural postures. Unnatural postures are accompanied by pain and symptoms of fatigue, and are felt to be uncomfortable. This led us to the conclusion that a seat profile is desirable which causes the least possible discomfort and pain to as many persons as possible. We may further assume that such a seat profile also constitutes a favourable precondition for the preservation of efficiency.

This led us to the present investigations. Starting from orthopaedically recommended seat profiles, we systematically developed an auditorium seat by virtue of the data obtained from test subjects. Apart from orthopaedic recommendations we also considered the functions of an auditorium seat.

The study was conducted in the following four stages.

Stage 1. Behavioural study on students in existing auditoriums.

Stage 2. Investigation of different seat profiles and development of a functional auditorium seat, using experimental subjects.

Stage 3. Comparative check of the developed auditorium seat in standardized sitting tests.

Stage 4. Comparative check of the developed auditorium seat under conditions obtaining during lectures.

2. Stage One: Behavioural Study on Students in Existing Auditoriums

2.1. *Method*

In four auditoriums we analysed the seated behaviour of students during lectures by means of the multi-moment technique and, with the aid of questionnaires, we carried out a survey of the subjective sensations and assessment of students when sitting in these auditoriums. The multi-moment process comprised 2798 observations; and 546 students, both male and female, completed the questionnaire. On the average, the students accounted for 92 per cent of the persons observed.

Figure 1 shows drawings of the four auditorium seats.

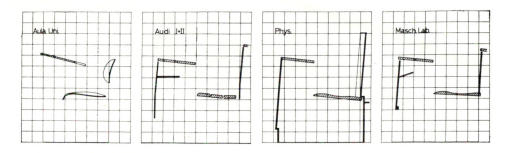

Figure 1. The seats of the four auditoriums.
A module area has a side length of 10×10 cm.

2.2. *Results*

Table 1 represents the most important results of the multi-moment process; we have dispensed with the results of auditorium 'Masch Lab' since the number of observations was too small there.

Table 1.　Results of the multi-moment technique

Number of observations : ' Aula ' $= 861$, Physics $= 963$, ' Audi ' I and II $= 974$.

Characteristic	Share of lecture time		
	Aula	Physics	Audi I and II
	(%)	(%)	(%)
Listens	22	33	28
Writes	65	54	61
Reads	5	3	1
Speaks	3	5	5
Attention diverted	5	5	6
Sleeps	2	0	1
Forwardly inclined	18	22	30
Sitting straight	63	67	60
Reclining	18	11	11
Resting against back	41	27	28
Not resting against back	58	73	73
Rests on table, right	93	86	83
Rests on table, left	87	75	84

The following results deserve particular mention.

　Students write for more than half the time.

　Students rest their lower arms on the writing surface for a large part of the lecture.

　Upright sitting postures are observed two-thirds of the time.

The frequency of leaning against the backrest is greater *in* auditorium than in others.

The observations reveal the significance of the design of the writ... and of the back rest.

Table 2. Assessments of students of seats in four auditoriums. The figures are p... related to the number of questionnaires completed per auditorium. The horiz... indicate pairs of values which are significantly different with $p < 0.05$ (Chi²)

Question	Answers	Aula university $n = 195$ (%)	Audi I and II $n = 198$ (%)	Physics $n = 131$ (%)	Ma... labor... n = ... (%)
How do you assess the seat when leaning back ?	Comfortable	30	11	18	12
	Medium	61	55	60	53
	Uncomfortable	9	34	18	35
How do you assess the seat when leaning forward ?	Comfortable	15	28	26	33
	Medium	52	50	56	47
	Uncomfortable	33	22	18	20
Did you feel any pain during the lecture ?	Yes	46	55	55	65
If so, where ?	In the back	20	37	31	35
	Buttocks	16	24	24	29
	Upper leg	9	8	10	15
	Lower leg	3	4	14	12
	At the back of the neck	10	12	6	12
	In the shoulders	5	10	2	8
Do you use the back rest ?	Always	43	6	11	17
	Frequently	45	64	63	61
	Rarely	12	30	26	22
Is the height of the writing surface	Good	73	92	69	81
	Too high	6	3	6	3
	Too low	21	5	25	16
Is the size of the writing surface	Good	11	27	23	33
	Too large	0	0	0	0
	Too small	89	73	77	67
Is leg-room	Good	57	47	24	43
	Insufficient	43	53	76	57

Table 2 shows the results of the 546 questionnaires. The following results may be particularly singled out.

In the case of the ' Aula ' seat, the back rest is used significantly more often and, again significantly, is more often thought comfortable. We presume that this is due to the closely adjacent table and the configuration of the back rest.

The same seat, however, obtained the lowest marks for sitting while leaning forward.

Complaints of pain are frequently indicated for all types, backache being to the fore. This reveals that particular attention must be given to the design of the back rest.

Complaints related to the lower leg are significantly more frequent in the Physics auditorium than in the ' Aula '. We believe that the limited room allowed for the legs is the cause; 32 cm between the front edge of the seat and the front end of the foot space is obviously insufficient. This is also shown by the reply made to the direct question concerning leg room.

It is best to dispense with evaluating the height of the writing surface in Auditoriums I and II. No definitive conclusion can be drawn since the judgment is determined not only by the height as such but also by the seat depth.

The table top surfaces are felt to be too small by the majority in all auditoriums. (Dimensions ranged between 30×57 cm and 30×60 cm.) This shows that, in principle, larger surfaces should be provided.

Leg room was said to be unsatisfactory by the majority. Table 3 compares the replies with the seat dimensions. This survey reveals that, if possible, more room in which to move one's legs should be allotted.

A seat depth of 38 cm is frequently thought inadequate in Auditoriums I and II. The other seats, 42 to 44 cm deep, are judged to be significantly better.

Table 3. The reply ' too little leg-room ' and the associated auditorium seat dimensions

Auditorium	Front edge of seat to front boundary (cm)	Overall depth of auditorium seat (back to foot) (cm)	' Too little leg-room ' (%)
Aula	Not fixedly delimited	Not fixedly delimited	43
Audi I and II	49·5	87·5	53
Masch Lab.	39	81	57
Physics	32	76	76

3. Stage Two: Study of Various Seat Profiles and Development of an Auditorium Seat, Using Test Subjects

3.1. *The Seats*

We produced five seats adjustable in the correlation between seat surface and back. The frames were made of wood and the two adjustable profile portions (seat surface and back) consisted of *Styropor* (expanded polystyrene) coated with sack-ticking and filter paper and then varnished. Figure 2 shows the five test seats.

Arranged in front of the seat was a small writing-table, the writing surface of which could be adjusted for height and inclination. The surface measured 40×80 cm.

The five seat profiles are seen in Figures 3 a and b.

The first ' seat-wedge ' profile complies with Schneider and Lippert (*op. cit.*) who wished to achieve lordosing of the lumbar vertebral column by means of the seat-wedge.

The second profile is an auditorium seat according to Keegan (*op. cit.*) characterized by low profiling.

The third 'Hy 300' profile is based on experience from the development of an easy chair (Grandjean and Burandt 1964, Grandjean *et al.* 1967). The seat is characterized by a long support for the upper leg and a long and concave back with a low-slung lumbosacral support.

The fourth profile 'Hy 200' is again derived from the shape of the previously developed easy chair (Grandjean and Burandt *op. cit.*, Grandjean *et al. op. cit.*). Its seat is provided with a reduced seat-wedge, the back with a pronounced lumbar wad while it curves concavely and provides a degree of support for the shoulders.

The fifth profile 'Hy 100' is a lighter version of the first 'seat-wedge' profile.

Figure 2. The five test seats of Stage Two of the study. From left to right: profile with seat-wedge; profile according to Keegan; Hy 300 profile; Hy 200 profile; Hy 100 profile.

3.2. *Subjects*

As subjects (Ss) we had available 27 male and 3 female students. The average age was 22·6 years (ranging from 20 to 27 years). Average body measurements were

Height (with shoes)		177·4 cm
Standard deviation		± 6·9 cm
Extremes	163·6 and	195·5 cm

The average height of the 30 Ss agrees well with corresponding figures of German and US recruits. We have no reason to assume that our subjects were not representative of students in Switzerland.

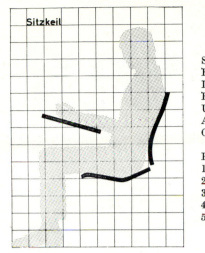

Sitzkeil

	uncomfortable		medium		comfortable	
	Ss	%	Ss	%	Ss	%
Shoulder	3	10	18	60	9	30
Back	9	30	8	27	13	43
Loin	12	40	9	30	9	30
Buttocks	6	20	18	60	6	20
Upper leg	10	33	13	43	7	21
Arms	0	0	10	33	20	67
General	11	37	16	53	3	10

Ranking	Ss	%
1st rank	1	4
2nd rank	0	0
3rd rank	4	16
4th rank	11	44
5th rank	9	36

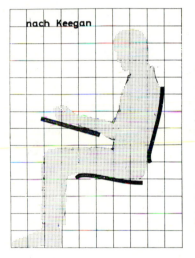

nach Keegan

	uncomfortable		medium		comfortable	
	Ss	%	Ss	%	Ss	%
Shoulder	2	7	14	47	14	47
Back	3	10	14	47	13	43
Loin	9	30	11	37	10	33
Buttocks	6	20	15	50	9	30
Upper leg	4	13	10	33	16	53
Arms	2	7	7	23	21	70
General	5	17	14	47	11	37

Ranking	Ss	%
1st rank	2	8
2nd rank	12	48
3rd rank	7	28
4th rank	3	12
5th rank	1	4

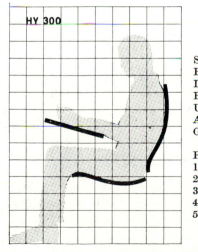

HY 300

	uncomfortable		medium		comfortable	
	Ss	%	Ss	%	Ss	%
Shoulder	3	10	9	30	18	60
Back	5	17	8	27	17	57
Loin	7	23	13	43	10	33
Buttocks	2	7	10	33	18	60
Upper leg	1	3	4	13	25	83
Arms	2	7	9	30	19	63
General	3	10	12	40	15	50

Ranking	Ss	%
1st rank	10	36
2nd rank	7	25
3rd rank	4	14
4th rank	2	7
5th rank	5	18

Figure 3. Assessment of the five test seats in Stage Two by 30 Ss.
(a) profiles : seat-wedge, Keegan and Hy 300.

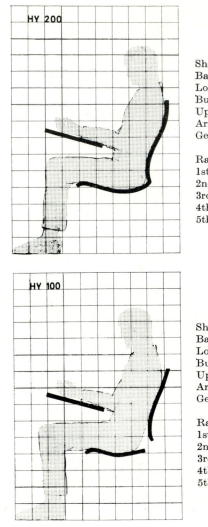

HY 200

	uncomfortable		medium		comfortable	
	Ss	%	Ss	%	Ss	%
Shoulder	2	7	5	17	23	78
Back	1	3	8	27	21	70
Loin	4	13	11	34	15	50
Buttocks	1	3	10	33	19	63
Upper leg	2	7	7	23	21	70
Arms	0	0	5	17	25	83
General	0	0	8	27	22	73

Ranking	Ss	%
1st rank	13	45
2nd rank	9	31
3rd rank	6	21
4th rank	1	3
5th rank	0	0

HY 100

	uncomfortable		medium		comfortable	
	Ss	%	Ss	%	Ss	%
Shoulder	4	13	14	47	12	40
Back	1	3	15	50	13	43
Loin	7	27	15	50	8	27
Buttocks	10	33	10	33	10	33
Upper leg	10	33	11	34	9	30
Arms	0	0	9	30	21	70
General	10	33	14	47	6	20

Ranking	Ss	%
1st rank	2	8
2nd rank	1	4
3rd rank	5	19
4th rank	8	31
5th rank	10	38

Figure 3 *cont.* Assessment of the five test seats in Stage Two by 30 Ss.
(b) profiles : Hy 200 and Hy 100.

3.3. *Procedure*

The 30 Ss came individually to the tests which lasted for 20 or 30 minutes. Initially, the relationship between back, seat surface, seat height and writing surface was adjusted until the Ss stated that the most comfortable posture had been achieved. When this optimum individual posture had been obtained with all five seats, we handed the Ss a questionnaire on which he or she gave a separate judgment for each seat.

Evaluation of the questionnaire yielded information on favourable and unfavourable shapes, inclinations and dimensions. By virtue of these results we then proceeded to develop two new variants ('Hy 210' and 'Hy 211').

3.4. *Results*

The results of the questionnaires evaluated are shown in Figure 3 a and b. The seats represented reflect the mean values of all individual optimum positions. Results show considerable differences between the five seats in terms of the assessment of discomfort caused to various parts of the body. This is the more striking as each S was in a position to select the relationship between the units most favourable to him or her.

For further development the following results were particularly significant.

1. The *seat surface* of 'Hy 300' and 'Hy 200' was clearly given top marks; the adjective 'uncomfortable' for buttocks and upper legs appears three to six times less frequently. The very bad judgment passed on the two 'seat-wedge' and 'Hy 100' variants is of particular interest; it confirms earlier studies performed by Burandt and Grandjean (1964) according to which a pronounced angle upwards in the seat surface according to the suggestions of Schneider and Lippert (*op. cit.*) is highly uncomfortable.

    For further development we considered the seat surface of 'Hy 300' since the epithet 'comfortable' for the buttocks and upper legs was there most frequently.

2. The *back rest* of 'Hy 200' was given highest, that of the 'seat-wedge' lowest, marks. The 'uncomfortable' to 'comfortable' ratios were most favourable with 'Hy 200' for the shoulders, with 'Hy 200' for the back, and with 'Hy 200' for the loins.

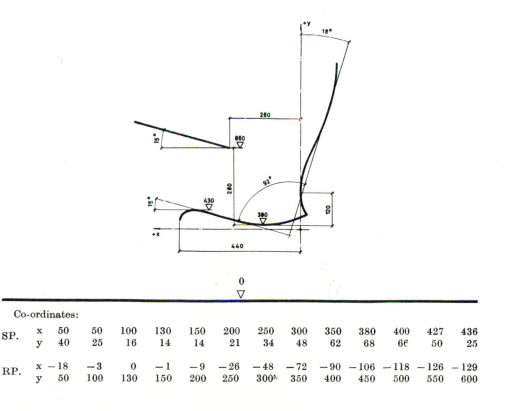

Co-ordinates:

SP.	x	50	50	100	130	150	200	250	300	350	380	400	427	436
	y	40	25	16	14	14	21	34	48	62	68	66	50	25
RP.	x	−18	−3	0	−1	−9	−26	−48	−72	−90	−106	−118	−126	−129
	y	50	100	130	150	200	250	300	350	400	450	500	550	600

Figure 4. Profile and dimensions of the developed auditorium seat Hy 211. Width of seat : 500 mm. Width of back rest : 500 mm. Curvature radius of back rest : 800 mm. The seat surface is of a concavity that increases towards the rear (cf. Figure 5). SP = seat surface profile. RP = back-rest profile.

From this we concluded that Ss give preference to a comparatively high back with a pronounced lumbar wad and with the upper half concave.

3. The *position of the writing surface* may be assessed by virtue of the data regarding sensations in the arms. No indications of discomfort were caused by ' seat-wedge ', ' Hy 200 ' or ' Hy 100 '. Further considering the indication ' comfortable ', ' Hy 200 ' obtained top rating.

From this analysis of discomfort caused to individual parts of the body it was found that the seat surfaces of ' Hy 300 ' and the back of ' Hy 200 ' would form the principal basis on which further development was to proceed.

Regarding also the general assessment the picture is analogous: ' Hy 200 ' is rated best, followed by ' Hy 300 ' and ' Keegan ', while the ' seat-wedge ' and ' Hy 100 ' are clearly considered worst.

3.5. *The ' Hy 211 ' Auditorium Seat Developed*

On the strength of the results discussed in the preceding chapter we developed the ' Hy 211 ' auditorium seat, formed from the parts of the seats tested which had most infrequently been rated ' uncomfortable '. In substance, the seat surface of ' Hy 300 ' and the back rest of ' Hy 200 ' were adopted.

As regards dimensions (seat height, seat width, etc.) we relied partly on the observations of Ss and partly on generally known anthropometric data.

The profile of the new ' Hy 211 ' seat and its dimensions are shown in Figure 4.

4. Stage Three: Comparative Check of the Developed Auditorium Seat in Standardized Sitting Tests

4.1. *The Seats Tested*

The following auditorium seats were compared.

1. Seat ' Hy 200 ' already studied in Stage Two.

2. Seat ' Hy 210 ' of which the longitudinal section is identical with ' Hy 211 '; however, it is not provided with a curvature of the seat surface and the back (in planes normal to the plane of drawing).

3. Seat ' Hy 211 ' as seen in Figure 4.

4. ' Eames Educational Seating ', a shell seat with associated folding table.

5. ' Aula ' seat as analysed in Stage One.

6. An auditorium seat according to Keegan, the table top being arranged substantially horizontally and relatively far (approx 40 cm) in front of the back rest for the purpose of these experiments. This corresponds to the usual auditorium seat arrangements.

The seat profiles are shown in Figure 5 a and b, the curvature being indicated by lateral delimiting lines (thin lines).

4.2. *Subjects*

As test subjects we had available 36 males and 4 females. The average age was 23 years (ranging from 17 to 28 years). Average body measurements were

Height (with shoes)	178 cm
Standard	± 7 cm
Extremes	158 and 198 cm

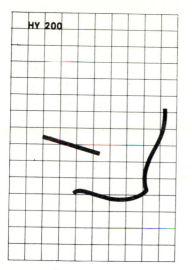

	uncomfortable		medium		comfortable	
	Ss	%	Ss	%	Ss	%
Shoulder	6	17	10	28	21	58
Back	4	11	13	39	18	50
Loin	5	14	20	53	12	33
Buttocks	5	14	12	33	19	53
Upper leg	5	14	18	50	13	36
Arms	3	8	14	39	19	53
General	3	8	21	58	12	33

Number of significant differences between preferences: 2 items.

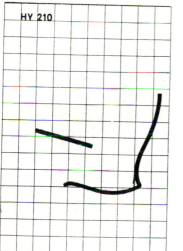

	uncomfortable		medium		comfortable	
	Ss	%	Ss	%	Ss	%
Shoulder	2	6	11	31	22	63
Back	6	17	16	44	14	39
Loin	11	30	17	47	8	22
Buttocks	4	11	16	44	16	44
Upper leg	10	28	12	33	13	39
Arms	1	3	18	50	17	47
General	4	11	23	64	9	25

Number of significant differences between preferences: 2 items.

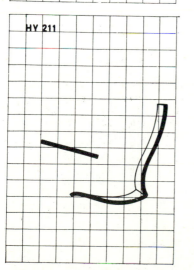

	uncomfortable		medium		comfortable	
	Ss	%	Ss	%	Ss	%
Shoulder	1	3	16	44	19	53
Back	1	3	12	33	23	64
Loin	7	19	15	42	14	39
Buttocks	1	3	13	36	22	61
Upper leg	6	17	15	42	15	42
Arms	2	6	14	40	19	54
General	1	3	20	57	15	42

Number of significant differences between preferences: 4 items.

Figure 5 a.

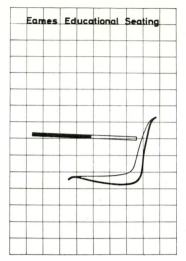

	uncomfortable		medium		comfortable	
	Ss	%	Ss	%	Ss	%
Shoulder	8	23	12	34	15	43
Back	3	9	13	38	18	52
Loin	4	11	16	44	16	44
Buttocks	2	6	8	22	26	72
Upper leg	5	14	11	30	21	58
Arms	13	36	15	42	8	22
General	6	17	16	44	14	39

Number of significant differences between preferences: 2 items.

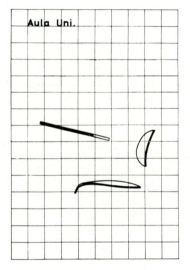

	uncomfortable		medium		comfortable	
	Ss	%	Ss	%	Ss	%
Shoulder	11	31	14	40	10	29
Back	8	23	13	37	14	40
Loin	6	17	15	42	15	42
Buttocks	11	31	17	49	8	22
Upper leg	9	25	17	49	10	28
Arms	21	58	10	28	5	14
General	14	39	18	50	4	11

Number of significant differences between preferences: 0 item.

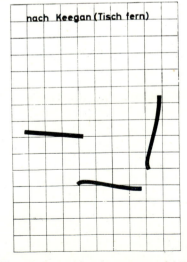

	uncomfortable		medium		comfortable	
	Ss	%	Ss	%	Ss	%
Shoulder	5	14	16	44	15	42
Back	4	11	15	42	17	47
Loin	12	33	15	42	9	25
Buttocks	10	29	18	51	7	20
Upper leg	8	22	18	50	10	28
Arms	19	53	9	25	8	22
General	9	25	22	61	5	14

Number of significant differences between preferences: 1 item.

Figure 5b.

Figures 5a and b. Assessment of ' Hy 211 ' in laboratory tests as compared to five other seat profiles. The thin lines indicate the profiles of the lateral borders while the heavy line shows the profile in longitudinal section.

4.3. *Procedure*

The 40 Ss came to the tests individually. In the first test, pair comparisons were made. The Ss had to test two seats each and mark the better one. In this manner every seat was compared with every other and judged in 15 pair comparisons. The choice was entered on forms for sequential analysis so that the statistical result could be read directly upon completion of all tests.

Subsequently, the Ss sat on each seat in haphazard order for five minutes and completed the same questionnaire that had been used in Stage Two.

4.4. *Results*

The results of the questionnaire test and pair comparisons are shown in Figure 5 (a) and (b). The following conclusions may be drawn.

1. With the backs some 5 cm higher ('Hy 210' and 'Hy 211'), 'uncomfortable' in the *scapular region* is clearly less frequent than with the others. The concave configuration of the upper half of the back rest is likely to have contributed to this result.
2. *Back and loin* give few causes for complaint in 'Eames', 'Hy 211' and 'Hy 200'. The other three profiles are clearly less successful. 'Hy 210' which has the same profile as 'Hy 211' but no curvature in back, was 'uncomfortable' a surprising number of times.
3. *Buttocks and upper legs*, in respect of 'Hy 211' and 'Eames', are seldom rated as 'uncomfortable'. 'Aula' and 'Keegan' on the other hand are rarely rated 'comfortable'. The two former seats are characterized by a dished configuration of the seat surface. We assume that this shape permits a more uniform distribution of the pressure throughout the buttocks and that this is why it is favourably commented upon.
4. The *arms* are seldom 'uncomfortable' on 'Hy 200', 'Hy 210' and 'Hy 211'. Here the writing surfaces are inclined at 15°, their front edges being located at 28 cm above the seat surface (lowest point) and 26 cm in front of the back (vertical tangent on lumbar wad).
5. The general assessments and results of the pair comparisons are compiled in Table 4. In general, 'Hy 211' is clearly favoured. It may here be mentioned that 'Eames' would be fairly near 'Hy 211' if the judgments regarding shoulders and arms had been more favourable.

Table 4. General assessment and results of pair comparisons in stage three (36 Ss)

Auditorium seat	Pair comparison (significant preferences)	General judgment as 'uncomfortable' (%)
Hy 200	2×	8
Hy 210	2×	11
Hy 211	4×	3
Eames	2×	17
Aula seat	0	39
Keegan	1×	25

5. Stage Four: Comparative Check of the Developed Auditorium Chair under Conditions Obtaining during Lectures

5.1. *Procedure*

The same seats were tested as in Stage Three, excepting the 'Aula' seat which had to be dispensed with for space reasons.

G

The tests were made during regular lectures. Prior to every lecture, five students (male or female) were requested to sit on the five experimental seats and to complete the questionnaire at the end of the lecture. Two hundred questionnaires were completed.

Table 5. Assessment of sitting behaviour and physical sensations caused by five auditorium seats under lecture conditions. The figures are percentages related to the number of questionnaires completed (about 40) per seat: 200 questionnaires in all. The horizontal lines connect pairs of values of which the differences are significant at $p < 0.05$

Question	Answers	Keegan (%)	Hy 211 (%)	Hy 210 (%)	Hy 200 (%)	Eames (%)
Do you use the back rest ?	Always	18	46	32	33	34
	Frequently	53	54	64	56	52
	Seldom	29	0	5	10	13
Do your buttocks slide forward readily ?	Yes	47	54	38	48	37
	No	53	46	62	52	63
If so, what is your reaction ?	Agreeable	16	10	15	10	0
	Indifferent	13	25	13	15	16
	Disagreeable	16	18	10	20	21
Did you feel acute discomfort during the lecture ?	Yes	46	30	51	45	41
	No	54	70	49	55	59
If so, where ?	Back of neck	8	3	3	13	8
	Shoulders	8	8	0	5	5
	Back	24	16	28	28	14
	Buttocks	14	5	0	8	8
	Upper leg	0	5	23	5	11
	Lower leg and feet	5	0	8	8	11
How do you qualify the seat when you lean back ?	Comfortable	29	63	53	48	43
	Medium	53	34	33	30	37
	Uncomfortable	18	3	15	22	20
How do you qualify the seat when you lean forward ?	Comfortable	20	26	18	24	31
	Medium	40	65	46	66	49
	Uncomfortable	40	9	36	10	20
Have you personally come across auditorium furniture which is more comfortable than this ?	Yes	20	3	14	15	17
	No	80	97	86	85	83

5.2. *Results*

The results are shown in Tables 5 and 6. The following conclusions may be drawn (for sitting during one hour).

1. The *back rest* is most used with 'Hy 211' and most favourably assessed for lean-back sitting postures.

2. 'Hy 211' is favourably assessed for sitting in a forward posture: comparatively few data are given in respect of pains in the buttocks and upper legs. This would indicate that the seat surface of 'Hy 211' is very acceptable and this is confirmed by the replies to the questions relating to inclination and depth of the seat surface (Table 6).

Table 6. Assessment of dimensioning of five auditorium seats under lecture conditions. The figures are percentages related to the questionnaires completed (about 40) per seat. 200 questionnaires in all. The horizontal lines connect pairs of values of which the differences are significant at $p < 0.05$

Question	Answers	Keegan (%)	Hy 211 (%)	Hy 210 (%)	Hy 200 (%)	Eames (%)
How do the following design features match your build ?						
Is the height of the seat surface above the floor	Good	75	70	75	72	68
	Excessive	3	0	10	0	24
	Too low	23	30	15	28	8
Is the backward inclination of the seat surface	Good	75	90	73	72	78
	Excessive	15	5	20	13	16
	Insufficient	10	5	7	15	5
Is the depth of the seat surface from front to rear edge	Good	70	84	78	76	71
	Excessive	13	5	7	11	11
	Insufficient	18	10	15	13	18
Is the backward inclination of the back rest	Good	59	72	59	74	54
	Excessive	28	18	28	23	22
	Insufficient	13	10	13	3	24
Is the height of the writing surface	Good	64	75	80	80	56
	Excessive	21	5	0	5	3
	Too low	15	20	20	15	42
Is the size of the writing surface	Good	62	62	50	44	11
	Excessive	3	3	0	0	0
	Insufficient	36	36	50	56	89
Is the inclination of the writing surface	Good	62	62	72	62	90
	Excessive	3	38	26	38	5
	Insufficient	36	0	3	0	5
Is the distance of the front edge of the writing surface from the upper body	Good	26	76	65	56	66
	Excessive	72	3	0	5	24
	Insufficient	3	21	35	39	11

3. The *seat height* of ' Eames ' was 47 cm; of the other seats, 43 to 44 cm. Although the differences are not significant, it may nonetheless be concluded that seat heights of 43 to 44 cm are somewhat too low for students.

4. The *inclination of the seat surface of* 15° (Figure 4) of ' Hy 211 ' is considered to be good.

5. The *seat depth* of 43 to 44 cm of all seats is largely described as good.

6. As regards the writing surface, the height (front edge to seat surface) of 28 cm is assessed to be good by the majority; however, ' too low ' is more frequently stated than ' too high '. The inclination of 15° (' Hy 211 ', ' Hy 210 ' and ' Hy 200 ') is largely described as good; ' excessive ' is more frequently scored than ' insufficient '. With ' Keegan ', the inclination of 5° is found to be insufficient; on the other hand, the majority finds ' Eames ' good.

6.   Discussion of Results and Recommendation for an Auditorium Seat

Comparative tests performed on various auditorium seats clearly revealed that ' Hy 211 ' obtained the best grades from the students.   It is particularly interesting that the two postures ' inclined forward ' and ' lean-back sitting '— the most frequent postures observed in lectures—caused the least number of ' uncomfortable ' ratings with ' Hy 211 '.

We attribute the favourable assessment of ' Hy 211 ' to the particular configuration of the seat surface and its relationship to the back.   The seat curvature of ' Hy 211 ' resembles a partial sphere which admits of several positions of equilibrium of the body and facilitates transition from one posture to another.   The back rest can perform its supporting function in any posture.   These considerations are illustrated in Figure 6.

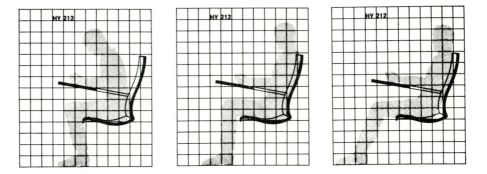

Figure 6.   The recommended ' Hy 212 ' auditorium seat with three typical seat arrangements. The thin lines indicate the profile of the lateral borders while the heavy line shows the longitudinal section at centre.

Figure 7.   The plaster model of the recommended auditorium seat ' Hy 212 '.

It was found in Stage Four that with the ' Hy 211 ' seat, the buttocks slide forward readily and that the angle of the writing surface is somewhat excessive. This induced us to modify slightly the ' Hy 211 ' auditorium seat: the inclination of the writing surface was reduced to $10°$ and the front edge of the table shifted forward by 4 cm (front edge of table—vertical tangent to lumbar wad = 30 cm). The latter solution will allow stouter persons to be seated comfortably.

In order to maintain a comfortable position of the arm during writing, we suggest a lateral arm-rest as a projection of the writing surface.

These modifications will very likely result in a greater forward inclination of the body during writing; this has caused us to lower the supporting surface of the upper legs by roughly 1 cm as compared with the central longitudinal section.

The slightly modified seat was designated as ' Hy 212 ' and is shown in Figures 6 and 7.

The present study was designed to enable the development of an auditorium seat which meets its specific functions while avoiding, to the maximum possible extent, discomfort. To this end, sitting behaviour was analysed during lectures by the multi-moment process. Subsequently, an auditorium seat was improved in stages by virtue of the results of seating tests. The final result is a recommended new auditorium seat which causes a minimum of discomfort in both reclining and forward inclined postures.

Cette étude est destinée à permettre la fabrication de sièges d'auditorium; ceux-ci doivent dans la mesure du possible éviter la sensation d'inconfort. Dans ce but les auteurs étudient le comportement de sujets assis et lisant, grâce à un système multimoment. Par la suite a été construit un siège qui possède les propriétés qui sont apparues nécessaires d'après les tests entrepris. Le résultat en est un siège d'un nouveau type qui est recommandé, car il produit un minimum d'inconfort en position penchée vers l'arrière ou vers l'avant.

Ziel der vorliegenden Untersuchung war die Entwicklung eines Hörsaalsitzes, der bei grösstmöglicher Vermeidung von Unbequemlichkeiten den spezifischen Funktionen gerecht wird. Zu diesem Zweck wurde zuerst während Vorlesungen das Sitzverhalten mit Multimomentaufnahmen analysiert. Hernach wurde auf Grund der Ergebnisse von Sitzversuchen mit Studenten ein Hörsaalsitz stufenweise verbessert. Das Schlussergebnis ist ein empfohlener neuer Hörsaalsitz, der sowohl beim Anlehnen an die Rückenlehne als auch bei vorgeneigtem Sitzen ein Minimum an Unbequemlichkeit hervorruft.

References

ÅKERBLOM, B., 1948, *Standing and Sitting Posture* (Stockholm : NORDISKA BOKHANDELN).

BURANDT, U., and GRANDJEAN, E., 1964, Die Wirkungen verschiedenartig profilierter Sitzflächen von Bürostühlen auf die Sitzhaltung. *Int. Z. angew. Physiol. einschl. Arbeitsphysiol.*, **20**, 441–452.

GRANDJEAN, E., BÖNI, A., and KRETZSCHMAR, H., 1967, Entwicklung eines Ruhesesselprofils für gesunde und rückenkranke Menschen. *Wohnungsmedizin*, **5**, 51–56.

GRANDJEAN, E., and BURANDT, U., 1964, Die physiologische Gestaltung von Ruhesesseln. *Bauen und Wohnen*, 233–236.

KEEGAN, J. J., 1962, Evaluation and improvement of seats. *Industr. med. Surg.*, **31**, 137.

OXFORD, H. W., 1966 a, Are you aware of the danger of sitting incorrectly ? *NAMCO, Australia*.

OXFORD, H. W., 1966 b, The problem of misfit furniture. *Department of Education, Sydney*.

SCHNEIDER, H. J., and LIPPERT, H., 1961, Das Sitzproblem in funktionell-anatomischer Sicht. *Med. Klin.*, **56**, 1164.

SCHOBERTH H,. 1962, *Sitzhaltung, Sitzschaden, Sitzmöbel* (Berlin : SPRINGER).

Die Verteilung des Körperdrucks auf Sitzfläche und Rückenlehne als Problem der Industrieanthropologie

Von H. W. Jürgens

Anthropologisches Institut der Universität, Kiel, Deutschland

Das Bemühen, die Beziehungen zwischen dem sitzenden Menschen und dem Gerät, auf dem er sitzt, aufzuklären, hat bereits zur Untersuchung zahlreicher Aspekte im Bereich der Anthropometrie, Anatomie und Physiologie geführt. Über eines der grundlegenden Probleme bei der Gestaltung eines Sitzes, die Verteilung des Körperdrucks auf die Unterstützungsfläche, wissen wir im Vergleich zu anderen Bereichen sehr wenig. Dieser Mangel an Wissen hat seine Ursache vor allem im methodologischen Bereich. Es erweist sich als eine komplizierte und langwierige Aufgabe, in diesem Bereich Daten zu gewinnen.

Auf der anderen Seite ist die Prüfung der Druckverteilung eine unabdingbare Voraussetzung für jeden Vergleich von Sitzmöbeln, der über subjektive Beurteilungen oder indirekte Erfassungsmethoden hinausgehen soll. Auch die Prüfung von Hypothesen über die optimale Gestaltung von Sitz und Rückenlehne und ihres Verhältnisses zueinander hängt von der Prüfung von Druckverteilungen ab.

Da diese Fragestellungen nicht neu sind, sondern in der Forschungspraxis immer wieder auftauchen, wurden in diesem Bereich verschiedene Geräte und Methoden entwickelt. Man stellte u.a. ein dem menschlichen Körper nachgebildetes Phantom her, dessen Oberfläche mit zahlreichen Geberelementen besetzt ist, die den Auflagedruck auf einem zu prüfenden Sitzmöbel anzeigen. Man entwickelte eine flexible Platte, die mit elektrischen oder hydraulischen Geberelementen besetzt ist und, zwischen Körper und Sitz gelegt, den Auflagedruck anzeigt (Möller u. Grygar 1957). O'Hara (1962) verteilte elektronische, Schoberth (1962) hydraulische Druckmesser auf einer starren planen Sitzfläche, um den Auflagedruck beim Sitzen zu messen.

Wir werden hier über Ergebnisse mit einem Druckmeßapparat berichten, mit dem wir seit einigen Jahren experimentiert haben (das Gerät wurde erstmals 1958 auf der 6. Tagung der Deutschen Gesellschaft für Anthropologie in Kiel vorgestellt). Das Gerät besteht aus zwei 176 × 176 großen Flächen, die aus starkem engmaschigem Drahtgeflecht bestehen, das in einen Stahlrohrrahmen gespannt ist. Diese beiden Platten werden—hochkant stehend—durch Stahlstreben in 98 cm Abstand voneinander parallel gehalten. Durch Stahlstäbe, die durch die Maschen der beiden Flächen geschoben werden, läßt sich jedes beliebige Sitzprofil (und Rückenlehnenprofil) konstruieren. Die Sitz- und Rückenflächen werden mit hydraulischen Druckaufnehmern belegt. Diese bestehen aus 5 × 5 cm großen, 0,5 cm dicken mit Manometerflüssigkeit gefüllten Kunststoffkissen, von denen ein Kunststoff-Druckschlauch zu einem Manometer führt. Die Manometer, die oberhalb der Meßskala einen Luftkessel haben, sind geschlossen; die Messung erfolgt also gegen den Druck der eingeschlossenen Luft. Diese Einrichtung hat den Vorteil, daß der Auflagedruck auf die Druckaufnehmer diese nur wenig verformt, so daß eine gleichartige

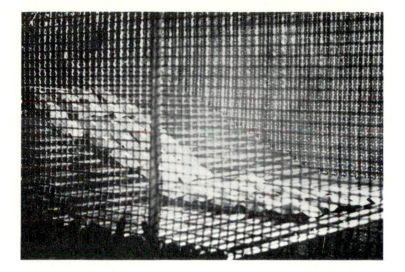

Abb. 1. Messapparat für die Druckverteilungsmessung (Blick ins Innere).

Oberfläche der Sitz- und Lehnenfläche erhalten bleibt. Die Registrierung der Manometerstände bei Druckbelastung im Sitzversuch erfolgt fotografisch (vgl. Abb. 1und 2). Die Füße ruhen während des Versuchs auf einer bzw. zwei Federwagen.

Mit Hilfe dieser Versuchsanordnung konnten wir bei 104 erwachsenen Personen beiderlei Geschlechts (Körpergewicht von 43 bis 92 kg) u. a. die Frage prüfen, wie sich zwei grundsätzlich verschieden gebaute Stühle bei Arbeits- und Ruhehaltung hinsichtlich der Druckaufnahme verhalten. Vor Beginn des Versuchs wurden die Versuchspersonen eingehend anthropometrisch erfaßt. Diese Messungen dienen dazu, die Auflagedruckzonen, die die manometrische Messung anzeigt, am Körper zu identifizieren. Es wurden daher

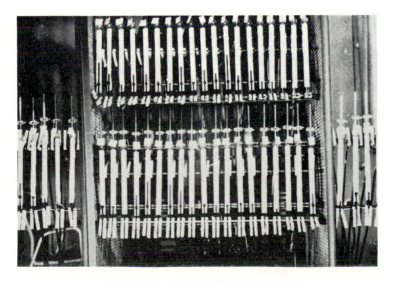

Abb. 2. Manometeranordnung am Messapparat.

u.a. erfaßt: Abstand der Tubera ossis ischii voneinander, Höhe des Becken-
kammes über dem Sitz, Höhe des 10. Brustwirbels über dem Sitz, Abstand
der Sehne des M. bizeps femoris bei rechtwinklig gebeugtem Bein über dem
Boden etc.

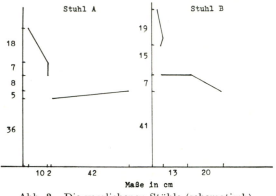

Abb. 3.   Die verglichenen Stühle (schematisch).

Die beiden verglichenen Stühle sind in Abb. 3 und 4 dargestellt. Bei
beiden Stühlen ist die Vorderkante der Sitzfläche 41 cm hoch. Bei Stuhl A
fällt sie dann kontinuierlich nach hinten ab. Bei Stuhl B steigt sie fast zwei
Drittel der Sitzfläche an und geht dann in eine horizontal gestellte Sitzfläche
über. Stuhl A, der sich an den von Åkerblom angegebenen Entwurf anlehnt,
ist als Arbeits- wie auch als Ruhestuhl (also für die vordere und die hintere
Sitzhaltung) vorgesehen. Stuhl B, der an Kutschersitze, Orgelbänke und die
von ihnen abgeleiteten Entwürfe von Zeller (1958) und Jürgens (1959) angelehnt
ist, ist im wesentlichen als Arbeitsstuhl, also für eine vordere Sitzhaltung,

Abb. 4.   Modelle der verglichenen Stühle.

gedacht. Die Tubera ruhen hier noch auf dem horizontal gestellten Bereich der Sitzfläche; der geneigte vordere Teil der Sitzfläche dient der Beinabführung. Während Stuhl A durch seine Sitzneigung eine Benutzung der Rückenlehne in beiden Haltungen fördern soll, ist bei Stuhl B die Rückenlehne weniger von Bedeutung (von Zeller wurde z.B. keine Rückenlehne vorgesehen).

Unsere Versuchsanordnung erfaßt (nach vorheriger Instruktion über die einzunehmenden Sitzhaltungen) bei allen Personen auf beiden Stühlen eine mittlere Sitzhaltung ohne Anlehnung, eine hintere Sitzhaltung mit voller Anlehnung und eine vordere Arbeitssitzhaltung mit Beckenrandanlehnung. Bei den Versuchen konnte die Sitzfläche auf Grund der mäßigen Sitztiefe und der relativ niedrigen Sitzflächenhöhe von allen Personen in gleicher Weise besessen werden. Die Rückenlehne wurde bei der hinteren Sitzhaltung vor allem im oberen Abschnitt, bei der vorderen Sitzhaltung im Unterabschnitt im Bereich des Lehnen-Knicks benutzt.

Vergleichen wir, ohne zunächst auf die Einzelheiten der Druckverteilung einzugehen, wie sich bei den beiden Stühlen und den verschiedenen Sitzhaltungen der Körperdruck auf Sitzfläche, Rückenlehne und Fußauflage verteilt (Tab. 1).

Tabelle 1

	Stuhl A			Stuhl B		
	Der Auflagedruck verteilt sich (in %) auf					
	Sitz-fläche	Rücken-lehne	Fußauf-lage	Sitz-fläche	Rücken-lehne	Fußauf-lage
mittlere Sitzhaltung (unangelehnt)	76	—	24	64	—	36
hintere Sitzhaltung (voll angelehnt)	68	15	17	63	10	27
vordere (Arbeits) Sitzhaltung (mit Beckenabstützung durch Lehne)	60	10	30	57	5	38

Zu dieser Verteilung ist anzumerken, daß als Folge der individuell recht verschiedenen Sitzhaltungen (auch der Variation der Sitzhaltung einer Person bei wiederholten Versuchen) das Verhältnis von Sitzdruck plus Lehnendruck zu Fußdruck erhebliche Schwankungen aufweisen kann. Dagegen ist das Verhältnis von Sitzdruck zu Lehnendruck bei den einzelnen Sitzhaltungen bei allen Versuchspersonen viel stabiler. Bei einzelnen Versuchspersonen wurde festgestellt, daß bei der vorderen Sitzhaltung mit Beckenabstützung die Addition von Sitz- + Lehnen- + Fußdruck das Körpergewicht etwas überstieg. Da Meßfehler ausgeschlossen werden konnten, ist anzunehmen, daß von diesen Probanden in der erwähnten Sitzhaltung ein aktiver Druck auf die Fußunterstützung bzw. Rückenlehne ausgeübt wurde. Bei längerem Sitzen bildete sich der Drucküberschuß zurück.

Vergleichen wir die in Tab. 1 dargestellte Druckverteilung auf den beiden Stühlen bei den verschiedenen Sitzhaltungen, dann ist zunächst kritisch anzumerken, daß eine mittlere Sitzhaltung auf beiden Stühlen wegen der Neigung der Sitzflächen keine praktische Bedeutung hat. Diese Werte wurden nur

der Vollständigkeit halber aufgenommen. Grundsätzlich unterscheidet sich
Stuhl A von Stuhl B dadurch, daß bei letzterem mehr Gewicht auf die Füße
verlagert wird. Aber auch hinsichtlich des auf den Stuhl ausgeübten Druk-
kes zeigt sich insofern ein Unterschied, als die Druckaufnahme der Rückenlehne
bei Stuhl B im Verhältnis zur Druckaufnahme der Sitzfläche kleiner ist als bei
Stuhl A. In Anbetracht der erheblichen Vorwärtsneigung der Sitzfläche des
Stuhles B erscheint aber beachtlich, daß selbst bei der vorderen Sitzhaltung die
Rückenlehne noch so weit in Funktion tritt, daß sie bei der Beckenrandab-
stützung 5% des gesamten ausgeübten Druckes aufnimmt.

Stuhl B ist unter dem Gesichtspunkt konstruiert, daß bei der vorderen
(Arbeits) Sitzhaltung (für die er im wesentlichen gedacht ist) Gewicht auf die
Beine verlagert wird. Es erscheint daher beachtlich, daß gerade in dieser
Haltung der Anteil des Gewichts, der auf die Füße verlagert wird, dem bei
Benutzung des Stuhles A am nächsten kommt. Die Unterschiede des Druckes
auf die Fußauflage bei den verschiedenen Sitzhaltungen sind bei Stuhl B
erheblich geringer als bei Stuhl A.

Die in Abb. 5 dargestellten Sitzdruckverteilungen geben in den Zahlenwerten
an, wieviel Prozent des auf die gesamte Sitzfläche bzw. Rückenfläche drück-
enden Körpergewichts auf die jeweilige Druckmesseinheit (d.h. also auf
$5,5 \times 5,5$ cm Auflagefläche) drücken. Auf dieser Basis ist eine Umrechnung in
den Druck je cm² ebenso möglich wie auch (mit Hilfe von Tab. 1) die Umrechnung auf Anteile am gesamten Körpergewicht. Für den hier von uns vorgenom-
menen Vergleich der beiden Stühle bei den verschiedenen Sitzhaltungen untereinander sind diese Umrechnungen jedoch nicht erforderlich.

Vergleichen wir zunächst die Druckverteilung auf der Sitzfläche, dann
zeigen sich bei den beiden Stühlen unabhängig von der Sitzhaltung typische
Verteilungsbilder: bei Stuhl B liegt eine Zone geringen Drucks an der vorderen
Kante des Sitzes. Dieser Bereich nimmt jeweils nur etwa 4% des gesamten
Sitzdrucks auf. Bei Stuhl A ist dagegen die Stuhlkante erheblich stärker
belastet. 9 bis 15% des Sitzdrucks finden sich hier. Eine Zone geringen
Drucks findet sich bei Stuhl A erst in der der Vorderkante folgenden Zone der
Sitzfläche. Dieser Befund bei Stuhl A überrascht. Die Höhe der Vorderkante
des Stuhls war mit 41 cm über dem Boden so gewählt, daß alle Versuchspersonen
(die Untersuchung wurde mit Schuhbekleidung vorgenommen) eine Unterschenkellänge aufwiesen, die der Sitzflächenhöhe entsprach oder sie übertraf.
Es entsteht aber dennoch ein Druckverteilungseffekt wie er für zu hohe Stühle
charakteristisch ist (vgl. Abb. 6, nach O'Hara *op. cit.*).

Die Auflagezone der Tubera ossis ischii ist bei allen Stühlen als Hauptdruckzone besonders herausgehoben. Sie liegt im hinteren Drittel des Sitzes.
Eine Ausnahme macht Stuhl A bei der hinteren (Ruhe) Sitzhaltung. Hier
wird von fast allen Benutzern das Becken nach vorn verlagert, um die Rücken-
lehnenabstützung voll ausnutzen zu können. Der hinterste Teil der Sitzfläche
wird bei dieser Sitzhaltung fast nicht belastet. Der Vergleich des anteiligen
Sitzgewichts, das von den Tubera getragen wird, zeigt, daß Stuhl B in diesem
Bereich stets mehr Druck aufnimmt als Stuhl A. Es wird hier ein Effekt
erzeugt, der an sich von Stühlen mit besonders niedriger Sitzflächenhöhe
bekannt ist. Die Sitzfläche von Stuhl B liegt aber gerade in dem Auflage-
bereich der Tubera besonders hoch (vor allem auch im Vergleich mit Stuhl A,
vgl. Abb. 3). Bei beiden Stühlen ist festzustellen, daß bei der hinteren

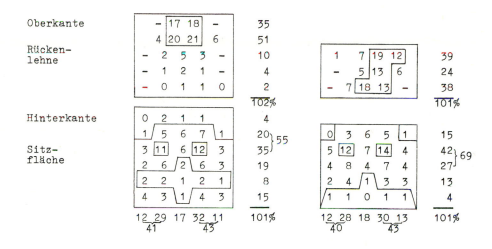

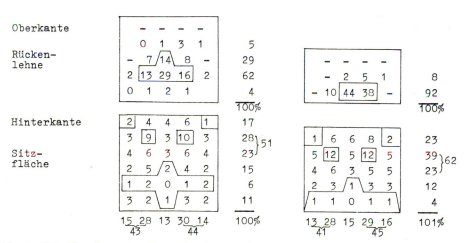

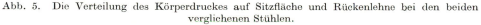

Abb. 5. Die Verteilung des Körperdruckes auf Sitzfläche und Rückenlehne bei den beiden verglichenen Stühlen.

Sitzhaltung, also bei der Ruhehaltung, mehr Gewicht auf die Tubera verlagert wird als bei der vorderen (Arbeits) Haltung.

Schließlich ist noch ein Hinweis auf die Asymmetrie der Sitzhaltung zu geben, auf die auch schon Schoberth (*op. cit.*) hinweis. Bei beiden Sitzhaltungen und auf beiden Stühlen—besonders betont jeweils bei Stuhl B—findet sich eine stärkere Belastung der linken Seite. Über die Ursache dieser Asymmetrie können wir aus unserem Untersuchungsgut keinen Anhalt gewinnen.

Sitz 5 cm höher als Unterschenkellänge

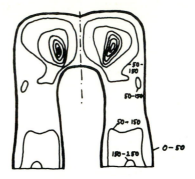

Verteilung des Auflagedrucks

Zahlenwerte je Feld von innen
nach außen in p/cm²

650	–	950
450	–	650
350	–	450
250	–	350
150	–	250
50	–	150
0	–	50

Abb. 6. Verteilung des Körperdruckes auf die Sitzfläche bei einem zu hohen Stuhl
(nach O'Hara, 1962).

Bei dem Vergleich der Rückenlehnen ist Tab. 1 zu berücksichtigen, die aufzeigte, daß Stuhl B—seiner Konstruktion entsprechend—weniger Druck auf seine Rückenlehne aufnimmt als Stuhl A. Das Druckverteilungsbild der Rückenlehne bietet bei beiden Stühlen keine Überraschung. Bei der hinteren Sitzhaltung verlagert sich die Hauptstützzone nach oben (besonders deutlich bei Stuhl A), während bei der Arbeitshaltung nur eine kleine Zone der Lehne im Bereich des Kreuzbeins und des oberen Beckenrandes Abstützung gibt. Bei Stuhl B ist die wirksame Stützzone in beiden Fällen kleiner als bei Stuhl A.

Eine Bewertung dieser zunächst rein deskriptiv dargestellten Befunde hängt von der vorliegenden Fragestellung bzw. von den über bestimmte Sitzhaltungen aufgestellten Hypothesen ab. Prüfen wir die Eignung der Stühle A und B für die Arbeitshaltung, dann ergibt sich—wenn wir nur die hier vorliegenden Druckverteilungsdaten verwenden—für Stuhl B (gegenüber Stuhl A) als Vorteil, daß mehr Gewicht auf die Region der Tubera verlagert wird und der Bereich der Oberschenkelunterseite an der vorderen Sitzflächenkante entlastet wird. Bei Stuhl B wird das Gewicht gleichmäßiger auf alle Teile der Sitzfläche verteilt. Als Nachteil ist bei Stuhl B bei der vorderen Sitzhaltung festzustellen, daß die Beckenabstützung nur in einem sehr kleinen Stützbereich wirksam wird. Hier ist Stuhl A günstiger.

Für die hintere Sitzhaltung erscheint von der Sitzfläche her Stuhl B ebenfalls günstiger, da ein Kantendruck an der Vorderkante des Sitzes vermieden wird. Die Rückenabstützung liegt jedoch bei Stuhl B in dieser Haltung ungünstig, weil sie auf relativ kleinem Bereich und im wesentlichen nur als Kreuzlehne wirksam wird. Hier bietet Stuhl A durch die gleichmäßigere Verteilung der Druckaufnahme auf eine weitere Fläche und durch die nach-oben-Verlagerung der Hauptdruckzone Vorteile.

Wesentlich erscheint es uns festzustellen, daß die Beurteilung eines Sitzes allein auf Grund der Druckverteilungen nicht möglich ist. Andererseits sollten aber Prüfungen dieser Art integrierende Bestandteile jeder Untersuchung einer Sitz- und Stützfläche für die menschliche Sitzhaltung sein.

Es wird eine Versuchseinrichtung beschrieben, mit deren Hilfe der Auflagedruck des sitzenden menschlichen Körpers auf Sitz und Rückenlehne jedes beliebig einstellbaren Sitzprofils geprüft werden kann. Am Beispiel des Vergleichs von zwei Stühlen mit unterschiedlich geneigter Sitzfläche werden die Möglichkeiten zum Einsatz des Meßverfahrens geprüft und die empirisch an 104 Personen gewonnenen Befunde dargestellt.

A test device is described by means of which the reaction pressure of the sitting human body to the seat and the back-rest in any reproducible sitting profile can be shown. By comparing two chairs of different seat inclination the results of this device are tested and the empirical values found for 104 persons are given.

Dans cet article on décrit un dispositif permettant l'évaluation de la pression sur le siège et le dossier du corps humain assis dans des postures diverses. La méthode d'évaluation a été appliquée à différentes inclinations du siège de deux chaises. Les résultats présentés s'appliquent à 104 sujets.

Literatur

Jürgens, H. W., 1959, Sitzprofil für Arbeitsstühle. Schautafel und Modelle. *Ausst. d. Muthesius-Werkschule, Kiel.*

Möller, E., und Grygar, O., 1957, Vorrichtung zum Messen von Drucken belasteter, gefederter Sitz- und Liegeeinrichtungen. *Deutsche Patentschrift* 1 001 832 *München.*

O'Hara, T., 1962, Zusammenhang zwischen Entwurf und Mensch (jap.). *Kenchiku-Bunka,* **194,** 138.

Schoberth, H., 1962, *Sitzhaltung—Sitzschaden—Sitzmöbel* (Berlin : Springer).

Zeller, M., 1958, Der Arbeitsstuhl und die Zivilisationskrankheiten. *Möbelkultur,* **10,** 133.

Über die Messung des Sitzkomforts von Autositzen

Von A. Rieck

Max-Planck Institut für Arbeitsphysiologie, Dortmund, Deutschland

1. Einführung

Die Beurteilung des Komforts von Sitzen basiert heute noch in erster Linie auf dem subjektiven Urteil. Die folgenden Messungen dienten dem Versuch, eine objektive Methode zur Beurteilung des Komforts von Autositzen zu finden, die mindestens eine Stützung der Resultate der subjektiven Befragung ermöglicht, vielleicht sogar erlaubt, ganz auf die Befragung zu verzichten. Die Entwicklung der dazu nötigen Messeinrichtung basierte auf folgenden Arbeitshypothesen.

1. Die Anzahl der auf einem Sitz unbewusst ausgeführten Bewegungen in der Zeiteinheit ist als ein objektives Maß für dessen Bequemlichkeit in der Weise anzusehen, dass wenige Bewegungen in der Zeiteinheit mit subjektiv positivem Komfortempfinden korrelieren.

2. Die Intensität der Bewegungen ist ebenfalls als ein objektives Maß für die Bequemlichkeit eines Sitzes anzusehen, und zwar in der Weise, dass geringe Intensität der Bewegungen in der Zeiteinheit mit subjektiv positivem Komfortempfinden korrelieren.

Die Intensität der Bewegung ist definiert als Weg des Schwerpunktes der Versuchperson (Vp.) von einer Ruhelage in die nächste.

2. Methode

Zur Messung der Bewegungen nach Anzahl und Intensität wurde die in Abb. 1 schematisch wiedergegebene Versuchseinrichtung entwickelt. Da es sich um die Untersuchung von Fahrzeugsitzen handelt, wurde versucht, die Sitzhaltung eines Fahrzeugführers und seine Tätigkeit durch eine—aus Versuchsgründen standardisierte—Aufgabe zu simulieren. Die Einrichtung besteht aus einer steifen Messplattform (800×1300 mm^2), die den zu testenden Sitz, das Pedal und das Steuerrad trägt und auf 4 Kraftmessdosen ruht. Alle Teile sind zur optimalen Anpassung an die Vp., wie in den in Abb. 1 angedeutet, stufenlos verstellbar. Die Registrierung der Bewegung erfolgte diskontinuierlich mittels Lochstreifen. Alle 5,12 sec (Gerätekonstante) wurden die Lagekoordinaten des Schwerpunktes (x und y) abgefragt, die als verstärktes elektrisches Signal einer Kraftmessung (DMS-Technik) mit den beiden in Abb. 1 dargestellten Wheatstonschen Brückenschaltungen erzeugt wurden. Die Auswertung erfolgte mittels Computer und einem entsprechenden Programm, das die Bewegungen nach Anzahl und Intensität berechnete.

Die vergleichende statistische Betrachtung der Sitze untereinander erfolgte anhand der erhaltenen Messwerte durch den parameterfreien Test für korrelierende Stichproben von Wilcoxon.

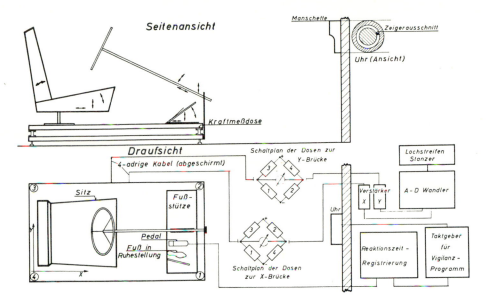

Abb. 1. Schematische Darstellung der Versuchseinrichtung.

Parallel zur Messung wurde die Befragung als Hilfsmittel der vergleichenden Beurteilung der Sitze herangezogen. Ausserdem sollte nach der Korrelation zwischen den Messwerten und der subjektiven Aussage gefragt werden. Der Fragebogen umfasste 7 Fragen (s. Tab. 1), die der Vp. aber in Form eines Heftes so vorgelegt wurden, dass auf jedem Blatt eine Frage allein stand, damit sie in ihrer Urteilsfindung von ihren voraufgegangenen Antworten möglichst unbeeinflusst blieb. Durch Anordnung der 7 Fragen in zwei verschiedenen Reihenfolgen und zwei Variationen der Anordnung der Polaritäten dieser Fragen wurden 4 Varianten des Fragebogens erstellt. Diese Varianten wurden gleich häufig auf die Einflussgrössen, wie Geschlecht der Vp. und Tageszeit, verteilt und sollten bewirken, dass denkbare Einflüsse einer schematisierten Beantwortung der Fragen theoretisch eliminiert wurden. Ihr Urteil hatte die Vp. in der Weise abzugeben, dass sie die 100 mm lange Strecke zwischen den Polaritäten an der Stelle durchkreuzte, die ihrem momentanen Urteil entsprach. Dabei sollte sie sich die Strecke zwischen den Polaritäten als lineare Skala vorstellen. Die Umsetzung der Urteile in quantitative Angaben, die der statistischen Betrachtung zugänglich sind, erfolgte durch Ausmessen der Strecke bis zur Urteilsangabe der Vp. von der Seite der positiv wertenden Polarität aus.

Die Standardisierung der Tätigkeit während des Versuches sollte bewirken, dass die Vp. weitgehend unbewusst sass und dennoch unter etwa ähnlichen Bedingungen an den Sitz '' gefesselt '' war wie der Lenker eines Fahrzeuges. Dazu hing in Augenhöhe, etwa 3 m vor der Vp., eine spezielle Uhr. Die Vp. hatte deren schrittweise vorrückenden Zeiger zu beobachten und auf dessen stochastisch auftretende Doppelschritte durch schnellen Pedaldruck zu reagieren. Gute Motivation und damit weitgehende Ablenkung vom eigentlichen Sitzen wurde durch diesen gleichzeitig ablaufenden Vigilanzversuch erzielt, der mit einem Prämiensystem gekoppelt war (Singer und Rutenfranz).

Tabelle 1. Der Fragebogen.—Art und Variation der Darbietung des Fragebogens s. Text

1.   wohl   Ich fühle mich auf diesem Sitz   nicht wohl

2.   bequem   Auf diesem Sitz sitze ich   nicht bequem

3.   angenehm   Die Form dieser Rückenlehne ist   nicht angenehm

4.   nicht angenehm   Die Form dieser Sitzfläche ist   angenehm

5.   härter   Ich wünsche diese Polsterung wäre   weicher

6.   zufrieden   Ich wäre mit diesem Sitz in einem Fahrzeug   nicht zufrieden

7.   schnell   Ich glaube, auf diesem Sitz ermüde ich   langsam

3. Die untersuchten Sitze

Aus Raumgründen muss hier auf die Beschreibung der konstruktiven Unterschiede der untersuchten 5 Sitze im Detail verzichtet werden. Nur so viel sei gesagt, dass es sich bei 3 von ihnen um kommerzielle Fahrzeugsitze mit deutlich verschiedener Profilierung und Härte der Polsterung handelte. Der 4. Sitz entstand durch Modifikation eines der drei mittels eines aufblasbaren Lenden-Stützkissens. Sitz Nr. 5 war mit einer hölzernen und ebenen Sitzfläche und ebensolcher Rückenlehne als Vergleichssitz in die Testreihe aufgenommen worden.

4. Versuchsdurchführung

Die Versuche sind mit 40 Vpn.—20 männlichen und 20 weiblichen Studenten —durchgeführt worden, wobei jede Vp. auf jedem der 5 Sitze über eine Testdauer von 2 Stunden gesessen hat. Die Darbietung der 5 Sitze erfolgte in 5 verschiedenen Reihenfolgen, die nach dem lateinischen Quadrat aufgebaut waren. Jeweils 8 Vpn.—je 4 männliche und 4 weibliche, von denen je 2 vormittags und 2 nachmittags erschienen—erhielten die Sitze in der gleichen Reihenfolge dargeboten. Auf diese Weise wurden mögliche Einflüsse auf das Verhalten theoretisch ausgeglichen, die durch die Gewöhnung an die Versuchssituation mit zunehmender Zahl der Versuche denkbar sind, sowie auch durch die Tageszeit.

Von der Vp. wurde verlangt, dass sie während des Versuches—vergleichbar mit den Umständen während einer gleichförmigen Fahrt auf der Autobahn— den rechten Fuss in der angegebenen Ruhestellung (Abb. 1) halten und nur im Fall einer erforderlichen Reaktion kurzzeitig auf das links daneben liegende Pedal und wieder zurücksetzen soll. Ferner war mindestens eine Hand am Steuer zu halten, sowie der ' Uhr ' ständig im Auge zu behalten.

Die Befragung der Vp. erfolgte zweimal im Sitzen. Das erste Mal direkt vor Versuchsbeginn, nachdem die Vp. schon etwa 10 Min. gesessen hatte (während dieser Zeit war die subjektiv optimale Anpassung von Sitz, Steuerrad und Pedalen vorgenommen worden), das zweite Mal unmittelbar nach Versuchsende.

5. Ergebnisse und Diskussion

Abb. 2 zeigt die Ergebnsse der Erst- und Zweitbefragung. Dargestellt wurden die Mittelwerte des subjektiven Urteils aller 40 Vpn. Die statistische Untersuchung des Materials zeigt, dass die 4 gepolsterten Sitze weder nach der Erst- noch nach der Zweitbefragung statistisch signifikant voneinander unterschieden wurden. Nur Sitz Nr. 5 wurde eindeutig negativ beurteilt und statistisch klar von den übrigen 4 unterschieden.

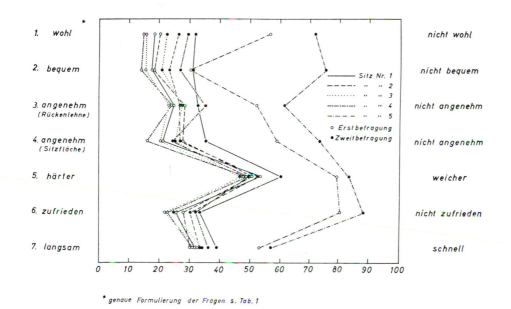

Abb. 2. Polaritätenprofile des subjektiven Urteils aus der Befragung.—Eingetragen sind die Mittelwerte aus den Antworten aller 40 Vpn., getrennt nach Erst- und Zweitbefragung (Vgl. auch Tab. 1).

Entsprechend den Arbeitshypothesen hätte auf Sitz Nr. 5, dem hölzernen Sitz, die grössere Anzahl von Bewegungen ausgeführt werden müssen. Abb. 3 zeigt aber, dass dies nicht der Fall ist. Hier wurde die Zahl der Vpn. über der Anzahl ihrer Bewegungen, die in Klassen zusammengefasst wurden, aufgetragen. Für den Fall, dass die Sitze nach der Anzahl der Bewegungen statistisch voneinander zu unterscheiden sind, hätte eine Häufung von Vpn. bei der Einteilung nach Bewegungsklassen auf den verschiedenen Sitzen in verschiedenen Klassen sichtbar werden müssen—für Sitz Nr. 5 besonders in den hohen Klassen.

Aber auch der Wilcoxontest sagte aus, dass keiner der 5 Sitze von den anderen—auch nicht der Holzsitz—statistisch signifikant unterschieden werden kann.

H

Ebensowenig kann mit diesem Test ein Unterschied der Sitze hinsichtlich der Intensität der auf ihnen ausgeführten Bewegungen nachgewiesen werden. Das ist einleuchtend, wenn man weiss, dass Intensität und Anzahl der Bewegungen mit $r = 0,96$ sehr hoch positiv und mit $P = 0,1$ gesichert miteinander korrelieren.

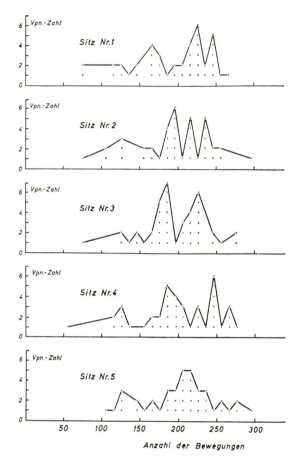

Abb. 3. Häufigkeitsverteilung der Vpn. auf die in Klassen unterteilte 'Anzahl der Bewegungen'.
Klassenbreite 10 Bewegungen.

Abb. 4, in der die Anzahl der Bewegungen der 40 Vpn. im Mittel über die Versuchsdauer aufgetragen ist, bestätigt durch die nahe beieinander verlaufenden Linien für die 5 Sitze optisch, was die Rechnung ergab, nämlich, dass kein Unterschied im Sitzverhalten des Versuchspersonen-Kollektivs in Abhängigkeit vom Sitz nachweisbar ist.

Schliesslich war auch keine signifikante Korrelation zwischen der Anzahl bzw. der Intensität der Bewegungen und dem subjektiven Urteil nachzuweisen. Dabei ging für das Urteil die Summe der Antworten auf die ersten 4 Fragen (s. Tab. 1) aus der Befragung nach Versuchsende ein, die u.E. allein eine wirklich wertende Aussagekraft besassen.

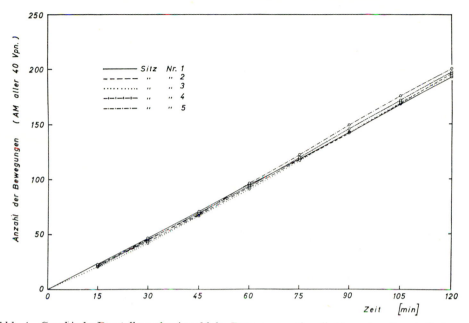

Abb. 4. Graphische Darstellung der Anzahl der Bewegungen über die Versuchszeit.—Aufgetragen ist das Mittel der Anzahl der Bewegungen aller 40 Vpn. für die 5 Sitze über die Versuchsdauer von 2 Stunden.

Es wurde die Hypothese aufgestellt, dass wenige, unbewusst im Sitzen ausgeführte Bewegungen pro Zeiteinheit und Bewegungen geringer Intensität mit subjektiv positivem Komfortempfinden korrelieren. Entsprechend dieser Hypothese haben 40 Vpn. je 5 verschiedene Fahrzeugsitze während jeweils 2 Stunden dauernder Versuche getestet. Neben den Messungen erfolgte zweimal eine Befragung der Vpn.

Die Messungen erlaubten keine statistisch signifikante Unterscheidung der Sitze untereinander; die Befragung liess nur den Sitz mit harter Sitzplatte und Lehne von den übrigen 4 statistisch signifikant trennen. Zwischen den Ergebnissen der Messung und der Befragung besteht keine statistisch signifikante Korrelation.

The hypothesis was set up that few unconsciously made movements per unit time as well as movements of only small intensity correlate with subjective positive feeling of comfort while sitting on a seat. According to this hypothesis each of 40 subjects tested five different car-seats for two hours. Twice during the test the subjects had to answer a questionnaire concerning their judgment of the seat.

There was no statistically significant difference between the seats according to the measurement. By means of the answers in the questionnaire one seat only could be separated as statistically significant from the other four seats. No statistically significant correlation exists between either the measurements or the subjective judgments of the subjects.

D'après l'hypothèse émise dans cette recherche, un certain nombre de mouvements inconscients faits par unité de temps, ainsi que des mouvements de faible amplitude seraient corrélés avec la sensation subjective de confort du subjet assis sur un siège. Pour éprouver cette hypothèse, on a fait tester à chacun des 40 sujets, 5 sièges de voiture différents pendant deux heures. Au cours de l'épreuve, les sujets devaient remplir, à deux reprises, un questionnaire se rapportant à l'appréciation de la qualité du siège.

Les résultats ne permettent pas de conclure à des différences statistiquement significatives entre les critères d'évaluation des sièges. D'après l'analyse des réponses au questionnaire, un siège seulement diffère significativement des quatre autres. Mais il n'y a pas de corrélations statistiquement significatives ni entre les mensurations effectuées, ni entre les appréciations subjectives fournies par les sujets.

Literatur

LIENERT, G. A., 1962, *Verteilungsfreie Methoden der Biostatistik* (Meisenheim an Glan : A. HAIN).

SINGER, R., und RUTENFRANZ, J., Untersuchung der Frage der Abhängigkeit der Vigilanzleistung von der Signalwahrscheinlichkeit bzw. Signalrate. *Im Erscheinen in „Studia Psychologica"*

Die Wirbelsäule von Schulkindern—Orthopädische Forderungen an Schulsitze

Von H. Schoberth

Orthopädischen Universitätsklinik Friedrichsheim, Frankfurt, Deutschland

Funktionell betrachtet, besteht die Wirbelsäule des Menschen aus einem unbeweglichen Mittelstück, der oberen Brustwirbelsäule und 2 beweglichen Endstücken, der Halswirbelsäule einerseits und der Lendenwirbelsäule mit der unteren Brustwirbelsäule andererseits. Die Lendenwirbelsäule wiederum schließt sich dem starren Sakralteil der Wirbelsäule, dem Kreuzbein an. Das Sakrum ist in den Iliosakralgelenken praktisch unbeweglich mit dem Darmbein verbunden. Sieht man von zyklusbedingten Lockerungen der Iliosakralfugen oder von traumatisch verursachten Stabilitätsminderungen ab, kann man die beiden Ossa coxae und das Sakrum als eine unbewegliche Einheit betrachten. Auf dem in der Symphyse geschlossenen Beckenring steht die Wirbelsäule nach einem Vergleich von Mollier wie der Mastbaum auf dem Deck eines Schiffes. Die Basis für den gesamten Wirbelsäulenaufbau ist die Deckplatte des Kreuzbeines. Ist diese nach ventral abfallend geneigt, wird die untere Lendenwirbelsäule zwangsläufig in eine Ventriflexion gezwungen. Zum aufrechten Stand muß dann die Rumpfwirbelsäule nach dorsal gebogen werden. So entsteht im Lendenteil die Lordose. Ist die Kreuzbeindeckplatte horizontal gestellt, oder nach dorsal abfallend, entwickelt sich kompensatorisch eine Steilstellung, oder eine Totalkyphose.

Grundsätzlich gelten die gleichen Bedingungen für die Form der Wirbelsäule im Sitzen.

Beim Hinsetzen werden die Knie- und Hüftgelenke gebeugt (Abb. 1), gleichzeitig erfolgt eine Schwerpunktverlagerung des Rumpfes nach ventral durch eine ergiebige Ventriflexion. Hat das Gesäß auf dem Sitzbrett seine Auflage gefunden, wird die Wirbelsäule zur aufrechten Haltung sekundär wieder gestreckt.

Infolge der Beugung in den Hüften und der Entspannung der ischiocruralen Muskeln durch die Flexion der Kniegelenke ist die Haltung des Beckens im Sitzen eine andere als im Stehen.

Gehen wir noch einmal von der Haltung der Wirbelsäule im Stehen aus. Das Becken dreht sich je nach Einsatz der ventralen oder dorsalen Muskelzüge um die frontale Hüftgelenksachse, vollführt also eine Beugung oder Streckung. Dabei wird das ganze Becken nach vorne oder nach hinten gedreht (Abb. 2). In diesem Zusammenhang spricht man von einer Kippung des Beckens, wenn im Stehen der Winkel zwischen Rumpfachse und Beinachse im Hüftgelenk kleiner wird als 180 Grad, d.h. eine Beugehaltung im Hüftgelenk besteht. Demgegenüber ist von einer Beckenaufrichtung die Rede, wenn das Hüftgelenk gestreckt oder gar überstreckt wird. Im Extremfall ist bei der Aufrichtung des Beckens die Hüfte überstreckt, mit anderen Worten, das Ligamentum iliofemorale maximal angespannt. Mit der Kippung des Beckens nimmt die

Abb. 1. Beugung der Wirbelsäule beim Hinsetzen. Man erkennt die ververmehrte Rundung der Wirbelsäule.

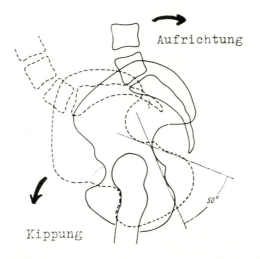

Abb. 2. Kippung und Aufrichtung des Beckens um die quere Hüftgelenksachse. Die Vorkippung führt zu einer Lordose der Lendenwirbelsäule, die Aufrichtung zu einer Streckung.

Lordose der Lendenwirbelsäule zu, mit der Rückdrehung oder Aufrichtung flacht sich die Lordose ab, die Lendenwirbelsäule steht dann steil. Auch im Sitzen kann man von einer Kippung und einer Aufrichtung des Beckens sprechen. Zum Verständnis der Terminus technicus richtet man den Blick auf die Stellung des Kreuzbeines und spricht dann von einer Aufrichtung, wenn die Kreuzbeinlängsachse sich der Vertikalen nähert, oder vertikal steht und von einer Kippung, wenn das Kreuzbein sich ventral senkt und der Horizontalen zustrebt.

Wir haben das Zusammenspiel von Beckenstellung und Wirbelsäulenform im Sitzen untersucht. Dabei haben wir gefunden, daß eine Abhängigkeit von der Neigung der Kreuzbeindeckplatte geegn den Horizont besteht.

Ist der ventrale Winkel zwischen Kreuzbeindeckplatte und Horizont größer als 16 Grad, dann ist eine Lordose der Lendenwirbelsäule zu beobachten. Zeigt die Kreuzbeindeckplatte dagegen einen dorsalen Abfall von mehr als 10 Grad, dann nimmt die Lendenwirbelsäule immer eine Kyphose ein. Diese Gesetzmäßigkeiten gelten freilich nur für den Fall, daß die Lendenwirbelsäule in ihren einzelnen Segmenten beweglich ist. Bestehen am lumbosakralen Übergang Segmentverschiebungen im Sinne einer Lumbalisation oder Sakralisation, oder finden sich Fixierungen in höhergelegenen Abschnitten, dann muß sich die Bewegungsbehinderung auch auf die Form der Wirbelsäule selbst auswirken.

Zusammengefaßt ergibt sich : Die Form der Wirbelsäule im Stehen wie im Sitzen hängt unmittelbar von der Stellung des Beckens, erkennbar an der Neigung der Kreuzbeindeckplatte, ab. Wir haben als Bezugsebene die Kreuzbeindeckplatte gewählt, weil sie die Wirbelsäulenform sicherer bestimmt als die Beckenstellung selbst. Wie bekannt, bestehen nämlich mitunter nicht unerhebliche Unterschiede im Einbau des Kreuzbeines in den Beckenring.

Im Röntgenbild des Beckens kann man im seitlichen Strahlengang den aufsteigenden Sitzbeinast mit der Spina ischiadica immer gut erkennen. Legt man an den Scheitelpunkt der Incisura ischiadica major und der Incisura

ischiadica minor eine Tangente, dann hat man damit den aufsteigenden
Sitzbeinast, den Sitzbalken nach Waldeyer erfaßt (Abb. 3). Mit der Sitzbeintangente ist die Stellung des Beckens im Raum definiert, wenn man diese mit
einer Hilfslinie, z.B. der Raumhorizontalen in Beziehung bringt. Wir haben
zur Festlegung der Wirbelsäulenform im Sitzen die Raumhorizontale gewählt,
weil sie der ausgeloteten Sitzfläche des Hockers entsprach, auf dem die
auszuwertenden Röntgenaufnahmen gefertigt worden sind.

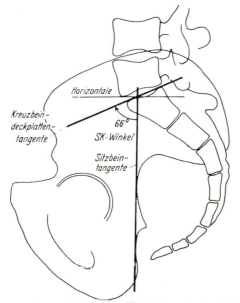

Abb. 3. Eine Tangente an dem aufsteigenden Sitzbeinast in der seitlichen Röntgenaufnahme
erlaubt die genaue Lokalisation der Beckenstellung im Raum.

Analysiert man auf seitlichen Aufnahmen der Wirbelsäule das Kreuzbein,
so lassen sich hinsichtlich seiner Krümmung große Unterschiede finden. Wir
haben sie nach Radlauer mit dem mittleren Sakralkrümmungsindex definiert.
So läßt sich die Kreuzbeinkrümmung exakt festlegen. Sie reicht von einer
fast gestreckten Form, deren Index unter dem von anthropoiden Affen liegt,
bis zu stark gekrümmten Formen im Sinne des Os sacrum arcuatum nach
Scherb. Bei den weniger gekrümmten Kreuzbeinformen mit einem kleinen
Index ist der Winkel zwischen Sitzbeintangente und Kreuzbeindeckplatte,
von uns SK-Winkel gennant, im allgemeinen bergrößert. Je nach ihrem
Krümmungsindex haben wir eine Kreuzbeinform A, B, C und D unterschieden,
wobei die Kreuzbeine vom Typ A einen Index unter 12 hatten, also sehr
gestreckt sind. Von einem Kreuzbein D haben wir von einer Indexzahl von
über 30 gesprochen. Den Zusammenhang zeigt das Diagramm. Kreuzbeine
mit geringer Krümmung zeigen stets einen geringeren Ventralabfall der Kreuzbeindeckplatte. Die Kreuzbeintypen A und B als die geringeren Krümmungen,
sind im Kindesalter eindeutig dominierend, während beim Menschen jenseits
des 40sten Lebensjahres die stark gekrümmten Typen C und D überwiegen.
Lassen Sie mich an das erinnern, was wir vorher festgestellt haben. Die
Form der Lendenwirbelsäule hängt von der Neigung der Kreuzbeindeckplatte
ab. Unter Berücksichtigung des eben Gesagten ergibt sich, daß bei einem

großen SK-Winkel, also bei gestrecktem Kreuzbein, die Ventralneigung der Kreuzbeindeckplatte zur Erreichung einer Lendenlordose nur durch eine verstärkte Kippung des Beckens erreicht werden kann.

Wir haben gesehen, daß beim jugendlichen Individuum das Kreuzbein eine relativ gestreckte Form aufweist. Die Kreuzbeinkrümmung des Erwachsenen kommt offensichtlich durch eine Ventral- und Kaudalverlagerung von S^1 im Laufe der postnatalen Entwicklung um das 8. bis 12. Lebensjahr zustande und nimmt beim Erwachsenen bis zum Greisenalter hin noch zu. Sie verläuft also ähnlich wie die Änderung des Schenkelhalswinkels, der unter dem Einfluß der Belastung beim alten Menschen auf Werte unter 120 Grad absinken kann. Wir glauben, daß die Belastung des Kreuzbeines bei Beckenkippung die Ventral- und Kaudalverlagerung von S^1 fördert, mit anderen Worten, die Ausbildung der Kreuzbeinkrümmung verstärkt. So sehen wir in der typischen Haltung des Kleinkindes, das im Stehen den Oberkörper auf dem Becken zurückneigt (Abb. 4), nicht etwa den Ausdruck einer krankhaften oder schlechten Haltung, sondern eine physiologische Notwendigkeit.

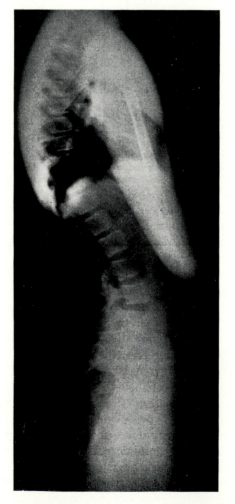

Abb. 4. Haltung der Wirbelsäule im Stehen. Der Oberkörper ist im ganzen auf dem Becken zurückgeneigt. Nur die unteren Lendensegmente stehen in Lordose.

Die angestellten Überlegungen bedürfen indessen einer Ergänzung. Wir haben bisher nur vom lumbosakralen Übergang und der Bedeutung der Beckenstellung für die Haltung gesprochen. Tatsächlich stellt sie den entscheidenden Faktor für die Gestaltung von Sitzmöbeln, im speziellen von Schulsitzen dar. Die Form der Wirbelsäule hängt aber ohne Zweifel von einer ganzen Reihe anderer Faktoren ab, die im einzelnen wenigstens andeutungsweise angesprochen sein sollen.

Wie Töndury gezeigt hat, ist das normale Wachstum der Wirbelkörper und damit auch die Ausbildung der Bandscheiben an die ungestörte Involution der Chorda dorsalis gebunden. Bleibt dies ungenügend, kann man im späteren Alter im hinteren Drittel der Wirbelkörper im seitlichen Röntgenbild Einbuchtungen der Grund- und Deckplatten erkennen, die perlschnurartig übereinandergereiht sein können. Sie sind stets Ausdruck einer tiefgreifenden Veränderung im Bewegungssegment. Funktionell ist in diesen Abschnitten der Wirbelsäule die Beweglichkeit behindert oder ganz aufgehoben. Ähnliche Befunde findet man auch bei andersartigen Ossifikationsstörungen der Wirbelsäule, die man heute unter dem Begriff der enchondralen Dysostosen zusammenfaßt. Daß es sich dabei um vererbbare Störungen handelt, zeigen die einschlägigen Untersuchungen von Mau, Rathke und anderen. Oft sind enchondrale Dysostosen der Wirbelsäule mit Dysplasien großer Körpergelenke, am häufigsten der Hüftgelenke, verbunden. In die gleiche Gruppe, wenn auch graduell oder kausal-genetisch prinzipiell verschieden, gehören die Ossifikationsstörungen im Adoleszentenalter, die unter der Bezeichnung der Scheuermann'schen Adoleszentenkyphose bekannt sind. Bei der Scheuermann'schen Krankheit handelt es sich im Prinzip um eine Ossifikationsstörung an den Wachstumszonen des Wirbels, den Grund- und Deckplatten. Sie hat mit einem Haltungsschaden nichts zu tun. In diesem Zusammenhang ist von Bedeutung, daß in allen Fällen die Bandscheibe mit betroffen ist. Sie zeigt am Ende eine fibröse Umwandlung, mit anderen Worten, sie hat ihre Elastzität eingebüßt. Funktionell sind die betroffenen Segmente dann versteift. Für die klinische Wertigkeit ist es nun von großer Bedeutung, an welchen Wirbelsäulenabschnitten die Versteifung sitzt und zu welcher Gesamthaltung sie führt. Am häufigsten findet man umschriebene Fixierungen der Wirbelsäule beim Jugendlichen im Dorsalabschnitt, seltener dorsolumbal, häufiger wieder am lumbosakralen Übergang. Die einfache Inspektion, auch die einmalige Röntgenaufnahme ist nicht in der Lage die wahren Verhältnisse zu klären. Aus diesem Grunde muß man die Beweglichkeit der Wirbelsäule im einzelnen feststellen und nach Möglichkeit objektivieren. Dazu eignet sich ohne Zweifel am besten die Wirbelsäulenganzaufnahme, die in den beiden extremen Sitzhaltungen, der maximalen Aufrichtung und der tiefsten Senkung durchgeführt wird. Sie ist immer notwendig, wenn man Einblick in die knöchernen Formabweichungen und die Lokalisation der Fixierung gewinnen will. Die technische Durchführung ist aber apparativ aufwendig und darum nicht überall möglich. Ersatzweise kann man zur groben Orientierung die Fotositzkurven heranziehen. Wir fertigen dazu je eine Aufnahme in aufrechter Haltung und in Ruhehaltung unter den gleichen Bedingungen im seitlichen Strahlengang. Aus dem Verlauf der Rückenkontur, die wir auf ein Millimeter-papier übertragen, lassen sich gewisse Schlüsse ziehen.

In Ruhehaltung zeigen alle Menschen, mit Ausnahme schwer fixierter
Deformierungen der Wirbelsäule, eine Totalkyphose.   Sie wird aus statischen
Gründen notwendig, weil durch die Entspannung der Ischiocruralmuskulatur
und die Beugung in den Hüftgelenken die Beckenaufrichtung viel ausgeprägter
ist als in Stehen.   Soll der Rumpfschwerpunkt über die Unterstützungsfläche
des Rumpfes, das Areal um die Sitzbeinhöcker gebracht werden, muß eine
maximale Ventriflexion der Gesamtwirbelsäule erfolgen.   Die Totalkyphose im
Sitzen ist nach der Gradausprägung recht unterschiedlich (Abb. 5).   Sie
reicht vom abgeflachten Rücken bis zur verstärkten Kyphose.   Besonders
bedeutungsvoll ist der Sitz des Krümmungsscheitels.   Beim normalen Jugendlichen fanden wir bei Schuluntersuchungen an 1035 Kindern im Alter von 6 bis
15 Jahren bei 668, d.h. in 64,5% der Fälle, eine kontinuierliche Rundung.   Bei
164 Kindern, d.h. in 15,9% war der Krümmungsscheitel im Bereich der Brustwirbelsäule gelegen.   In diesen Fällen ist zunächst an eine fixierte Brustkyphose
zu denken, auf alle Fälle ist die weitere orthopädische Analyse notwendig.   Bei
203 Kindern, d.h. in 19,6% der Fälle, lag der Krümmungsscheitel der Kyphose
an Dorsolumbalübergung, bzw. im Bereich der Lendenwirbelsäule.

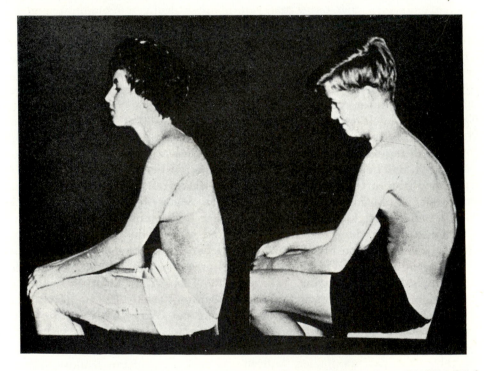

Abb. 5.   2 verschiedene Kyphoseformen im entspannten Sitzen, (*a*) Flachrücken : Die Wirbelsäule
         kann nicht weit genug gebeugt werden, (*b*) Rundrücken : Im Sitzen vermehrte
         Beweglichkeit der Gesamt-WS.

Auch bei der Aufrichtung bestehen im seitlichen Bild erhebliche Unterschiede.
Erfolgt die Aufrichtung aus mittlerer oder hinterer Sitzhaltung, verläuft sie
immer synchron mit einer Kippung des Beckens.   Weil diese bei Menschen mit
versteiftem Hüftgelenk nicht möglich ist, ändert sich wie bei diesem 12 jährigen
Mädchen die Wirbelsäulenform auch durch die Streckung nicht.   Hüftversteifte

müssen die fehlende Beckenkippung durch Inanspruchnahme besonderer Sitzhilfen kompensieren. Erfolgt die Aufrichtung der Wirbelsäule aus vorderer Sitzhaltung, genügt die Streckbewegung alleine, da die Beckenkippung schon in der Ruhehaltung vorhanden ist.

Von den untersuchten 1035 Schulkindern zeigte die Hälfte auch in voller Aufrichtung nur eine angedeutete Lordosierung der Wirbelsäule, so daß man insgesamt den Eindruck einer Streckung oder Steilhaltung der Rumpfwirbelsäule gewinnt. Bei 30,5% fanden wir eine Lordose oberhalb der Beckenkämme und bei 19,4% war die lordotische Einziehung in dem Bereich der unteren und mittleren BWS verschoben.

Diese Ergebnisse haben wir an 125 Jugendlichen im Alter zwischen 18 und 25 Jahren, bei 102 Mädchen und 23 Männern nachgeprüft. Während in Ruhehaltung etwa die gleichen Ergebnisse feststellbar waren, zeigten sich größere Unterschiede in der Lordoseform. Bei 41,5% der Untersuchten war der Scheitelpunkt der Lordose im Bereich der oberen Lendenwirbelsäule gelegen, bei einigen fand sich wie in diesem Beispiel eine Lordose in der mittleren BWS.

Eine genaue Analyse ist bei den Fällen einer verstärkten Kyphose am Dorsolumbalübergang notwendig (Abb. 6). Nach dem klinischen Aspekt handelt es sich dabei immer um einen sog. Sitzbuckel, der oft fälschlicherweise mit der Rachitis in Zusammenhang gebracht wird. Am häufigsten ist der Sitzbuckel durch eine extreme Beweglichkeit am dorsolumbalen Übergang hervorgerufen. Man erkennt das an der verminderten Ausbiegung nach dorsal in Ruhehaltung und der verstärkten Einziehung, bzw. Lordose in maximaler Aufrichtung. Die Fixierung der Brustwirbelsäule in Streckhaltung, z.B. beim Flachrücken, ruft eine vermehrte Beweglichkeit am Übergang zu den beweglicheren Abschnitten der unteren Rumpfwirbelsäule hervor. So findet man auf den Wirbelsäulenganzaufnahmen die Kyphose der Lendenwirbelsäule und die Streckhaltung, d.h. die Minderung der Krümmung der Brustwirbelsäule vermehrt. Nur in seltenen Fällen ist die dorsolumbale Kyphose durch effektive Versteifungen am Brust-Lenden-übergang hervorgerufen. In diesen Fällen bleibt bei der Lordosierung die dorsolumbale Kyphose bestehen und erst darüber läßt sich eine lordotische Bewegung feststellen. Auch Versteifungen im Bereich der Lendenwirbelsäule können eine Sitzkyphose hervorrufen. Die größte Bedeutung kommt dabei der Sakralisation zu. Durch Einbeziehung des letzten Lendenwirbels in das Kreuzbein wird der Bewegungsraum der Lendenwirbelsäule von 5 auf 4 Segmente reduziert. Zum Ausgleich muß dann entweder die Exkursionsfähigkeit der verbliebenen Segmente verstärkt, oder die Lordosierung in die unteren Brustsegmente verlagert werden. Bei übermäßiger Beanspruchung einzelner Bewegungssegmente folgt der Überlastung bald die Degeneration der Bandscheibe, die zu deformierenden Arthrosen der kleinen Wirbelgelenke oder zu universellen Lockerungen einzelner Segmente führt. In beiden Fällen ist der vorzeitige Funktionsabbau die unausbleibliche Folge. Im Bereich der mittleren und oberen Brustwirbelsäule sind Bewegungs-einbußen nicht allzu bedeutungsvoll, weil hier ja schon physiologischerweise die Beweglichkeit recht gering ist. So besitzen die hochsitzenden umschriebenen Fixierungen nur dann Bedeutung, wenn sie mit einer verstärkten Kyphosierung der oberen Brustsegmente einhergehen. Das ist nicht nur beim totalrunden Rücken der Fall, sondern betrifft auch jene Flachrückenformen, die im oberen Brustabschnitt umgrenzte Kyphosen aufweisen. Der Grund ist

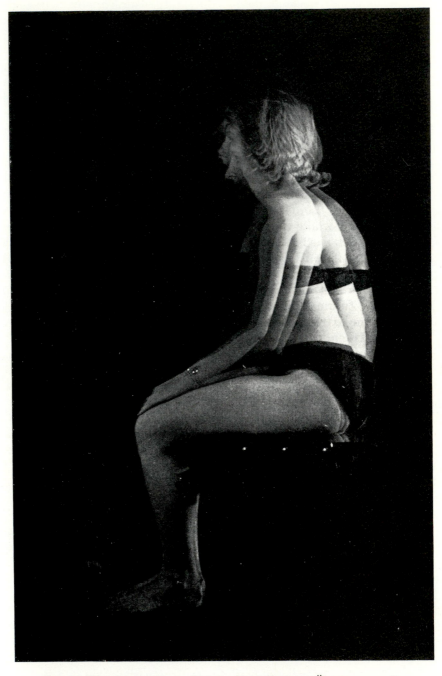

Abb. 6. Wirbelsäulen-Ganzaufnahme, Versteifung am Übergang von der Brust- zur Lendenwirbelsäule, mangelnde Aufrichtung im Sitzen.

leicht einzusehen und statisch genauso wie die vermehrte Beckenkippung zu bewerten. In beiden Fällen handelt es sich im Prinzip darum, daß die Basis des beweglicheren und damit dynamisch beanspruchten Teilstückes der Wirbelsäule eine vermehrte Ventralneigung zeigt.

Im Sitzen spielt die Lendenlordose überhaupt keine Rolle. Sie ist durch die Drehung des Beckens, also passiv, auszugleichen und in Ruhehaltung an sich nicht vorhanden. Wie Äkerblom gezeigt hat, wird die Ventriflexion, mit anderen Worten, die weitere Kyphosierung durch Anspannung der Ligamenta flava begrenzt. Der Erector trunci ist weitgehend erschlafft. Das zeigt eindrucksvoll das elektromyographische Bild (Abb. 7). Beim Sitzen mit totalrundem Rücken finden wir eine geringe Aktivität der Brust- und Lendenmuskulatur. Die geringe Zahl der Entladungen, erkennbar an der Abnahme der Frequenz und Amplitude läßt den Schluß zu, daß die stoffwechselfordernde Haltearbeit geringer ist. Daraus könnte der Schluß gezogen werden, daß das Sitzen mit Totalkyphose ökonomisch besonders günstig sei und darum anzustreben wäre. Tatsächlich sind die Ruhesessel ja auch nach diesen Gedankengängen konzipiert und konstruiert. Hier besteht aber ein Widerspruch zur täglichen Erfahrung. Langes Sitzen mit totalrundem Rücken führt bekanntermaßen zu Schmerzen im Kreuz, die dem Autofahrer nach längerer Fahrzeit durchaus geläufig sind. Diese Schmerzen sind indessen primär nicht myogen ausgelöst, sondern arthrogen durch die Dehnung der Gelenkkapseln entstanden, die in Kyphose über längere Zeit auf Zug beansprucht werden. Sekundär entstehen in diesen Fällen dann erst die Muskelschmerzen. Sie werden durch Elastizitätseinbußen infolge lokaler Verspannungen und die Überlastung der Sehnenansätze an Beckenkamm und Wirbelfortsätzen im Sinne von Tendopathien ausgelöst. Aus diesem Grunde ist das Sitzen mit totalrundem Rücken ohne Stützmöglichkeit irgend welcher Art schädlich und führt zwangsläufig zu Beschwerden. An der Brustwirbelsäule sind die arthrogenen Schmerzen durch Kapseldehnung bedeutungslos, weil in den fraglichen oberen Brustabschnitten die Bewegungsausschläge an sich gering sind, bzw. völlig fehlen. So spielt hier die myogene Schmerzentstehung die führende Rolle.

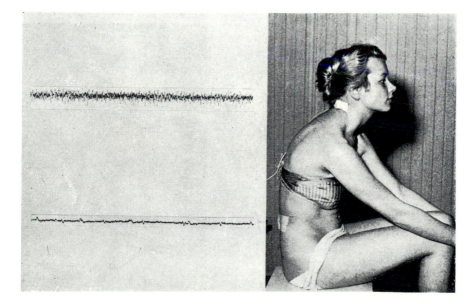

Abb. 7. Elektromyographische Kurve im lockeren Sitzen. Der lange Rückenstrecker im Rumpfbereich ist erschlafft, erkennbar an der verminderten Frequenz und der geringen Amplitudenhöhe.

Ganz anders liegen die Verhältnisse am Zervikalübergang. Je mehr die obere Brustwirbelsäule kyphotisch gekrümmt ist, umso stärker muß die Halswirbelsäule zum Blick geradeaus lordosiert werden. Dazu ist aktive Haltearbeit der Nackenmuskeln notwendig.

Eine Kopfhaltung ohne nennenswerte Muskelanstrengung ist nur möglich, wenn es gelingt die Halswirbelsäule so auszubalancieren, daß der Schwerpunkt des Schädels in die Unterstützungsfläche der Atlaskondylen fällt. Das ist aber nur denkbar, wenn auch die Rumpfwirbelsäule im ganzen einen gestreckten Verlauf zeigt. Die passive Begrenzung der Beckenaufrichtung erlaubt diese Streckhaltung ohne nennenswerte Muskelarbeit. Darum hat die Lehnengestaltung am Schul und Arbeitssitz eine so große Bedeutung. Eine Entlastung der Nackenmuskulatur kann auch durch das Auflegen der Arme auf einen Tisch oder auf Armlehnen erreicht werden, weil damit die Haltearbeit der Schultergürtelmuskeln, die an der Halswirbelsäule entspringen, reduziert ist. Der Schultergürtel wird bekanntlich im wesentlichen vom absteigenden Teil des Trapezius, vom Levator scapulae und dem Sternocleidomastoideus getragen. Trapezius und Levator scapulae spielen aber zugleich bei der Streckung der Halswirbelsäule eine wichtige Rolle. Sie sind darum bei der Hyperlordose vermehrt in Aktion, was sich elektromyographisch belegen läßt (Abb. 8). Die Folge einer derartigen Doppelbelastung ist nicht selten das Cervicalsyndrom, das nach unseren Untersuchungen weit häufiger durch unphysiologische Sitzhaltung, als durch die sicher überwertete Osteochondrose der cervicalen Wirbelsäule hervorgerufen wird. Die Brachialgien von Stenotypistinnen jugendlichen Alters und von Schulkindern um das 16. und 17. Lebensjahr finden so eine Erklärung. Stenosen im Bereich des Foramen intervertebrale, etwa durch exostosenartige Bildungen oder Lockerungen im Bewegungssegment sind in diesem Alter ohne nennenswerte Bedeutung.

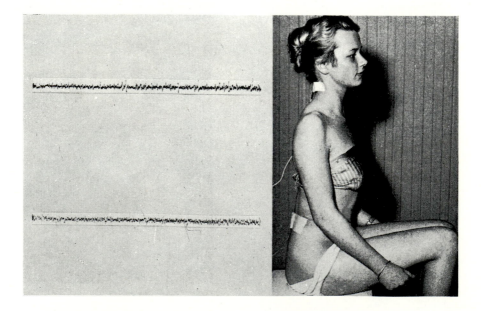

Abb. 8. Aktivität der Hals- und Rücken-Muskulatur im aufrechten Sitzen,
vermehrte Entladungen im Bereich der Rumpfstreckmuskualtur.

Damit schließt sich der Kreis. Aus den angestellten Untersuchungen lassen sich Folgerungen ziehen, die beim Bau von Schulsitzen berücksichtigt werden müssen.

Beim Schulanfänger mit 6 Jahren ist die Form der Wirbelsäule noch nicht fertig ausgebildet. Dubois fand, daß im 7. Lebensjahr noch bei 60% aller Kinder die Lendenlordose fehlt. Sie entwickelt sich im Zusammenhang mit

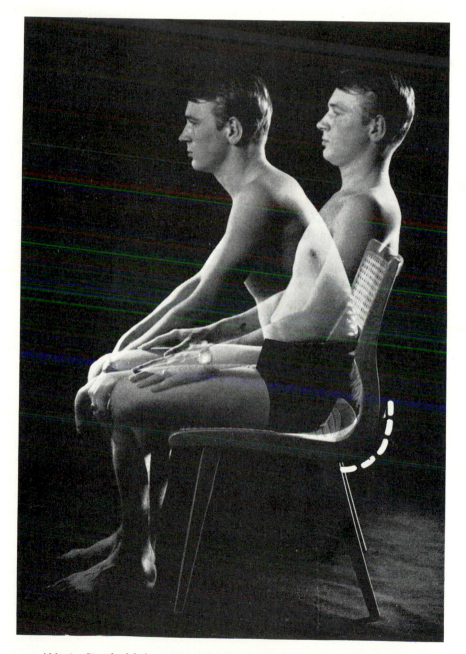

Abb. 9. Durch richtige Abstützung des Beckens Lehnenkontakt in vorderer und hinterer Sitzhaltung.

der Kaudal- und Ventralverschiebung des I. Sakralwirbels im Alter von 8 bis 12 Jahren, die zur Krümmung des Kreuzbeines führt. Beim lockeren entspannten Sitzen ist das Becken aus statischen Gründen soweit aufgerichtet, daß in hinterer Sitzhaltung die Kreuzbeindeckplatte horizontal steht oder gar nach dorsal hin abfällt. So bildet sich eine großbogige Kyphose der gesamten Rumpfwirbelsäule aus. Auf der nach ventral geneigten Deckplatte des 1. Brustwirbels baut sich in dieser Haltung die vermehrt lordotisch gekrümmte Halswirbelsäule auf. Sie muß durch zusätzliche Muskelarbeit vonseiten der Schultergürtelmuskulatur gehalten werden. Zur Vermeidung von vorzeitigen Ermüdungserscheinungen ist die Begrenzung der Beckenaufrichtung notwendig. Sie kann nur durch Kräfte erfolgen, die am Becken selbst angreifen. Ein Keil im hinteren Anteil des Sitzes ist beim Schulsitz ungünstig, weil dadurch ein Ventralschub des Beckens zustande kommt, mit anderen Worten, das Kind vom Sitz nach vorne herunterrutscht. Die Abstützung kann darum nur am oberen Beckenrand und an der Rückfläche des Kreuzbeines, das anatomisch etwas tiefer und ventral von den beiden hinteren Beckenkämmen gelegen ist, erfolgen (Abb. 9). Durch Ausbildung eines Lehnenwulstes, der das Becken faßt, ist in der Schreib- und der Lesehaltung ein Lehnenkontakt gewährleistet. Auf dem dadurch fixierten Becken kann die Wirbelsäule frei ausbalanciert werden. So ist die ökonomische Sitzhaltung gewährleistet. Die Höhe des Lehnenwulstes richtet sich nach der Höhe des Beckens und liegt zwischen 10 und 22 cm über dem Sitzbrett je nach Lebensalter. In der hinteren Sitzhaltung, der Höhrhaltung entsprechend, kann der obere Lehnenteil beansprucht werden, der nach dorsal hin antsteigt. Er darf aber nur temporär benutzt werden, weil in dieser Sitzhaltung die Ruhekyphose stets maximal ausgebildet ist.

Die Sitzhöhe über dem Boden richtet sich nach der Länge der Unterschenkel. Sie sollen auf jeden Fall senkrecht stehen, während die Oberschenkel dem Sitzbrett horizontal aufliegen müssen. Genaue Abmessungen ergeben sich nach anthropologischen Reihenuntersuchungen und sind je nach Rasse unterschiedlich. Wichtig ist auf jeden Fall, daß dem Schulkind genügend Bewegungsfreiheit bleibt. Aus diesem Grunde muß auf die Höhe des Schultisches, die notwendige Beinfreiheit und den seitlichen Abstand zum Sitznachbarn besonders geachtet werden.

Der gute Schulsitz hat aber nur dann einen Sinn, wenn er vom Schüler auch richtig benutzt wird. Trotz allem bleibt die ausschließlich sitzende Lebensweise eine Belastung. Durch körperliche Betätigung in Spiel und Sport wird der notwendige Ausgleich geschaffen, der zur vollen Ausreifung der endgültigen Wirbelsäulenform erforderlich ist. Wer sich um die Verbesserung der Schulmöbel bemüht, sollte darüber nicht vergessen, daß er damit auch die Verpflichtung übernimmt, der Bewegungsarmut unseres Zeitalters entgegenzuwirken. So verstanden hat das Wort von Spitzy heute noch seine besondere Aktualität: Eine Schulbank ist umso besser, je weniger das Kind darin sitzt.

Die Form der Wirbelsäule wird durch die Stellung des Beckens entscheidend bestimmt. Weil die Beckenaufrichtung im Sitzen viel ausgeprägter ist als im Stehen wird die Kyphose vollkommen. So ist es möglich, die maximale Kyphosierbarkeit durch die Ruhesitzhaltung zu erfassen. Die Form der Lendenwirbelsäule hängt von der Neigung der Kreuzbeindeckplatte ab.

Sitzanalysen zeigen darüber hinaus, daß Bewegungsbehinderungen in den Lenden- und Brustsegmenten sich auf die Gesamtform auswirken. Am Lumbosacral-Übergang spielen neben der Sacralisation vor allem Krümmungsanomalien des Kreuzbeines eine Rolle. In höheren Segmenten sind Bandscheibendifferenzierungsstörungen und Abweichungen der normalen Wirbelkörperverknöcherung von besonderem Interesse.

Die Wirbelsäulenganzaufnahme in 2 extremen Haltungen, nämlich der Ruhehaltung und der angespannten, aufrechten Haltung vermittelt den exakten Überblick über die Wirbelsäulenkapazität, worunter mit Schede der maximale Bewegungsumfang verstanden wird. Einen guten Überblick vermitteln aber auch die Foto-Sitz-Kurven, welche die Form des Rückens und damit die Wirbelsäulenkapazität wiedergeben. Während die Lendenlordose im Sitzen ohne Bedeutung ist, verdient die Halslordose besondere Beachtung. Sie ist durch eine vermehrte Ruhekyphose der BWS oft verstärkt. Dadurch werden die Schultergürtelmuskeln statisch überlastet und Ursache hartnäckiger Nacken-Arm-Schmerzen.

Zur Vermeidung von vorzeitigen Ermüdungserscheinungen muß die Beckenrotation begrenzt werden. Die Abstützung erfolgt durch einen Lehnenwulst, der den oberen Beckenrand und das Kreuzbein abstützt. Eine Abstützung in der unteren Brustwirbelsäule ist in Ruhehaltung notwendig.

Zur Vermeidung von bleibenden Sitzschäden ist auf einen genügenden Ausgleichssport besonderer Wert zu legen.

The position of the pelvis determines the shape of the spine. The lumbar lordosis is of little importance in the sitting posture. In order to avoid fatigue it is necessary to support the iliac crest and the sacrum. A support of the lower part of the thoracic spine is necessary for a rest position.

Orthopaedic damage due to sitting posture can be avoided by proper physical activity in sports and games.

La position du bassin détermine la forme de la colonne vertébrale. La lordose lombaire est de peu d'importance dans la position assise. Pour éviter la fatigue il convient de soutenir la crête iliaque et le sacrum. Pour une position de repos, il convient de soutenir la partie basse de la colonne thoracique.

Les troubles orthopédiques dus à la position assise doivent être évités par une activité physique appropriée, dans le sport ou les jeux.

Entwicklung eines Verschiebbaren Rückenprofils für Auto-und Ruhesitze

Von M. Rizzi

Rheumaerkrankungen FMH, Schenzengraben 23, 8002 Zürich

Bei der Beurteilung der Sitzschäden der Wirbelsäule ist anatomisch gesehen folgendes zu beachten:

Zwei Segmente sind äusserst beweglich: die Hals- und die Lendenwirbelsäule.

Ein Segment ist nur bedingt beweglich und zwar die Brustwirbelsäule.

Das Kreuzbein ist völlig unbeweglich.

Die Wirbelkörper, verschieden in der Grösse und in der Form, sind voneinander getrennt durch die Bandscheiben. Diese letzten zeigen je nach der Höhe der Wirbelsäule-Segmente gewisse Charakteristika, die der Funktion entsprechen.

Zwei Wirbelkörper und die dazwischen liegende Bandscheibe bilden nach Junghans ein *vertebrales Segment*.

Jedes Segment gilt als funktionelle Einheit. Das Synchrone-Gesamtspiel der verschiedenen Segmente gestattet eine harmonische Funktiontüchtigkeit der Wirbelsäule. Jede Störung der Funktion einer einzelnen Einheit provoziert die reflexogene Blockade der oben liegenden. (Das Symptom ist röntgenologisch verifizierbar und geht unter dem Namen Guntz'sches Phaenomen).

Neben diesen Elementen spielen die Bänder, die Kapseln und die angrenzende Muskulatur mit den Sehnen in der Anatomie und Pathologie der Wirbelsäule eine sehr grosse Rolle.

Aeusserst wichtig für die Interpretation des Schmerzphaenomens sind die Kenntnisse der eigenen Innervation der Bänder, die ein Segment umgeben sowie die topographischen Verhältnisse zwischen Bandscheiben und Nervenwurzel bezw. Rückenmark.

Die Ligamenta longitudinalia anteriora et posteriora sowie das Periost werden durch nicht eingekepselte Endungen versorgt, ebenso die intervertebralen Gelenke. Die freien Fasern wurden in den äussersten Schichten der Anuli fibrosi der Bandscheiben sowie im Bereich des Ligamentum longitudinale posterius beschrieben.

Die gesamte Wirbelsäule zeigt bekanntlich ein S-förmiges Profil. Das quasi starre Segment der Brustwirbelsäule in kyphotischer Stellung bildet die Basis für die cervicale Lordose; das statische Kreuzbein gilt als Basis für die Lendenlordose. Diese beiden Uebergangsstellen sind besonders zu berücksichtigen, sei es wegen der verschiedenen Strukturen, sei es wegen der physiologischen Belastungen, die ein bewegliches Segment zu die andern, weniger beweglichen, erleidet.

Die Schmerzen in diesen Uebergangszonen speziell bei anhaltenden gleichen Lagen sind nur durch die Verschiedenheit der Strukturen erklärbar.

Die letzten zwei cervicalen und die letzten zwei lumbalen Bandscheiben verfallen der Osteochondrose in besonderem Masse.

Der Morbus Scheuermann, eine typische Krankheit des thorakalen Niveaus entwickelt sich auf Höhe der 1. und 2. Lendenwirbelkörpers am ungünstigsten.

Jede Aenderung des S-förmigen Profils erfolgt durch Zug- und Druckkräfte, die die Form der Bandscheiben unmodellieren.

An den beiden lordotischen Wirbelsäulensegmente, namentlich den lumbalen wird ersichtlich, wie in der Lordose durch zentripetalen Kräfte der Gallertkern im Bereich des mittleren und des vorderen Drittels einer Bandscheibe gehalten wird.

Jede Aenderung der trapezoidalen Form (hinten schmäler und vorne mehr eröffnet) begünstigt einen erhöhten Druck der Gallertmasse in Richtung des Ligam. long. posterius.

Dies wird bestätigt durch die Untersuchungen des intradiscalen Druckes. Dieser wird im Sitzen um ca. 30% erhöht. (Nr. 1).

Diese Verschiebung bewirkt wohl eine Kyphosierung, jedoch auch eine vermehrte Spannung im Bereich des hinteren Ligamentum. Je länger eine solche Postur beibehalten wird, desto grösser wird der Druck zwischen Discus und Ligamentum.—Das Microtrauma des Schüttelns in kyphotischer Lage erhöht die Anfälligkeit. Dies erklärt die durch die Motorisierung bedingte Zunahme der Rückenleiden.

Ein weiterer erschwerender Faktor ist die Umkippung des Kreuzbeines um 30°, die stets eine Umformung der lumbosacralen Bandscheibe zur Folge hat.

Die Wirbel-Gelenkskapseln, sowie die Bänder werden bei jeder Aenderung der Form gespannt, ebenso die Rücken- und die Bauchmuskulatur. Jedes Andauern der sitzenden Haltung führt zu einer statischen Arbeit der Muskelmassen. Nachteilige biochemische Veränderungen, vorwiegend zirkulatorisch bedingte, bewirken eine starke Erhöhung der Müdigkeitserscheinungen. Abgesehen von der kapillären Stase häufen sich metabolische Produkte, an die nur durch eine Entspannung beseitigt werden können.

Die Sitze jeglicher Art, so wie sie heute auf dem Markt vorhanden sind, weisen physiologisch gesehen, die drei folgenden negative Momente auf:

(a) Die Umkippung des Sacrums um 30° und die sekundäre Umformung der lumbosacralen Bandscheibe.

(b) Die primäre Umformung im Bereich der übrigen lumbalen vertebralen Segmente.

(c) Die Ueberdehnung des ligamentären Systems der Uebergangszone zwischen der Brustwirbel—und Lendenwirbelsäule.

Verschiedene Methoden wurden zur physiologischen Beurteilung der sitzenden Haltung angewandt:

Die Psychotechnische und zwar durch die direkte oder indirekte Befragung mehrere Personengruppen über den Sitzkomfort.

Die Elektromyographische die über den Zustand der Muskulatur Aufschlüsse gibt.

Die Röntgenologische, die eine nähere Beurteilung der Bandscheibenumformung ermöglicht.

Die Photochemische, die über die Kontaktfreudigkeit des Körpers zur Sitz-bezw. Rückenlehne, insbesonders über die Druckverhältnisse Aufschluss gibt. (Nr. 2).

Keine der oben erwähnten Methode allein ist jedoch im Stande einen sicheren Bescheid zu geben.

Der ideale Sitz soll psychisches Wohlbefinden und eine einwandfreie anatomisch-physiologische Lage möglichst der gesamten vertebralen Struktur gewährleisten.

Das Wort Sitzen bedeutet, physiologisch gesehen, eine Kompromisslösung zwischen dem Stehen und dem Liegen. Schobert hat bereits zwei Arten von Sitzen beschrieben: die vordere und hintere. Zwischen diesen Extremen sind alle Varianten möglich. (Nr. 5).

Arbeitsmedizinisch wird zur Entlastung unseres Bewegungsapparates die sitzende Haltung bei der Arbeit mehr und mehr empfohlen, postur-physiologisch gesehen sollten jedoch die *Arbeitssitze* und die *Ruhesitze* getrennt beurteilt werden.

Eine dritte Kategorie sind die *Vehikel-Sitze* (Auto, Flugzeug, Zug). Diese bilden eine Zwischenlösung, weil sie teils als Arbeitssitztyp, teils als Relaxsitz concipiert wurden. Sie charakterisieren sich durch eine vehikelsbedingte Instabilität.

Unsere Kritik der heutigen *Arbeitssitze*, gestützt auf eine umfangreiche klinische Erfahrung hält folgende Mängel fest: (Nr. 3).

Der Sitz ist nicht genügend nach hinten geneigt, sodass ein nach-vorne-rutschen der Sitzknochen jederzeit möglich ist.

Die Rückenlehne ist wohl vertikal verstellbar, begleitet jedoch nicht die gelegentlichen Körperbewegungen, weil die Federung nicht für Excursionen in der Horizontal konzipiert wurde. Der hintere Lendenbausch übt keinen genügenden Druck aus, weder auf dem Sacrum noch auf die Brust-wirbel-Lendenwirbel-Uebergangzone. Auch während dem Anlehnen bedingt die hintere Stütze eine ganz unbequeme Haltung. Das Resultat ist eine nicht ausgenützte Sitzlänge und eine durchgehende. Kyphosierung die von dem occipitalen bis zum sacralen Bereich geht.

Die Lotlinie verläuft vom Gehörorgan aus vor den Schultern- und vor dem Hüftgelenk durch. Dies bedingt eine Stauung der thorakalen und abdominellen Höhle, eine Gesamtkyphosierung und eine anhaltende, statische Muskelarbeit.

Der *Relaxsitz* erwirkt durch die Erhöhung der hinteren Extremitäten auf Femurkopfhöhe wohl eine Entspannung der Bauchmuskulatur, was bei den Arbeitssitze nicht möglich ist. Dafür sollte der Wirbelsäule in ihrer Gesamtheit die bestmögliche Anlehnung und Stützung durch eine individuell anpassbare Lehne angeboten werden. Die heutigen Ruhesitze besitzen nicht die genügende Höhe zur Anlehnung der Halswirbelsäule bezw. des Kopfes und nur einige Typen ermöglichen eine Winkeländerung zwischen Sitz und Lehne, jedoch keine individuelle Abstufung zur Stützung der Brustwirbelsäule-Anteile.

Die Sitzhaltung auch auf dieses Stuhlart ist gekennzeichnet durch eine durchgehende Kyphosierung.

Ueber die *Autositze* wurden unzählige Studien und Publikationen veröffentlicht. Es ist erstaunlich zu sehen, und hier werden auch die Flugzeugfabriken mit inbegriffen, dass keine Erzeugnisse den ärztlichen Empfehlungen Rechnung tragen. (Nr. 4).

Der Winkel zwischen Sitz und Lehne ist wohl verschiebbar, die Höhe der Sitze nur in einigen teuren Modellen. Die Lehne berücksichtigt keines der anatomischen Postulate. Die Formgestaltung bleibt eine universelle und trägt der individuellen Körpergrösse bezw. der individuellen Eigenschaften der Wirbelsäule keine Rechnung.

Das Bedürfnis, ein neues Rückenlehneprofil zu entwickeln, kam spontan während der arbeitsmedizinischen Beurteilung einiger Arbeitergruppen, die ihre Arbeit ausschliesslich sitzend durchführen.

Die folgenden Postulate wurden bei der Entwicklung massgebend: (Abb. 1).

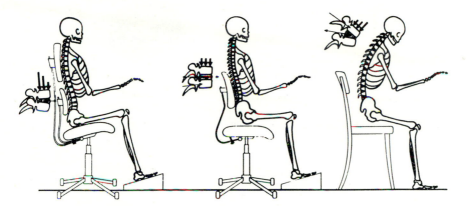

Abb. 1. Die drei Bilder zeigen die Druckbelastung der Lendenbandscheiben bei drei vecshiedenen Sitzhaltungen. Das vorgeschlagene Profil mit zwei Stützen gestattet eine vollkommene Anlehnung, begünstigt durch die besondere Neigung des Sitzes.

Der Sitz sollte von vorne nach hinten eine Neigung von 15% aufweisen um die Sitzknochen auf seine unterste Partie zu fixieren. Die Kniehüftgelenkslinie wird bei den Arbeitsstühlen den gleichen Abfall zeigen mit der Hilfe eines Fusschemels.

Die Rückenlehne soll zwei Stützen aufweisen, vertical anpassbar, die untere für die Lendenwirbel-Sacralzone, die obere für die Brustwirbel-Lendenwirbel-Uebergangspartie. Diese Stützen sollen ' Dynamisch ' konstruiert werden, d.h. mit einer eigenen horizontalen Krafteinwirkung, die ein aktives Mitmachen zur totalen Unterstützung der einzelnen Segmenten bei jeder Körperbewegung ermöglicht.

Dieses Sitzende wird somit ohne Möglichkeit zu verrutschen, die ganze Sitzlänge ausnützen und eine wirksame Unterstützung durch die zweiteilige dynamische Rückenlehne geniessen.

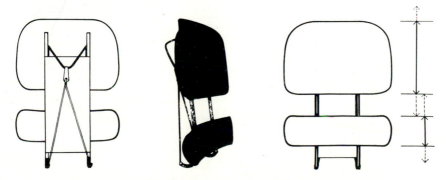

Abb. 2. Die Rückenlehne, wie sie für die Vehikelsitze konzipiert wurde.

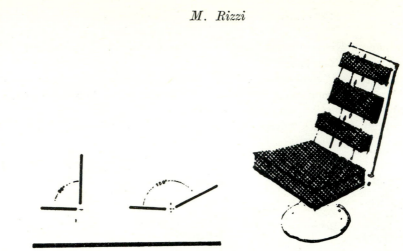

Abb. 3. Beim Relaxsitz berücksichtigen die drei verschiebbaren Rückenteile die Eigenschaften der Wirbelsäulesegmente.

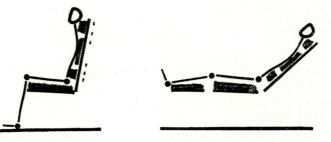

Abb. 4. Die Verstellbarkeit der Rückenlehne gestattet eine individuelle Anpassung in jeder gewünschten Lage.

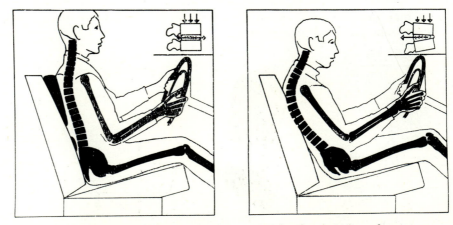

Abb. 5. Die Rückenlehne auf einem vorbestehenden Autositz aufgesetzt.

Das patentierte Profil* besteht für die Arbeitssitze aus zwei Teilen, bei den Relax und Vehikelsitzen aus drei Teilen.   (Abb. 2, 3).   Die Rückenlehne als Ganzes kann in ihrem Winkel zu der Sitzfläche verstellt werden.   Die verschiebbaren Rückenteile können zweckmässig durch eine speziell konzipierte Führungsfläche individuell so angebracht werden, dass jede körperliche Grösse die gewünschte Stütze an der richtigen Stelle sofort findet. (Abb. 4).

Man kann die Rückenlehne sowohl als Bestandteil des Sitzes ausbilden oder, wie bei den Autos, auf bestehende Sitze aufsteckbar fixieren.   (Abb. 5).

Abb. 6.   Bei der korrigierten Rückenlehne fällt die Lotlinie vom Gehörorgan auf das Hüftgelenk.

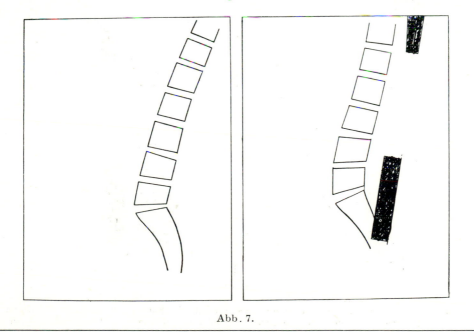

Abb. 7.

---

* Patent-Nummer : 446637.

Die so konzipierte Rückenlehne gewährleistet eine vollkommen gleichmässige Auslastung der Wirbelsäule. (Abb. 6).

Wir haben uns vorläufig mit zwei Testungen bemüht, die Richtigkeit unseres Vorschlages zu überprüfen.

In einer Rundfrage bei 200 Wirbelsäulen-Patienten wurde eine subjektive Besserung von 90% des Komforts beim Autofahren angegeben.

Röntgenologisch wurde die Lage der unteren Wirbelsäule auf Bandscheiben-Umformung untersucht. Die folgenden Bilder geben Rechenschaft über die erzielten Resultate. (Abb. 7, 8, 9).

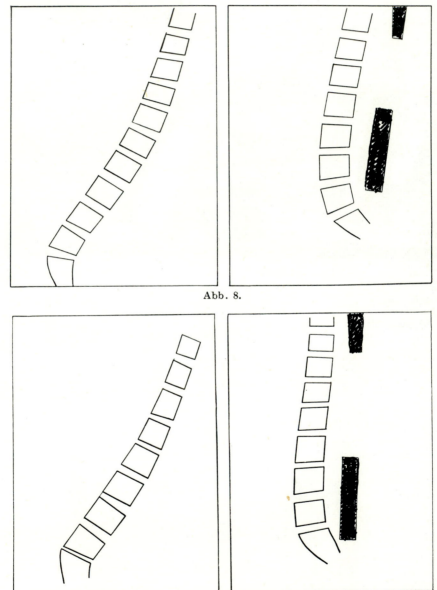

Abb. 8.

Abb. 7, 8, 9 Röntgenaufnahme eines gesunden Menschen beim Fahren *ohne* und *mit* der Rückenlehne in verschiedenen Autotypen. Die Rückenlehne verhindert eine Lendenkyphosierung und eine Umformung der Lendenbandscheibe womit eine physiologische Haltung gewährleistet wird.

Auf Grund klinischer Erfahrungen schlagen wir eine Rückenlehne mit neuartigem Profil vor : Sie besteht aus zwei oder drei gegeneinander und gegenüber der Sitzfläche verschiebbaren Rückenteilen. Gleichzeitig soll die Lehne als ganzes in ihrem Winkel zur Sitzfläche verstellbar sein.

Diese individuelle Anpassung der Polsterung an die anatomischen Gegebenheiten der Wirbelsäule schafft neben dem Gefühl von Komfort auch eine physiologisch ausgewogene Auslastung der Wirbelsäule.

Kontrollröntgenbilder bestätigen die einwandfreie Wirbelsäulenhaltung und Bandscheibenform und damit auch die Richtigkeit der anatomischen und physiologischen Ueberlegungen, von denen wir uns leiten liessen.

Clinical experience led to the proposal of a backrest with two or three mobile elements. The angle between this backrest and the seat surface must be variable. The seat will ensure an individual adaptation of the backrest profile to the anatomical conditions of the spine. X-ray studies confirmed the proper position of the spine with this seat.

L'expérience clinique suggère la conception d'un dossier comportant deux ou trois éléments mobiles. L'angle entre un tel dossier et la surface du siège doit être variable. Ce siège permet une adaptation individuelle du profil du dossier à la conformation anatomique de la colonne vertébrale. Les études faites à l'aide des rayons X ont confirmé la vraie position de la colonne vertébrale dans la cas de ce siège.

Literatur

BRUSSATIS, F., 1967, XI. *Vortragstagung ACS—Lausanne* (Bern : VERLAG STÄMPFLI), pp. 50–62.

KOHARA, J., und HOSHI, A., 1966, Fitting the seat to the passenger. *Industrial Design*, 54–65.

RIZZI, M., und GARTMANN, H., 1967, Arbeitsphysiologische Aspekte vertebraler Syndrome und ihre Prophylaxe durch technische Anpassung der Arbeitsplätze bei einer Fluggesellschaft. *Zeitschrift Präventivmed.*, **12**, 191–203.

SALVISBERG, M. D., und RIZZI, M., 1967, Occupationally caused rheumatic infections of air crew and air personnel. *XVI. Cong. Int. Med. aéron. et spaciale, Lisbonne*.

SCHOBERT, H., 1962, Sitzhaltung-Sitzschäden. *Sitzmöbel* (Berlin: VERLAG SPRINGER).

Körperhaltung und Sitzgestaltung des Kraftfahrzeugführers

Von G. Preuschen und H. Dupuis

Max Planck-Institut für Landarbeit und Landtechnik, Bad Kreuznach, Deutschland

Der Fahrersitz im Kraftwagen ist ein Arbeitssitz. Damit unterscheidet er sich grundsätzlich von Transportsitzen, die Ruhesitze sind, auch wenn diese aus optischen Gründen im Personenkraftwagen gleich dem Fahrersitz gestaltet werden. Es gelten somit grundsätzlich alle Anforderungen, die an einen Arbeitssitz gestellt werden müssen und die auch schon in anderen Beiträgen behandelt wurden.

1. Die ' Fesselung ' des Kraftfahrers

Die Besonderheit des Arbeitssitzes im Kraftfahrzeug ist die starke Fesselung der Person mit ihren Extremitäten und ihrem Kopf in ganz bestimmten Stellungsbereichen, die zusammen mit der Bereitschaft zu schneller Reaktion eine völlige Entspannung unmöglich macht. Diese Fesselung ist stärker als an jedem stationären Arbeitsplatz. Hohe Geschwindigkeiten und steigende Verkehrsdichte verlangen eine ununterbrochene Beobachtung der Fahrbahn und ihrer Flanken sowie die Bereitschaft, die Bedienteile unverzüglich zu benutzen. Es fehlt also selbst bei Überlandstrecken ein Entspannungszustand mit der Möglichkeit, den Kopf zu drehen, die Landschaft zu betrachten oder gar, wie in früheren Zeiten, mit Handgas zu fahren. Im Gegensatz dazu können Arbeitsplätze in der Industrie, die eine gleich starke Fesselung erzwingen, in beinahe beliebigen Abschnitten vorübergehend aufgegeben, kurzzeitig verlassen oder sogar vertauscht werden. Dadurch wird im Regelfall nur die Arbeitsleistung beeinträchtigt, nicht aber die Sicherheit. Eine weitere Besonderheit des Fahrerplatzes ist die fast ausschließlich statische Muskelbeanspruchung, die nur noch durch wenig dynamische Muskelarbeit unterbrochen wird. Die technische Entwicklung und die Verkaufspsychologie führen zu weiterer Verringerung dynamischer Arbeit (Servolenkung, Getriebeautomatik). Die für eine Entspannung der statisch verspannten Muskeln notwendige Dynamik würde daher häufige Fahrtunterbrechungen mit Verlassen des Wagens und Bewegungsarbeit erfordern. Da dies erfahrungsgemäß aber nicht geschieht, kann bei größeren Fahrstrecken die notwendige Entlastung nur durch geringe Bewegungen des Körpers auf dem Sitz herbeigeführt werden.

Fahrtunterbrechungen können weder durch Vorschriften, wie z. B. für Bus- und Lastwagenfahrer, noch durch Ermahnungen erzwungen werden. Das Wiederanfahren bedeutet für den Körper nämlich eine neue Arbeit und ist deswegen mit dem bekannten überhöhten Anfangsanstieg der Herzschlagfrequenz verknüpft. Sie dauert beim Berufsfahrer oft nur wenige (2-6) Minuten an, beim Nichtberufsfahrer, also der Mehrzahl aller Personenwagenfahrer, jedoch 10-15 Minuten. Diese Kreislaufänderung, die nicht durch erhöhten Sauerstoffbedarf erzwungen wird, wird gefühlsmäßig gescheut und damit auch Fahrtunterbrechungen.

2. Der Sitz als Schwingungstilger

Der Fahrersitz muß aber nicht nur im Hinblick auf die Haltungsfesselung und die Möglichkeit, die Körperstellung zu ändern, betrachtet werden, sondern muß auch den Schwingungen Rechnung tragen, die jedes Fahrzeug erzeugt und mehr oder weniger stark auf den sitzenden Menschen überleitet. Da jedes Fahrzeug nieder-frequente Schwingungen erzeugt, zum Teil durch Fahrbahneinflüsse, zum Teil aber auch durch Eigenfrequenzen am Fahrzeug bedingt, besteht grundsätzlich in jedem Fahrzeug die Gefahr, daß Schwingungen zwischen 1 und 10 Hertz auf den Oberkörper des Fahrers treffen. Der aufrecht stehende Mensch verfügt im Fuß, im Sprunggelenk und im Knie, ein wenig auch im Beckengelenk, über die Möglichkeit, Schwingungen durch ein stark gedämpftes Federsystem abzufangen oder zu mindern. Da die Federungs- und Dämpfungssysteme in weiten Bereichen und sehr schnell verstellbar sind, kommt es beim aufrechten Menschen sehr selten zu Erregungen der Eigenfrequenz des Körpers oder von Körperteilen. Der sitzende Mensch hat jedoch keine direkte Dämpfungsmöglichkeit vertikal eingeleiteter Schwingungen. Der Schwingweg der Bandscheiben ist sehr gering und entspricht im wesentlichen nur den Winkeländerungen zwischen zwei Wirbelkörpern, setzt also eine geschwungene Wirbellinie voraus (Abb. 1). Nur bei sehr stark entwickelter Rückenmuskulatur kann diese durch Verspannung dämpfend wirken. Aber gerade der Zivilisationsmensch pflegt eine sehr gering entwickelte Rückenmuskulatur zu haben. Die Eigenfrequenz des Oberkörpers liegt um 4 Hertz, mancher Organe (z. B. Magen) bei 4 bis 5 Hz und dem Vielfachen (8 bis 10 Hz). Treffen solche Frequenzen auf den Oberkörper, gerät er in Eigenschwingung, die sowohl lange andauern, als auch im System, also z. B. vom Gesäß bis zum Kopf, zu einer Beschleunigungsvergrößerung führen kann.

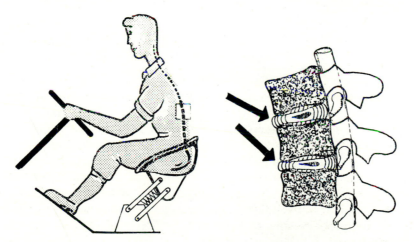

Abb. 1. Zu weite Entfernung zwischen Sitz und Lenkrad führt zu lokaler Beanspruchung und eingeschränkter Federwirkung der Bandscheiben (Beispiel: Traktoren—und Baumaschinenarbeit).

Solche Resonanzschwingungen werden als besonders belastend empfunden. Sie können bei langdauernder Einwirkung auch zu Skelett- oder Organschäden führen. Fahrzeugschwingungen können nur durch kombinierte Federdämpfungssysteme oder Stabilisatoren zwischen Fahrzeug und Achsen oder Fahrzeug

und Körper vermindert bzw. vernichtet werden.   Entsprechend der niedrigen
Schwingungsfrequenz und den relativ großen Amplituden muß bei federnden
Sitzträgern (in Lastwagen und Omnibussen) für ausreichende Schwingungs-
verminderung ein größerer Schwingweg in Kauf genommen werden, wodurch die
Körperstellung zu den Fesselungspunkten verändert wird.   Da die vertikalen
Schwingungen die unangenehmsten sind, muß vorwiegend mit einer Änderung
der Vertikalzuordnung gerechnet werden.   In Gummimetallteilen aufgehängte
Rückenlehnen wirken ebenfalls schwingungsisolierend, jedoch vorwiegend
im Frequenzbereich 15-30 Hz, ohne daß dabei eine Stellungsveränderung des
Körpers zu den Fesselungspunkten merkbar wird. (Abb. 2).

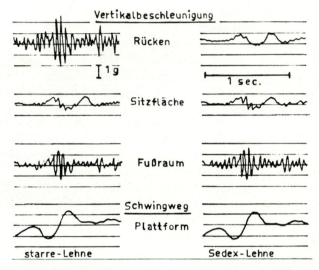

Abb. 2.   Eine federnd aufgehängte Rückenlehne ist in der Lage, die höheren Frequenzen der
Vertikalschwingungen zu absorbieren (Sedex-Lehne).

3. Sitzausformung und Bezug

Fahrersitze werden vorwiegend für längere Zeiten eingenommen.   Dement-
sprechend müssen sie als Dauersitze ausgebildet sein.   Form und Polsterung
dürfen weder zu einer zu großen gleichzeitig belasteten Hautfläche noch zu einem
zu großen Druck je Flächeneinheit führen.   Mit Rücksicht auf die schon
erwähnte Notwendigkeit, zur Muskelentspannung die Sitzposition ändern zu
können, bei gleichzeitiger Fixierung der Extremitäten auf bestimmte Punkte,
muß eine genügende Bewegungsmöglichkeit im Becken- und Rückenraum
gegeben sein.   Diese Notwendigkeit wird häufig vergessen, weil in der Ruhesitzhaltung die Körperbewegung weniger mit Becken und Rücken als vielmehr
durch Veränderung der Beinlage erreicht wird—aber gerade diese Änderung
ist im Kraftfahrzeug nicht möglich.   Aber auch aus Gründen des notwendigen
Wechsels der Hautbelastung darf nur ein bestimmter Teil der Gesäßhaut
gleichzeitig mit dem Sitz in Berührung kommen.   Daher sind Schalensitze
abzulehnen.   Eine seitliche Anlage für Oberschenkel und Rücken ist zur
Aufnahme von extremen Gegenkräftem aus der Steuerbewegung oder der
Kurvenbeschleunigung notwendig.   Mittlere bzw. vorausschaubare Kräfte
werden bei genügender Reibung zwischen Sitzbezug und Bekleidungsstoff

meist durch Andrücken des Körpers an den Sitz aufgefangen. Zwischen den seitlichen Anlagen und der Normalsitzposition muß ein gewisser Wanderungsweg vorhanden sein, der sowohl zur Änderung der Sitzposition benötigt wird wie auch zum Erreichen mancher Bedienteile (Abb. 3). Diese Positionsänderung darf auch nicht durch Sicherheitsgurte beeinträchtigt werden.

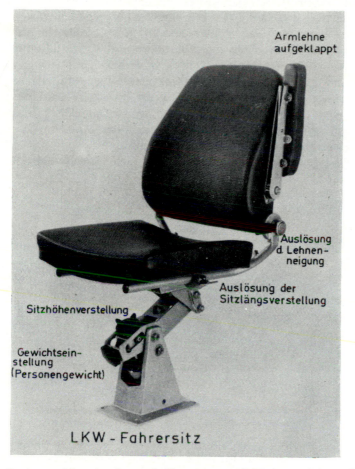

Abb. 3. Lastwagenfahrersitz, der große Bewegungsmöglichkeiten und dennoch sichere Abstützung bietet.

Sitz und Rückenlehne müssen gepolstert sein, um Schwingungen hoher Frequenzen und geringer Amplituden aufzufangen und hohe Flächenpressungen auf die Haut zu vermeiden. Der Bezugsstoff soll rauh sein, so daß Beschleunigungskräfte, insbesondere seitlich wirkende, die Haftung zwischen Anzugsstoff und Sitz nicht rasch überwinden können. Durch leichtes Andrücken des Rückens gegen die Lehne können dann sogar hohe Kräfte aufgenommen werden, so daß größere seitliche Anlagen entfallen können. Glatte Stoffe, wie Leder und Kunststoff, sind ungeeignet. Darüber hinaus muß die Sitzpolsterung einschließlich Bezugsstoff einerseits isolierend wirken, andererseits Feuchtigkeit und Wärme abführen können, also ein Textilverhalten entsprechend dem von Wolle zeigen.

4. Anpassung an verschiedene Körpergrößen

Die letzte und entscheidende Schwierigkeit des Fahrersitzes kommt durch die Tatsache der Serienproduktion auf der einen Seite und der Verschiedenheit der Körpermaße der Benutzer auf der anderen Seite. Da die Fahrerstellung durch Auge, Hand und Fuß gefesselt ist, genügt ein einfacher Höhenausgleich, wie er für nur fuß- oder handgefesselte Sitze ausreichend ist, ebensowenig wie Kombinationsverstellungen (Abb. 4). Für den Größenausgleich wäre es notwendig, sowohl eine Längs- und eine Höhenverstellung des Sitzes als auch eine Lenkradverstellung zu haben, dazu eine Veränderung der Sitztiefe durch Vor- und Zurückverstellen der Rückenlehne. Aus Gründen des Positionswechsels muß außerdem die Rückenlehne in der Neigung verstellbar sein, was wiederum eine Veränderung der Sitzflächenneigung nötig macht. Aus Gründen der passiven Sicherung sollen alle Verstellungen sicher arretierbar sein. Schließlich kann auch erwogen werden, sofern der Fahrer am Sitz angeschnallt ist, den Sitz selber als stoßverzehrendes Element zu gestalten.

Für die Gesamthaltung des Fahrers ergeben sich folgende Grundbeziehungen:

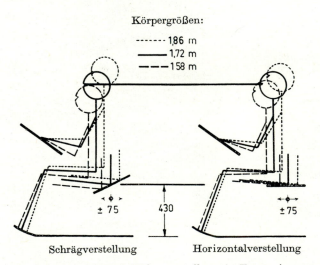

Abb. 4. Bei kombinierter Längs- und Höhenverstellung in Form einer nach vorn abwärts gerichteten Schrägverstellung werden die Sichtverhältnisse für den kleinen Fahrer wesentlich verschlechtert. (Beispiel : Traktorsitz)

5. Kopfhaltung

In Abbildung 5 ist die Zuordnung von Sehstrahl zu Kopfhaltung und Rückenneigung dargestellt. Nach den bisherigen Erfahrungen muß man annehmen, daß eine verkrampfte Kopfhaltung für den Fahrer besonders gefährlich ist, weil sie zur Minderung der Blutzufuhr zum Gehirn führen kann. Als verkrampft darf eine Stellung von mehr als 20° der Senkrechten durch Kopf und Hals zur Mittellinie des Rückens angesehen werden. Optimal im Hinblick auf die Kopfhaltung dürfte die Stellung A sein, B stellt den noch zulässigen Grenzfall dar und C muß für den normalen Fahrer abgelehnt werden, auch wenn diese " Sportwagen-Position " heute sehr modern ist. Die in Abbildung 5 dargestellten Verhältnisse gelten nur für den schnellen Überlandverkehr. Im Stadtverkehr, der einen dichter vor dem Fahrzeug liegenden

Auftreffpunkt des Sehstrahls erfordert, können zusätzliche Kopf- und Augapfelbewegungen in Kauf genommen werden, sofern sie nur kurzzeitig auftreten.

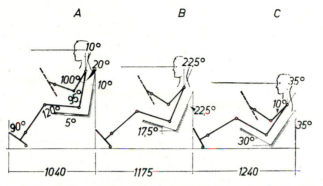

Abb. 5. Körperhaltung bei verschiedener Sitzhöhe und -neigung (Schema).

6. Rückenlehnenneigung

Der Sehstrahl begrenzt damit über die Kopfneigung die mögliche Neigung der Rückenlehne des Fahrersitzes. Eine stärkere Neigung der Rückenlehne ist auch aus schwingungstechnischen Gründen abzulehnen, da bei Neigungen von mehr als 20° Schwingwegvergrößerungen vom Gesäß bis zum Kopf auftreten können und somit auch stärkere Schwingbelastung der Kopf-Halspartie als bei mehr senkrechter Lehnenstellung. Bei geringen Sitzschwingungen und Langstreckenfahrten kann jedoch die Lehnenneigung vorübergehend bis auf 30° erhöht werden.

7. Hüftwinkel

Wenn in der Praxis, speziell bei Kleinwagen, die Fahrer eine stark geneigte Lehne anstreben, so aus einer unwillkürlichen Reaktion auf eine zweite Winkelstellung, die echte Grenzwerte hat. Das ist der Winkel zwischen Rumpf und Oberschenkel.

Die Empfehlungen hierfür sind, wie Tabelle 1 zeigt, außerordentlich unterschiedlich, was durch die verschiedenen Untersuchungsobjekte und -methoden begründet ist.

Ein Winkel von 90° kann nicht dauernd eingehalten werden. Die größte Annehmlichkeit scheint ein Winkel von 105 bis 115° zu bringen, weil diese Winkelstellung von der Mehrzahl der Menschen unbewußt gesucht wird. Das ist der Grund dafür, daß alle Rückenlehnen nach hinten geneigt sein müssen, weil sonst ja die Sitzfläche nach vorn abfallen mäßte. Das ist aber im Fahrzeug sowohl wegen der Gefahr des Abrutschens als auch vor allem wegen der Stellung der Beine zu den Pedalen unmöglich. Je mehr die Sitzfläche nach hinten fällt, um so stärker muß auch die Rückneigung der Rückenlehne sein. Mit Rücksicht auf die oben angegebene Grenze für die Kopfhaltung ist eine Rückwärtsneigung der Sitzfläche von 17,5° eben noch bequem; alle stärker geneigten Sitze zwingen den Fahrer, entweder die Rückenlehne stärker zu neigen und damit eine verkrampfte Kopfhaltung in Kauf zu nehmen oder das Becken vorzuschieben und einen fast vollkommenen Rundrücken auszuformen.

Tabelle 1. Empfohlene Winkel zwischen Rumpf und Oberschenkel (Hüftwinkel)

Literatur–Autor	empfohlener Hüftwinkel	Begründung	Untersuchungs– bzw. Anwendungs- objekt	Untersuchungs- methode
Keegan	115°	Beste Annäherung an " normale " Krümmung der LWS	KFZ-Sitze	Ermittlung der Krümmung der WS bei versch. Körperhaltung durch Röntgenverfahren.
Akerblom	115°	in Verbindung mit "Akerblom-Knick" als Stützung der LWS und BWS	Stühle	orthopädische Gesichtspunkte
Dreyfuß	115°	wie bei Akerblom	Passagiersitze in Flugzeugen	o. A.
Versace	135°	Annähernder Ausgleich zwischen Beuge- und Streckmuskeln, wenn Kniewinkel auch 135°	Sitze in Raumfahrzeugen	o. A.
Picard/ Wisner	85-100°	Abweichungen von diesem Winkelbereich führen zu Krampferscheinungen	KFZ-Sitze	o. A.
Rebiffé	95-135°	o. A.	KFZ-Sitze	erwähnte mögliche Kriterien, deren Anwendung jedoch nicht angegeben ist: (*a*) subjektive Meinungen (*b*) anatomisch-pathologische Gesichtspunkte (*c*) Muskelkontraktion (Myographie) (*d*) Gelenkarbeitsweise
Lehmann	134°	maximale Entspannung der Muskulatur	Versuchspersonen im Leigen	Unterwasserphotographie bei entspannter Körperhaltung
Grandjean/ Böni/ Kretzschmar	101-104° 105-108°	—	beim Lesen beim Ruhen	subjektiv bequemste Einstellung des Profiles der Sitzmaschine

8. Kniewinkel

Die Neigung der Sitzfläche wird heute meist im Zusammenhang mit der Stellung des Gespedals und der Länge der Extremitäten in Abhängigkeit von der Sitzhöhe ausgewählt. Auch hier gibt es Grenzwinkel, die eingehalten werden müssen. Primär muß aber von der Gesamtkörperhaltung ausgegangen werden. Die Annehmlichkeitsstellung des Kniewinkels (Tabelle 2) kann mit 110 bis 120° angenommen werden. Diese Stellung läßt sich ermüdungsarm über lange Zeit einhalten. Andererseits muß gefordert werden, daß der Fuß für den Beschleunigungsgeber (' Gaspedal ') so auf dem Pedal aufstehen kann, daß bei Leerlauf ein Winkel von rund 85°, bei Vollgas von 105° im Sprunggelenk entsteht.

9. Armhaltung

Wie schon Abbildung 1 zeigte, wirkt sich aber auch die Anordnung und Stellung des Steuerrades auf die Sitzhaltung des Fahrers aus. Abbildung 5 zeigt die günstigsten Ellenbogenwinkel, die die gerade zur Entspannung des Hand-Arm-Systems dringend notwendigen Lageänderungen am Steuerrad ebenso zulassen wie eine rasche Reaktion und ausreichende Kraftübertragung auf das Steuerrad. Auf langen Autobahnstrecken kann ein etwas größerer Ellenbogenwinkel zugelassen werden. Fahren 'mit gestreckten Armen' ist jedoch aus physiologischer Sicht bei üblicher Fahrzeuglenkung abzulehnen.

Tabelle 2. Empfohlene Kniewinkel

Literatur/autor	Empfohlener Kniewinkel	Begründung	Untersuchungs- bzw. Anwendungs- objekt	Untersuchungs- methode
Keegan	115°	wie Hüftwinkel	KFZ-Sitze	o. A.
Versace	135°	Annähernder Ausgleich zw. Beuge- und Streckmuskel, wenn Hüftwinkel auch 135°	Sitze in Raumfahrzeugen	o. A.
Picard/Wisner	100-120°	Abweichungen führen zu Krampferscheinungen	KFZ-Sitze	o. A.
Rebiffé	95-120°	o. A.	KFZ-Sitze	wie bei Tab. 1
Müller, A. E.	130-150°	Bereich größter Kraftausübung	nur für kraftbetonte Pedale	Maximalkraftmessungen
Kroemer/ Coermann	110-120°	Gesichtspunkt der Kraftausübung	KFZ-Sitze	o. A.
Lehmann	133°	maximale Entspannung der Muskulatur	Versuchs-Personen im Liegen	Unterwasserphotographie bei entspannter Körperhaltung
Dupuis/ Preuschen/ Schulte	110-120°	günstige Kraftausübung	Schlepper	Maximalkraftmessungen

Die zweckmäßige Neigung des Steuerrades und der Durchmesser bestimmen einerseits die Sitzverstellung, andererseits die Kopfhaltung des kleinen Fahrers. Die Sitzhöhenverstellung wird durch den beim Einsteigen für die Oberschenkel nötigen Freiraum zwischen Sitzfläche und Steuerradunterkante beschränkt. Andererseits kann die Oberkante des Steuerrades den Sichtbereich kleiner Personen so einengen, daß sie gezwungen sind, in aufgerichteter Position zu fahren.

Zeichnet man nun zu den hier angegebenen Grenzwinkeln die anthropometrischen Extremitätenwerte ein, und zwar für die üblicherweise angenommene Population von 95% aller Individuen eines Verkaufsgebietes, dann ergeben sich daraus sehr rasch Mindestwerte, die aber in vielen heutigen Personenkraftwagen nicht zu finden sind. Vorschläge für optimale Sitzmaße und Zuordnung der Bedienteile zeigen die Abbildungen 6, 7 und 8.

K

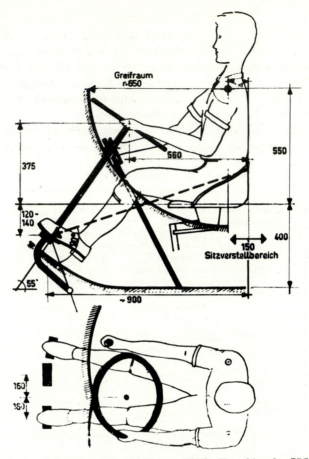

Abb. 6.   Führerstand für Traktoren (VDI—Vorschlag für ISO).

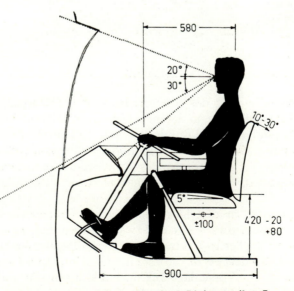

Abb. 7.   Arbeitsplatz im Standard-Linienomnibus I.

Der Fehler der meisten Personenkraftwagen ist eine zu niedrige Sitzhöhe über Flur, dadurch eine zu steile Anwinkelung des Knies und die Tendenz, die Rückenlehne sehr weit nach hinten zu neigen, um noch einen erträglichen Beckenwinkel zu finden. Geringe Sitzhöhen wären dann erträglich, wenn die Fahrzeuge so lang wären, daß auch große Fahrer einen sehr flachen Kniewinkel einhalten können. Das würde aber eine Verlängerung des Innenraumes gegenüber dem heute üblichen von 30 bis 40 cm bedeuten, damit hinter dem Fahrerplatz noch ein wirklicher Sitzplatz verbleibt. Andererseits soll das Fahrzeug kurz sein, um es bequem parken zu können. Vielfach ist aber auch die Längenverstellung der Sitze unzureichend, weil man zwar aus modischen Gründen (und mit Rücksicht auf den Luftwiderstand) das Fahrzeug und damit die Sitze niedriger gemacht hat, die Konsequenz aus der nunmehr anderen Beinstellung aber nicht zieht.

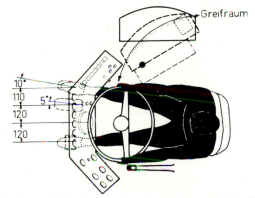

Abb. 8.   Arbeitsplatz im Standard–Linienomnibus II.

Günstiger ist die Situation in Lastwagen und Omnibussen. Die hier notwendige größere Höhe der Karosserie für den Zu- und Abgang der Fahrgäste erlaubt eine sehr viel größere Sitzhöhe für den Fahrer. Der Fahrersitz ist so hoch, daß die Rückenlehne nur wenig geneigt zu sein braucht. Die dadurch bedingte Beinstellung ist für die Ausübung größerer Kräfte zwar ungünstiger, was aber durch zunehmende Verwendung von Servoeinrichtungen nicht mehr so wichtig ist. Allerdings fehlt diesen Fahrzeugen häufig eine genügende Längsverstellung der Sitze, weil ein großer Nutzraum des Fahrzeugs wichtiger erscheint als die ausreichende Anpassung an große Fahrer. Wesentliche Bedeutung dürfte schließlich die Kopfstütze als Schutz bei Auffahrunfällen erlangen, über deren optimale Konstruktion jedoch noch keine ausreichenden Grundlagen vorliegen.

Zusammenfassend darf festgestellt werden, daß der Mensch keine Möglichkeit hat, sich so anzupassen, daß er in einer bestimmten Sitzform stundenlang mit gleicher Aufmerksamkeit die verantwortliche Arbeit des Fahrzeugführers leisten kann. Wesentliche Abweichungen von den genannten Grenzwinkeln bringen unnötige Körperbelastung, mindern die Wachsamkeit und können zu gesundheitlichen Störungen oder Dauerschäden führen. Die Optimierung der Fahrersitze ist vor allem aber im Hinblick auf Straßenverkehrssicherheit eine wichtige Forderung.

Alle Einzelheiten der Sitzgestaltung sollten auf folgende Grundsätze Rücksicht nehmen:

1. Der Sitz muß durch seine Form das Skelett optimal unterstützen und darf gleichzeitig Organe, Gewebe und Haut nur minimal belasten.

2. Er muß eine Veränderbarkeit der Körperhaltung ohne und mit Verstellvorrichtungen in bequemer und einfacher Wiese ermöglichen.

3. Der Sitz muß eine optimale Unterstützung der Arbeitsausführung erlauben.

Der Sitz des Kraftfahrers muß als ein Arbeitssitz angesehen werden, weil der Fahrer viele Arbeitsaufgaben wahrzunehmen hat. Die Besonderheit dieses Arbeitssitzes stellt die starke Fesselung der Person mit den Extremitäten und dem Kopf dar, die die manuellen, peduellen und visuellen Aufgaben notwendig machen. Dadurch wird der Umfang ermüdender statischer Muskelarbeit vergrößert, während auf der anderen Seite technische Entwicklungen (Servolenkung, Schaltautomatik) weniger dynamische Muskelarbeit erfordern.

Die Federungs- und Dämpfungssysteme der meisten Automobile schließen nicht aus, daß der menschliche Körper und seine Teilbereiche in Resonanz geraten. Hierdurch sind nicht nur Belästigungen, sondern auch gesundheitliche Schädigungen zu erwarten. Die technischen Hilfsmittel zur Schwingungsisolierung müssen daher verbessert oder besser genutzt werden.

Die Anforderungen an die äußere Gestaltung, Lage und Abmessung der Fahrersitze werden von den Arbeitsaufgaben, den anthropometrischen und arbeitsphysiologischen Voraussetzungen abgeleitet. Um einerseits Sitze an verschiedene Körpergrößen anzupassen und zum anderen in bestimmten Bereichen einen Haltungswechsel zu ermöglichen, sind ausreichende Verstellmöglichkeiten vorzusehen. Beispiele praktischer Fahrersitzgestaltung erläutern die theoretischen Grundlagen.

In contrast to an easy chair a driver's seat has to be seen as a work-seat. In particular the driver's seat has to have the man with his extremities and head in a fixed position in order to fulfil the manual, pedal and visual tasks of driving. Therefore on the one hand the amount of fatiguing static work increases while on the other technical improvements, such as servo-steering and automatic shifting, will demand less dynamic muscle work.

The spring and damping systems of many vehicles make it possible that the human body and parts of it go into resonance. This means not only discomfort, but also it may be followed by injuries to health. Technical improvements for vibration isolation need to be developed and applied.

The position and dimensions of drivers' seats depend on the different work tasks and the data from anthropometry and work-physiology. Further, the need to adapt to different body dimensions and to change the body position means that automobile seats must have sufficient adjustments. These theoretical ideas are illustrated by examples of practical seat designs.

Contrairement à la chaise de repos, le siège du conducteur doit être considéré comme un siège de travail où les membres et la tête du conducteur se trouvent dans une position fixe, en vue de l'exécution de la tâche visuelle et des manoeuvres de pilotage. C'est ainsi que, d'une part, la quantité de travail statique fatigant augmente et que, d'autre part, le travail musculaire dynamique diminue grâce aux améliorations technologiques telles que la direction assistée ou les boîtes de vitesse automatiques.

Les systèmes ressorts-amortisseurs de beaucoup de véhicules peuvent entraîner des phénomènes de résonnance se localisant dans tout le corps ou dans certaines de ses parties. Ce qui est à l'origine, non suelement de sensations d'inconfort, mais retentit également sur la santé du conducteur. Des améliorations techniques visant l'isolement des vibrations devraient être étudiés et appliquées.

La position et les dimensions du siège du conducteur dépendent des différentes tâches à accomplir, ainsi que des données anthropométriques et physiologiques. En outre, la nécessité de se conformer aux diverses dimensions corporelles et aux diverses postures possibles implique la conception d'un siège ajustable. Ces considérations théoriques sont illustrées à l'aide d'exemples serapportant à la conception des sièges.

Literatur

AKERBLOM, B., 1951, Gesundes Sitzen—Bequemes Sitzen. *Möbel-Kultur*, **3**, 1–4.

DREYFUß, H., 1960, The measure of man. *Whitney Libr. of Design, New York*, 22.

DUPUIS, H., PREUSCHEN, R., und SCHULTE, B., 1955, Zweckmäßige Gestaltung des Schlepper-führerstandes. *Heft 20 der Schriftenreihe Landarbeit und Technik*, 177 S., Bad Kreuznach.

GRANDJEAN, E., BÖNI, A., und KRETZSCHMAR, H., 1967, Physiologische Grundlagen für die Konstruktion von Motorfahrzeugen. ' Schädigungen der Wirbelsäule, verursacht durch das Motorfahrzeug.' *Vortragstagung des Automobil-Clubs der Schweiz*, **11**, 84–97.

KEEGAN, J. J., Das medizinische Problem der Abflachung der Lendenwirbelsäule in Kraftfahr-zeugsitzen. *SAE 838 A.*

KROEMER, K. H. E., und COERMANN, R., 1965, Die Gestaltung der Insassenkabine von Kraftfahr-zeugen. ' *Prüfliste* ' *und Bibliographie. Zbl. Verkehrs-Med.*, **11**, 213–223.

LEHMANN, G., 1940, Zur Physiologie des Liegens. *Arbeitsphysiologie*, **11**, 253–258.

MÜLLER, E. A., 1936, Die günstigste Anordnung im Sitzen betätigter Fußhebel. *Arbeits-physiologie*, **9**, 125.

PICARD, F., und WISNER, A., 1961, How can the physiologist contribute to the improvement of automobile seats ? 1961 *SAE International Congress, Detroit.*

REBIFFÉ, R., 1966, Ergonomische Untersuchung über die Anordnung der Fahrstellung in Personen-Kraftwagen. *IME-Symposium, London.*

VERSACE, J., 1960, Human engineering the seat. *ASBE 15. Annual technical convention.*

Le Siège du Conducteur: Son Adaptation Aux Exigences Fonctionnelles et Anthropometriques

Par R. Rebiffé

Laboratoire de Physiologie et de Biomécanique de la Régie Nationale des Usines Renault,
Rueil-Malmaison, France

1. Introduction

Ce qui fait la qualité d'un siège d'automobile, c'est avant tout sa plus ou moins bonne adaptation au poste de conduite dans lequel il s'intègre.

Il résulte de cette affirmation que l'étude du siège ne peut être envisagée isolément; le siège doit être considéré comme un des éléments d'une structure complexe qu'il convient d'aménager pour l'adapter à la structure encore plus complexe du corps humain.

De ce fait, la réalisation du siège optimal ne peut se faire en prenant uniquement en considération des données anatomo-physiologiques classiques; il faut à chaque fois, se poser la question de savoir quels rapports existent entre le siège et les différentes fonctions assurées au poste étudié.

Ce n'est pas un hasard si les spécialistes des problèmes d'ergonomie des véhicules ont perçu très tôt la nécessité d'une approche synthétique pour l'étude du poste de conduite ou de pilotage. En effet, dans ces postes, les exigences de la tâche imposent à opérateur une posture fixe et précise maintenue pendant un temps prolongé; l'agencement des différents éléments du poste qui fixent cette posture ne peut être déterminé par une série de recommandations séparées. Il existe des interrelations entre les différentes dimensions du poste qu'il faut connaître et respecter. C'est de l'application de ces principes déjà énoncés par Morant (1947) il y a plus de 20 ans que naît le bon poste de conduite.

Cette notion de base que nous avons développée (Wisner et Rebiffé 1963) s'est révélée particulièrement féconde pour l'étude des postes de conduite et des sièges d'automobiles et, également, pour l'étude des postes de travail en général.

A la suite de ces généralités, nous considèrerons pour l'étude d'un siège de véhicule, quel qu'il soit, trois stades successifs:

1. Analyse de la tâche effectuée au poste de conduite; dans cette analyse on met l'accent principalement sur la définition des exigences visuelles.
2. Détermination de la posture qui permet de satisfaire au mieux les exigences de la tâche.
3. Définition des caractéristiques fonctionnelles et dimensionnelles du siège qui assure le support optimal du corps de l'opérateur, dont la posture a été précédemment déterminée en fonction de l'analyse du travail.

C'est selon cette démarche méthodologique que nous nous proposons de considérer maintenant l'étude du siège conducteur des voitures automobiles particulières.

2. Analyse de l'Activite de Conduite

Du point de vue où nous nous plaçons, les points importants que l'analyse du travail de conduite nous révèle sont les suivants:

(a) Attitude à peu près fixe du conducteur conservée pendant de nombreuses heures.

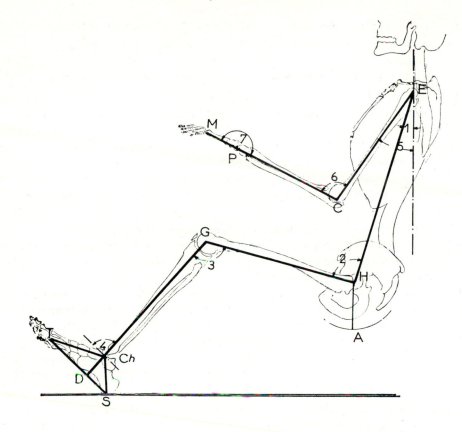

ANGLES DU CONFORT

$$20° < A_1 < 30°$$
$$95° < A_2 < 120°$$
$$95° < A_3 < 135°$$
$$90° < A_4 < 110°$$
$$10° \text{ à } 20°_{(1)} < A_5 < 45°_{(2)}$$
$$80° < A_6 < 120°$$
$$170° < A_7 < 190° \quad \text{Flexion. ext.}$$
$$\text{Inclinaison radiale } 180° < A_7 < 190° \quad \text{Inclinaison cubitale}$$

(1) Selon le dossier
(2) Avec appui de la main

DISTANCES°
INTER ARTICULAIRES

CHAINON	5%	50%	95%
TAILLE	157	170	183
E.C	26,5	29	31,5
C.P	24	26	28
P.M	6,5	7	7,5
H.G	38	41,5	45
G.Ch	37	40,5	44
Ch.D *	10,5	11,5	12,5
D.S *	5,5	6	6,5
D.T *	13	14,5	16
E.H(droit)	42	45,5	49
E.H(normal)	38	41,5	45
H.A	8,5	9,5	10,5

° en cm
* sujet chaussé

Figure 1. Modèle biomécanique du corps humain; angles du confort et distances interarticulaires.

(*b*) Nécessité d'une observation continue de l'environment extérieur.

(*c*) Ajustement permanent et simultané de la direction et de la vitesse du véhicule.

Chacun de ces points nous suggère un certain nombre de remarques:

(*a*) Une attitude à peu près fixe ne peut être conservée pendant un temps prolongé que si la posture est correcte. Cela signifie que les angles articulaires des différents segments corporels doivent être compris à l'intérieur de limites précises qu'il convient de définir. Si ces angles articulaires ne sont pas respectés, les douleurs musculaires et ligamentaires qui en résultent rendent la position rapidement inconfortable voire intolérable. La figure 1 représente la valeur des angles articulaires que nous proposons pour réaliser une posture correcte.

(*b*) La densité des informations visuelles à prélever à l'extérieur est très élevée. Cette prise d'informations nécessite donc, d'une part une observation continue de l'environment et, d'autre part, un champ de visibilité le plus vaste possible.

En ce qui concerne l'observation continue, on remarquera qu'elle n'exige pas une grande acuité visuelle; l'accommodation est nulle et le conducteur regarde droit devant lui; pour ce faire, le conducteur conserve la tête droite.

Pour obtenir un champ de visibilité étendu, il faut que les yeux soient convenablement situés par rapport aux superstructures du véhicule pour que celles-ci, ainsi que les accessoires de l'habitacle, ne masquent qu'au minimum le champ de vision normal du conducteur.

(*c*) Le guidage du véhicule, relativement simple, est effectué à l'aide du volant; le réglage de vitesse se fait par l'action du pied droit sur l'accélérateur et sur le frein; le pied gauche agit sur la pédale de débrayage.

Ces deux ajustements permanents ne comportent pas habituellement de grandes exigences de précision mais demandent un maniement aisé à cause de leur utilisation continue.

3. Determination de la Posture au Poste de Conduite

L'analyse du travail fait apparaître trois fonctions essentielles pour la tâche de conduite:

(*a*) Prélèvement des informations visuelles dans l'environment extérieur.

(*b*) Réglage de la vitesse.

(*c*) Guidage du véhicule.

Pour réaliser ces trois fonctions, le conducteur—nous l'avons vu plus haut— place ses yeux dans une position déterminée par rapport au pare-brise; ses mains reposent sur le volant, ses pieds actionnent les différentes pédales. On conçoit aisément que la position des yeux, des mains et des pieds étant ainsi fixée à l'intérieur de l'habitacle, la posture du conducteur se trouve du même coup assez étroitement déterminée; il ne reste, de ce fait, qu'une marge de liberté extrêmement réduite pour placer à son tour le siège dans cette structure.

Si, du fait d'un mauvais agencement des différenst éléments à l'intérieur du poste, la posture du conducteur n'est pas correcte, il n'est plus possible de trouver pour ce poste un siège qui améliore la posture; quelles que soient les qualités intrinsèques de ce siège, il sera considéré comme mauvais et ceci parce que la posture, telle qu'elle est déterminée par l'architecture du poste, ne peut être que mauvaise; le siège ne peut dans ce cas apporter aucune amélioration. Ce n'est pas le siège qui détermine la posture; il n'est là que pour supporter et maintenir une certaine posture qui doit être au préalable correcte.

Nous constatons fréquemment que les gens portent des jugements négatifs sur les sièges qu'ils rendent responsables de l'inconfort d'un poste; c'est souvent une erreur de diagnostic car l'inconfort provient, dans bien des cas, plus d'un aménagement dimensionnel défectueux du poste que du siège lui-même.

Nous considèrerons maintenant plus en détail la façon dont l'exécution dans de bonnes conditions des différentes fonctions de conduite détermine la posture du conducteur à son poste.

3.1. *Prélèvement des Informations Visuelles et Posture*

Si l'on classe par ordre d'importance le rôle des différentes fonctions énumérées ci-dessus dans la détermination de la posture, c'est la fonction visuelle qu'il faut mettre en premier.   Tout opérateur placé à un poste de travail cherche en effet à satisfaire d'abord les exigences visuelles du poste.   Pour cela, il adapte sa posture de façon à placer ses yeux dans la position qui lui permet de prélever, dans les meilleures conditions, les informations visuelles nécessaires à l'exécution de la tâche.   Plus ces exigences sont sévères, plus la position des yeux est déterminée, plus la zone dans laquelle ils doivent se placer est réduite et étroitement circonscrite.

Or, définir une position précise des yeux, une direction donnée du regard, revient à définir une position précise de la tête par rapport à la tâche: en effet, les yeux ne peuvent s'écarter d'une façon prolongée de leur position d'équilibre. Si la position de la tête est fixée, tout au moins dans un plan vertical, on ne dispose alors que de peu de liberté pour fixer celle du reste du corps; la position du buste en particulier et, corrélativement, celle du siège se trouvent donc déterminées.

Pour l'étude dimensionnelle d'un poste de travail, il est pratique de traduire les exigences visuelles en termes de: distance oeil-tâche, direction du regard, champ de vision balayé, etc. . . .

En ce qui concerne la tâche de conduite, c'est la direction du regard et le champ de vision que nous considèrerons.

3.2. *Réglage de la Vitesse et Posture*

La deuxième fonction importante à assurer au poste de conduite est le réglage de la vitesse du véhicule par action des pieds sur les différentes pédales. Cette action est continue au moins en ce qui concerne l'accélérateur; l'emplacement de cette pédale par rapport au siège doit donc être défini avec soin.

Nous avons vu plus haut que l'obtention d'une posture de conduite correcte était liée à la valeur des angles que faisaient les segments corporels entre eux.

La position des pédales par rapport au siège détermine en grande partie, au moins pour le membre inférieur, la valeur de ces angles articulaires: si la pédale est trop éloignée, les angles articulaires sont trop ouverts; dans le cas contraire ils sont trop fermés; dans l'un et l'autre cas, les articulations sont sollicitées au-delà des limites de confort. Il importe donc de connaître la distance optimale des pédales par rapport au siège. Il est facile de démontrer que cette distance optimale n'est pas fixe mais qu'elle est directement liée à la hauteur de l'assise et à l'inclinaison du dossier (Wisner et Rebiffé *op. cit.*)

A partir d'une construction géométrique simple on peut, connaissant la valeur optimale des angles intersegmentaires et les distances interarticulaires, établir la relation qui donne, en fonction de la hauteur d'assise, l'éloignement des pédales par rapport au siège, la hauteur de la pédale audessus du plancher ainsi que l'inclinaison de cette pédale (Rebiffé, 1966).

En ce qui concerne le siège plus particulièrement, on déduit, à partir de cette même construction géométrique, l'amplitude optimale de réglage antéro-postérieur du siège. De la même façon, on détermine, à partir de l'inclinaison des cuisses du conducteur ,la meilleure inclinaison du coussin du siège susceptible de procurer un support correct de la cuisse.

Ce problème est important car nombreuses sont les récriminations des conducteurs qui se plaignent, soit d'avoir la cuisse ' dans le vide ' soit, au contraire, d'avoir la cuisse comprimée par le bord antérieur du siège. Il est, en fait, assez difficile de trouver l'inclinaison optimale qui satisfasse l'ensemble des conducteurs, compte tenu des différences de taille importantes qui existent au sein de la clientèle.

Nous verrons, dans les chapitres suivants, comment le problème se complique du fait des modifications de l'inclinaison de la cuisse liées aux mouvements du pied sur les différentes pédales. Le rôle du garnissage du bord antérieur du coussin comme élément de support de la cuisse sera assez longuement analysé compte tenu de retentissement indirect de ce facteur sur la posture adoptée au poste de conduite. Nous pensons qu'il s'agit là d'un excellent exemple pour montrer les relations complexes qui existent entre les différents éléments du poste.

3.3. *Guidage du Véhicule*

La posture commence à s'esquisser assez précisément quand la position des yeux, des pédales et du siège est fixée. C'est dans une dernière étape que l'on disposera les commandes manuelles: volant, changement de vitesses, etc... par rapport au siège dans l'espace de travail du conducteur, pour que l'atteinte de celles-ci ne nécessite pas de mouvements du buste qui détériorent la qualité de la posture précédemment fixée.

Les commandes manuelles peuvent aider à fixer la posture, c'est le cas pour le volant, mais en aucun cas elles ne doivent la déterminer.

La localisation des commandes manuelles peut se faire par l'utilisation d'abaques qui donnent les zones d'atteinte maximale et les zones de confort dans des plans situés à différentes distances de l'opérateur.

4. **Determination des Caracteristiques Fonctionnelles et Dimensionnelles du Siège**

Nous venons de voir comment les exigences de la conduite, telles que nous les révèle l'analyse du travail, imposaient une posture parfaitement déterminée du conducteur à son poste. En ce qui concerne le passager, le problème est assez différent; dans ce poste, les exigences sont extrêmement peu sévères car nous avons affaire à un poste de repos plus qu'à un poste de travail. Dans ce cas, la définition des caractéristiques optimales du siège peut se faire presque indépendamment du poste dans lequel le siège s'intègre: plus on est libre au poste de travail, plus les qualités propres du siège comptent.

On remarquera cependant que cette indépendance du siège passager, vis-à-vis du reste du poste, n'est pas aussi grande qu'il apparaît en première analyse; en effet, dans ses propriétés dynamiques, on redécouvre que le siège passager n'est pas isolé mais constitue un élément d'un système plus complexe où les masses suspendues du véhicule et du corps humain interviennent (Wisner *et col* 1965).

Ces considérations théoriques étant faites, nous allons voir maintenant comment, d'une façon pratique, il est possible de fixer les principales caractéristiques du siège.

Pour faciliter la démonstration, nous considèrerons un véhicule dans lequel seules les superstructures sont fixées ; nous entendons par superstructures les éléments de la carrosserie tels que le plancher, le pare-brise et le pavillon.

Il est évident que, dans la pratique, le problème ne se présente pas toujours ainsi et l'itinéraire choisi pour l'étude d'un poste de conduite peut être un peu différent de celui emprunté pour l'aménagement d'un poste existant où un certain nombre de choix ont déjà été faits. L'esprit de la méthode que nous proposons n'en reste pas moins inchangé.

4.1. *Détermination de la Hauteur d'Assise et de l'Inclinaison du Dossier*

La position du siège à l'intérieur de l'habitacle doit être définie de façon à ce que les yeux des conducteurs de tailles diverses se situent dans une zone telle que les exigences visuelles de la conduite soient satisfaites.

On peut, d'une façon simplifiée, définir les exigences visuelles en précisant que l'axe de vision doit être horizontal et que les conducteurs doivent apercevoir la chaussée à 4 mètres au plus à l'avant du véhicule ; on considèrera un angle de vision de 15° minimum vers le haut comme satisfaisant.

Ces exigences visuelles se matérialisent par les droites AA′ et BB′. Pour que la vision de l'environnement extérieur soit correcte, les yeux des conducteurs doivent se placer à l'intérieur de l'angle formé par ces deux droites (figure 2).

Connaissant la distance verticale œil-siège correspondant à différentes inclinaisons du buste, il est facile de déterminer la hauteur de l'assise (H) qui permettra aux conducteurs de placer leurs yeux à l'intérieur de l'angle formé par les droites AA′ et BB′.

On choisira généralement une inclinaison de dossier telle que le buste soit à une inclinaison moyenne de 25° ; on admettra, selon les cas, une tolérance de $\pm 5°$ autour de cette valeur en sachant que :

(*a*) Une inclinaison du buste inférieure à 20° rend difficile le maintien d'un angle tronc-cuisse au moins égal à 95° et ceci d'autant plus que la cuisse est déjà fortement inclinée sur l'horizontale du fait d'une distance siège-pédale insuffisante. Par ailleurs, de telles inclinaisons n'offrent pas un support efficace pour le buste dont le poids est supporté presque uniquement par la région fessière.

(*b*) Une inclinaison du buste supérieure à 30° est difficilement compatible avec le maintien de la tête verticale imposé par les exigences de vision.

La hauteur de l'assise étant fixée, on définit également les positions extrêmes arrière du siège : si le siège recule au-delà du point M, les yeux du conducteur à grand buste passent au-dessus de la droite AA' ; si le siège recule au-delà du point N, les yeux du conducteur à petit buste passent en dessous de la droite BB'. Il faudra tenir compte de ces limitations pour la détermination du réglage antéro-postérieur du siège.

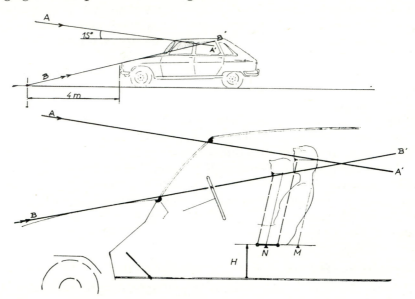

Figure 2.   Détermination de la hauteur d'assise en fonction des éxigences visuelles.

4.2. *Détermination de la Position du Siège dans l'Habitacle*

Nous venons de définir deux caractéristiques importantes du siège :
—sa hauteur d'assise par rapport au plancher.
—l'inclinaison de son dossier.

Nous allons définir maintenant la position relative du siège et des pédales.

Pour maintenir les angles articulaires de la hanche, du genou et de la cheville dans les limites de confort que nous nous sommes fixées, nous avons vu qu'il fallait respecter certains rapports dimensionnels entre la position du siège et des pédales.

La figure 3 montre la construction géométrique qui nous permet d'établir la relation dont nous parlions plus haut entre l'éloignement des pédales et la hauteur d'assise du siège. Les quadrilatères curvilignes $T_1 T_2 T_3 T_4$ et $t_1 t_2 t_3 t_4$ représentent les zones d'emplacement optimal des pédales pour les conducteurs grand et petit, lorsque l'un et l'autre sont assis sur un siège fixe sans réglage d'avant en arrière ; il n'existe dans ce cas aucun recouvrement : il est alors difficile de déterminer une position de pédale qui satisfasse l'ensemble des conducteurs ; on trouvera un compromis en plaçant la pédale entre les deux zones (ligne en tirets). On remarquera cependant que ce compromis n'est pas satisfaisant surtout pour les hauteurs d'assise faibles (partie supérieure des zones) : pour agir sur la pédale, le conducteur petit s'avance sur son siège et n'utilise plus le dossier, il met sa jambe en extension ; le grand, au contraire, est 'recroquevillé' sur son siège, les angles articulaires exagérément fermés.

On voit d'emblée la nécessité d'introduire un réglage du siège d'avant en arrière chaque fois que cela est possible. La figure 4 montre le chevauchement

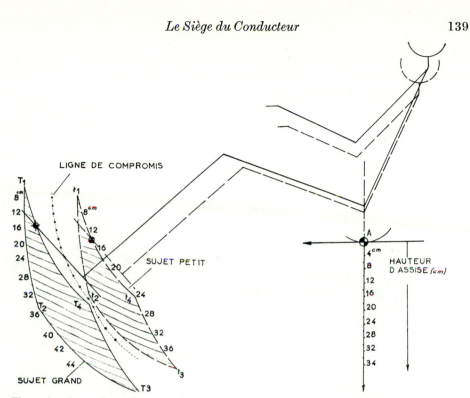

Figure 3. Zones de confort pour l'emplacement des pédales avec un siège non réglable.

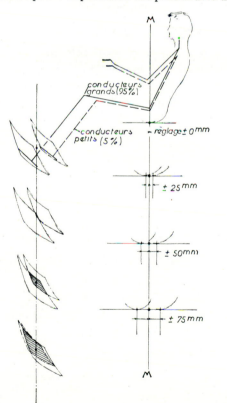

Figure 4. Influence de l'amplitude du réglage sur la zone commune de confort pour les pédales.

des zones de confort obtenu pour différents réglages du siège. On constate que le recouvrement maximal est obtenu pour un réglage de ± 75 mm autour de la position moyenne du siège.

La figure 5 montre la représentation, sous forme d'abaques directement utilisables, de la relation donnant la distance siège-pédale en fonction de la hauteur d'assise.

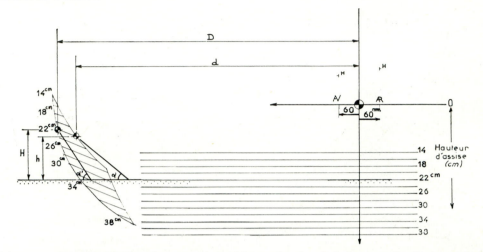

Figure 5.   Normes pour l'emplacement des pédales en fonction de la hauteur d'assise (exemple: haut. assise = 22 cm).

4.3. *Détermination de l'Inclinaison du Plateau de Siège*

L'inclinaison du plateau de siège peut être obtenue, nous l'avons vu plus haut, à l'aide de la construction géométrique de la figure 3.

On détermine par cette méthode l'inclinaison moyenne de la cuisse des conducteurs de taille extrême pour différents réglages du siège. C'est de cette inclinaison moyenne de la cuisse qu'il faut tenir compte pour déterminer l'inclinaison optimale du plateau de siège.

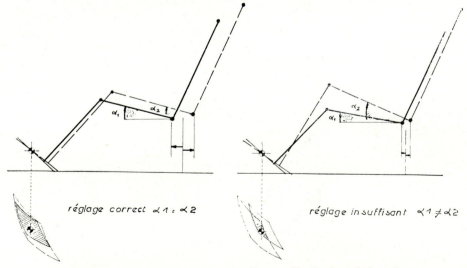

Figure 6.   Influence du réglage du siège sur l'inclinaison de la cuisse.

On remarquera par ailleurs, sur la figure 6, qu'un réglage de grande amplitude du siège permet d'obtenir des inclinaisons de cuisse à peu près identiques : $\alpha 1 = \alpha 2$. Si le réglage du siège est insuffisant le compromis est plus difficile à réaliser : $\alpha_1 \neq \alpha_2$.

4.4. *Détermination du Galbe du Dossier dans le Sens Longitudinal*

Pour la détermination du galbe du dossier dans le sens longitudinal, nous retiendrons les travaux d'Akerblom (1948) et de Keegan (1953, 1962) qui concluent à la nécessité de maintenir la lordose sacro-lombaire en position assis.

Sans nous étendre davantage sur cette question, nous tenons à faire une remarque très importante à ce sujet :

Le maintien de la concavité lombaire ne peut être obtenu uniquement par une configuration appropriée du galbe du dossier. En d'autres termes, il ne suffit pas de prévoir un renflement lombaire dans le dossier pour que la colonne vertébrale trouve un support adéquat. Il faut avant toute chose que l'angle que fait la cuisse avec le tronc soit suffisamment ouvert pour que ce renflement lombaire joue son rôle. Pour que cet angle soit suffisamment ouvert, il faut, et c'est ce que nous avons démontré précédemment, que les pédales soient à bonne distance du siège.

Si cette distance n'est pas respectée, l'angle tronc-cuisse est trop fermé, ce qui entraîne nécessairement une bascule du bassin vers l'arrière par l'action des muscles extenseurs de la cuisse et de la jambe ; le renflement du dossier ne peut empêcher cette bascule. La cyphose qui en résulte est inévitable ; elle est à l'origine de la pathologie lombaire bien connue des automobilistes, pathologie aggravée par l'action des secousses qui se transmettent à travers le siège et ceci d'autant plus que la voiture est mal suspendue.

5. Remarques sur les Qualités du Garnissage du Siège

Les considérations précédentes nous ont permis de définir avec précision les caractéristiques géométriques les plus importantes du siège.

Nous aborderons maintenant le problème de la consistance statique du siège en voyant comment le garnissage du siège intervient dans le confort en distribuant correctement les pressions sur les aires de support.

5.1. *Distribution des Pressions entre le Séant et le Coussin*

Les nombreuses études faites sur ce problème débouchent sur les mêmes conclusions, à savoir la nécessité de laisser aux tubérosités ischiatiques leur rôle de support du tronc. Les coussins mous et profonds qui répartissent de façon homogène les pressions entre le séant et son support donnent au premier abord une impression de confort ; en fait, cette répartition trop uniforme favorise la congestion des organes pelviens et gêne la circulation sanguine. La figure 7 montre un exemple de relevé des pressions réalisé sur un siège à l'aide de capteurs de pressions miniatures de 1 mm d'épaisseur. Cette mesure est réalisée sur un sujet assis en position de conduite. Ce siège est considéré comme bon ; on observe les pressions les plus fortes sous les ischions (90 g/cm^2) autour de cette région les pressions diminuent progressivement pour atteindre les valeurs les plus faibles à la périphérie et, ce qui est important, au niveau des cuisses ($P < 30 \text{ g/cm}^2$).

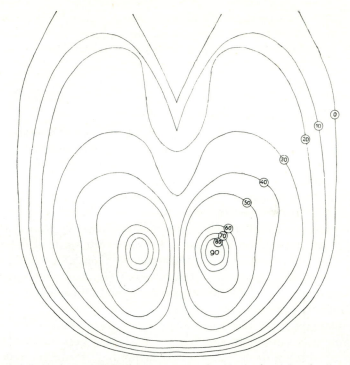

Figure 7. Carte des pressions s'exerçant entre le siége et la région fessière (g/cm²).

5.2. *Distribution des Pressions sous les Cuisses*

Le point important que nous voulons traiter dans ce dernier paragraphe concerne la consistance à donner au garnissage du siège dans sa partie antérieure pour procurer un support correct de la cuisse.

Akerblom (*op. cit.*) a montré, à l'aide de radiographies, comment un bord dur pouvait comprimer douloureusement la face postérieure du 1/3 inférieur de la cuisse. L'examen de l'anatomie de cette région montre, en effet, que les vaisseaux sanguins et le nerf sciatique peuvent être facilement comprimés entre le bord dur du siège et le fémur, sans être protégés par les muscles peu importants à ce niveau.

Pour soutenir la cuisse dans cette région, il faut donc un matériau suffisamment souple qui s'affaisse sous une faible pression unitaire. Il faut, par ailleurs, tenir compte du fait que ce n'est pas une cuisse immobile qu'il faut soutenir mais une cuisse en mouvement du fait de l'action permanente des pieds sur les pédales. Si nous soulignons ce point, c'est parce qu'à notre avis il n'est pas suffisamment pris en considération dans l'élaboration du siège.

Les figures 8 et 9 montrent comment, dans le mouvement d'extension du pied droit pour accélérer ou du pied gauche pour débrayer, la cuisse pivote autour de la hanche et diminue son inclinaison sur l'horizontale. Ce mouvement d'angulation de la cuisse peut être important pour certaines configurations de pédalier : il dépend de la position des pédales par rapport au siège, de la course des pédales ainsi que de leur trajectoire. On voit, dans l'exemple présenté, que l'angulation de la cuisse varie de 5° pour le mouvement d'accélération et de 10° pour le mouvement de débrayage, entraînant un abaissement de la région poplitée de respectivement 30 mm et 60 mm.

L'importance de la compression des parties molles de la cuisse peut être objectivée par la mesure des pressions qui s'exercent entre le bord antérieur du siège et la région poplitée.

La figure 10 montre l'augmentation de cette pression en fonction de l'enfoncement de la pédale d'accélération, obtenu par une extension du pied de 20°, et de l'enfoncement de la pédale de débrayage sur une course de 14 cm.

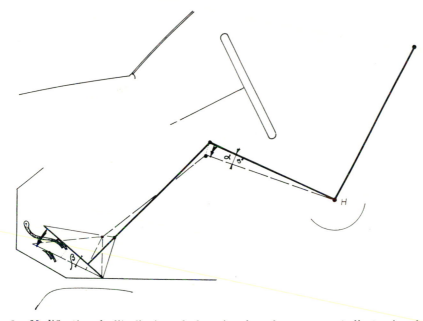

Figure 8.   Modification de l'inclinaison de la cuisse lors du mouvement d'extension du pied pour accélérer.

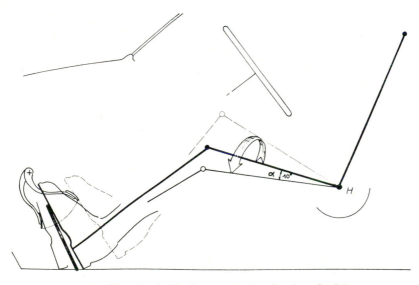

Figure 9.   Modification de l'inclinaison de la cuisse lors du débrayage.

L

Les mesures ont été réalisées en respectant la distance siège-pédale déterminée selon nos normes à partir de la hauteur d'assise. En ce qui concerne l'accélérateur, on constate que la pression de départ (accélérateur libre) est égale à 35 g/cm², ce que nous considérons comme tolérable ; en fin de course elle atteint 65 g/cm², ce qui est nettement trop élevé. Les mêmes constatations peuvent être faites au sujet de la pédale de débrayage où les pressions mesurées varient de 16 g/cm² au départ (la cuisse décolle presque du coussin) à 120 g/cm² en fin de course.

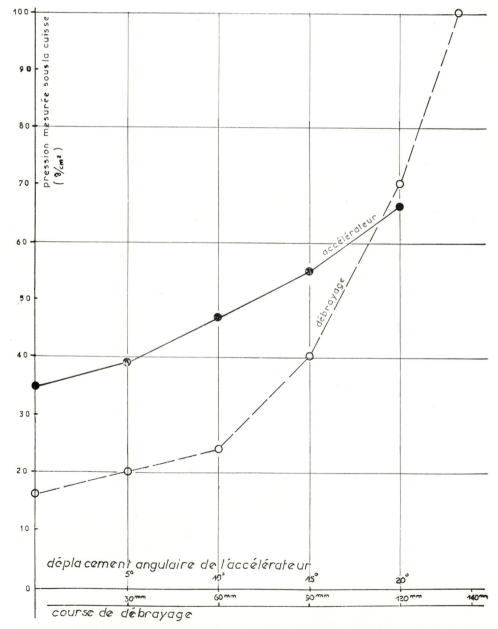

Figure 10. Modification des pressions qui s'exercent sous la cuisse quand le conducteur agit sur les pédales.

Ces pressions sont manifestement trop importantes pour être supportées par le conducteur. Nous pensons, en effet, qu'une pression sous la cuisse, supérieure à 30 g/cm², ne peut être supportée d'une façon continue sans apparition rapide de gêne douloureuse dans les membres inférieurs.

Dans le cas où, du fait d'une action intermittente sur la pédale, la compression n'est pas permanente, on peut tolérer des valeurs plus élevées ; 60 g/cm² nous paraît cependant une limite à ne pas dépasser.

Pour éviter la compression exagérée de la face postérieure de la cuisse, le conducteur cherche à réduire le contact avec le coussin ; pour cela, il augmente l'obliquité de sa cuisse en avançant son siège vers les pédales ; la réduction de l'angle tronc-cuisse, qui résulte de la diminution de la distance siège-pédale, a pour conséquence—nous l'avons vu précédemment—de favoriser la bascule du bassin et l'effacement de la concavité lombaire.

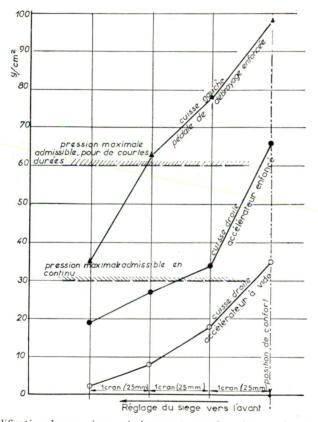

Figure 11. Modification des pressions qui s'exercent sous la cuisse en fonction de la distance siège – pédale.

La figure 11 montre la façon dont les pressions mesurées sous la cuisse diminuent quand on rapproche le siège des pédales. Si l'on se réfère aux valeurs des pressions admissibles proposées ci-dessus, à savoir 30 g/cm² du côté de l'accélérateur et 60 g/cm² du côté du débrayage, on constate que :

(*a*) lorsque le siège est réglé à bonne distance des pédales et que le pied repose sur l'accélérateur sans l'affaisser, la pression qui s'exerce sous la cuisse est correcte (33 g/cm²) ;

(b) lorsque, dans cette même position du siège, le conducteur accélère
" à fond " la pression s'élève à 65 g/cm² ; elle s'élève à 120 g/cm² quand il
débraye.

Pour éviter cela, le conducteur avance son siège d'un cran : la pression du
côté de l'accélérateur devient alors correcte (34 g/cm²) mais reste trop élevée
du côté du débrayage, il lui faut alors avancer d'un cran pour que la pression
s'abaisse à 63 g/cm². Cette réduction de la distance siège-pédales s'accompagne
d'une fermeture de l'angle tronc-cuisse de 6° environ.

Un autre fait mérite également d'être souligné : à la gêne due à la compression
de la cuisse, vient s'ajouter une fatigue musculaire supplémentaire liée à
l'effort de contraction des muscles extenseurs du pied et de la jambe qui
agissent pour affaisser la pédale—ce qui est normal—mais également pour
vaincre la résistance opposée par le bord antérieur du siège.

La solution la plus simple qui vient d'abord à l'esprit pour améliorer cette
situation, consiste à assouplir le plus possible la partie antérieure du siège.
Il faut evidémment agir dans ce sens, tout en respectant certaines limites :

(a) Un siège trop souple ne procure pas un support satisfaisant de la cuisse,
or la cuisse a besoin d'être soutenue :

— pour empêcher, autrement que par une contraction active, le mouvement
de dévers de la cuisse vers l'extérieur du corps quand le pied ne repose
que par la pointe du talon sur le plancher ;

— pour éviter que les muscles fléchisseurs, et principalement ceux du pied,
n'aient à agir seuls—par une contraction permanente—pour maintenir
la position sur l'accélérateur. En effet, l'accélérateur—nécessairement
souple—ne permet pas de limiter le mouvement d'extension du pied qui
survient naturellement sous l'effet de la force de pesanteur. Cette force
de pesanteur, lorsqu'elle n'est pas contrebalancée par le coussin du
siège, s'exerce selon une composante qui agit sur la cheville et fait
pivoter le pied autour du talon.

(b) La réalisation d'un siège dont le bord antérieur est très souple pose des
difficultés de construction, notamment en ce qui concerne sa bonne tenue à
l'usage.

En définitive, le diagnostic d'une compression exagérée de la cuisse par le
siège n'implique pas nécessairement de modifier la consistance statique du
coussin ; dans bien des cas, c'est au niveau du pédalier qu'il convient d'agir
en étudiant, par exemple, une disposition différente des pédales les unes par
rapport aux autres, ou encore en jouant sur la course et la trajectoire de ces
pédales.

A titre d'exemple, on peut voir à l'aide des figures 10 et 11 qu'une diminution
de 30 mm de la course de la pédale de débrayage ou un recul de 50 mm vers
l'arrière du véhicule de cette même pédale rameneraient les pressions mesurées
sous la cuisse à des valeurs voisines de celles admises ci-dessus.

6. Conclusions

Les considérations auxquelles nous nous sommes livrés au cours de cette
étude font apparaitre que le problème de la conception du siège n'est pas un
sujet facile.

Encore faut-il remarquer que seuls les aspects concernant le confort statique
du conducteur ont été abordés. Il va de soi que l'étude du siège doit être

envisagé dans une perpective plus large où les problèmes du confort dynamique et de sécurité jouent un rôle prépondérant.

Il est intéressant de souligner qu'en ce qui concerne ces deux autres aspects du problème, la démarche méthodologique que nous proposons s'applique sans restriction. La définition des caractéristiques dynamiques d'un siège ne peut se concevoir autrement qu'en considérant celui-ci comme un élément d'un système plus complexe où le véhicule et le corps humain interviennent comme des système de masses suspendues; de même la définition des caractéristiques d'un siège sous l'angle de la sécurité ne peut être faite qu'à partir de l'étude préalable du comportement au choc des structures de tôle qui constituent la carrosserie.

Dans ce travail on considère le siège essentiellement dans ses rapports avec les différentes fonctions à assurer au poste de conduite. Selon ce point de vue on envisage trois étapes pour l'étude du siège :
1. Analyse de la tache de conduite.
2. Détermination de la posture qui permet de satisfaire au mieux les exigences de la tache.
3. Définition des caractéristiques du siège qui procure le support optimal du conducteur dont la posture est déterminée par les exigences de la tache.
Les principales caractéristiques du siège obtenues par l'application de cette méthode sont :
— la hauteur d'assise,
— l'emplacement et l'amplitude de la zone de réglage,
— l'inclinaison du dossier,
— l'inclinaison de l'assise,
— la consistance statique du coussin.

In this paper we consider the seat essentially in its relationship with the different functions to be carried out in the driver's cab. According to this point of view we suggest the following three stages in the study of the seat.
1. Analysis of the driver's function.
2. Determination of the best body posture to satisfy in the best way the requirements of the function.
3. Definition of seat characteristics giving optimum support to the driver's posture as determined by the necessities of the function.
The main characteristics of the seat obtained by this method are : seating height, localization and extent of adjustment zone, seat back inclination, cushion inclination, and static consistency of the cushion.

In dieser Arbeit wurde der Sitz hauptsächlich in seiner Beziehung zu den verschiedenen Funktionen betrachtet, die in der Fahrerkabine auszuführen sind. Unter diesem Gesichtspunkt sehen wir drei Stadien für das Studium des Sitzes vor :
1. Analyse der Funktionen des Fahrers.
2. Bestimmung der für diese Funktionen günstigsten Körperhaltung.
3. Definition derjenigen Sitz- Eigenschaften, welche am besten die für die Fahrerfunktionen günstigste Körperhaltung unterstützen.
Die Haupteigenschaften der Sitze, die mit dieser Methode erhalen wurden, sind: Sitzhöhe, die Lage und die Ausdehnung der Anpassungszone, die Sitzrückenneigung, die Neigung des Sitzkissens und die statische Konsistenz des Kissens.

Bibliographie

AKERBLOM, B., 1948, *Standing and Sitting Posture* (Stockholm : NORDISKA BOKHANDELN).

KEEGAN, J. J., 1953, Alterations of the lumbar curve related to posture and seating. *J. Bone and Joint Surgery*, **35**, 589.

KEEGAN, J. J., 1962, Evaluation and improvements of seats. *Industrial Medicine and Surgery*.

MORANT, G. M., and RUFFELL-SMITH, H. P., 1947, Body measurements of pilots and cockpit dimensions. *Flying Personnel Research Committee*, 689.

REBIFFÉ, R., 1966, An ergonomic study of the arrangement of the driving position in motor cars. Symposium in *London J. Inst. mech. eng.*

WISNER, A., et REBIFFÉ, R., 1963, L'utilisation des données anthropométriques dans la conception du poste de travail. *Le Travail Humain*, **XXVI**, 193–217.

WISNER, A., DONNADIEU, A., and BERTHOZ, A., 1965, A biomechanical model of man for the study of vehicle seat and suspension. *The Intern J. of Production Res.*, **3**, 285–315.

Die Gestaltung des Führersitzes auf SBB-Lokomotiven und Gabelhubtraktoren

Von A. Serati

Chefarzt SBB, Bern, Schweiz

Der Uebergang von der Dampf- zu der elektrischen Traktion von Eisenbahnzügen brachte mit sich, dass sämtliche Führerstände unserer ersten Lokomotivserien für eine stehende Bedienung konzipiert wurden.

Während Jahrzehnten war man der Meinung, dass die senkrechte Körperhaltung des Führers eine unbedingte Voraussetzung für die Betriebssicherheit darstelle. Die Fortschritte, die im Bau und in der Bedienung elektrischer Triebfahrzeuge erzielt wurden, die Einführung tauglicher Sicherheitsvorrichtungen, die bei akut auftretenden Gesundheitsstörungen des Führers die Ausschaltung der Stromzuleitung zu den Triebmotoren unterbrechen und eine Schnellbremsung einleiten (Totmannpedal, induktive Zugssicherung, Wachsamkeitskontrolle) sowie die dringende Notwendigkeit einer physiologischen Arbeitsplatzgestaltung führten zur Erkenntnis, dass eine Lokomotive ebenso sicher in sitzender Stellung bedient werden kann.

Die Palettierung und die Mechanisierung des Güterumschlages verlangt den Einsatz von Gabelhubtraktoren. Innerhalb weniger Jahre musste deren Zahl sowohl in den grösseren als auch in den mittleren Bahnhöfen wesentlich vermehrt werden, so dass die Probleme der Gestaltung des Führersitzes an Bedeutung zunahm.

Durch das immer grösser werdende Kontingent von Führern, die Lokomotiven, Triebfahrzeuge und Gabelhubtraktoren bedienen, war es uns möglich, umfassende Erfahrungen inbezug auf die nachteiligen Wirkungen von Vibrationen auf den menschlichen Organismus zu sammeln.

Von seiten der Patienten und der behandelnden Aerzte häuften sich die Klagen über das Auftreten von vertebratogenen Syndromen, Verdauungsstörungen, Oppressionsgefühlen, Beinschmerzen, erhöhter Ermüdbarkeit. Wir selber sind beeindruckt von der Häufigkeit der Feststellung deformierender Thorakalspondylosen, die wir anlässlich der periodischen Untersuchungen von Lokomotivführern röntgenologisch diagnostizieren können (systematische Kontrolle der Thoraxorgane im Schirmbildverfahren in drei Strahlenrichtungen p.a., I. und II. schräger Durchmesser). Die Mehrzahl dieser Spondylosen der Brustsegmente bleiben allerdings symptomlos. Sie erlauben jedoch, gewisse Schlussfolgerungen zu ziehen inbezug auf die Häufigkeit von Spondylarthrosen und von lumbalen Spondylosen, die die Unverfügbarkeit des Personals namhaft beeinflussen. Wir sind überzeugt, dass Mikrotraumatisierungen, die auf Vibrationen zurückzuführen sind, als krankmachende Noxen nicht bagatellisiert werden müssen.

Gestützt auf die neuesten Erkenntnisse von Arbeitsphysiologen, Medizinern und Ingenieuren, die sich mit diesen Problemen unter den verschiedensten Perspektiven befasst haben, sowie auf eigene Erfahrungen (Einführung der Viking-Bostrom-Sitze in den Mowag-Frontlenkern für Städtetransporte der

SBB Führersitz   Führersitz Bostrom Viking
                     Original Ausführung

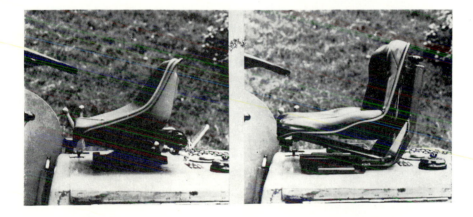

Führersitz Bostrom Baltic   Führersitz Grammer

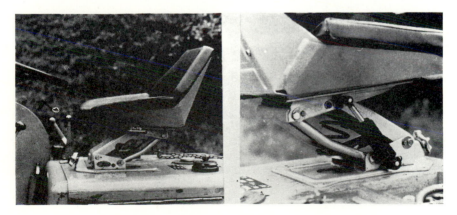

Abb. 1. Führersitz Bremshey FA 414.

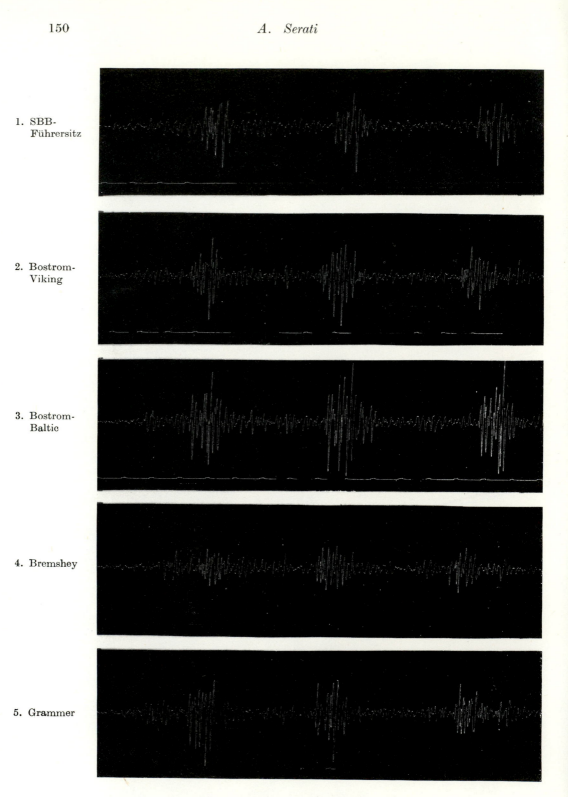

1. SBB-
   Führersitz

2. Bostrom-
   Viking

3. Bostrom-
   Baltic

4. Bremshey

5. Grammer

Abb. 2.   Schwingungsdiagramme.

Postbetriebe, Schwebesitze der PTT-Cars) suchten wir in Verbindung mit der Abteilung für Zugförderung und Werkstättedienst der Generaldirektion SBB die geeignetsten Lösungen, um unsere neuen Triebfahrzeuge und Gabelhubstapler mit Sitzen zu dotieren, welche unter den verschiedensten Betriebssituationen eine genügende Schwingungsdämpfung garantieren können.

Der Gestaltung der Sitzschale schenkten wir besondere Aufmerksamkeit: Sie muss den arbeitsphysiologischen Kriterien entsprechen, die freie Beweglichkeit des Führers ermöglichen (Fluchtweg bei Betriebsgefährdungen oder ausserordentlichen Vorkommnissen) und seiner Wachsamkeit nicht entgegenwirken.

Wir waren darüber einig, dass in Uebereinstimmung mit den Anforderungen von Oehmen und Herzog die Federkonstante des Sitzes auf das Gewicht des jeweiligen Fahrers einstellbar sein muss und dass der Sitz von Parallellenkern an ein Federsystem getragen werden muss, der durch einen Dämpfungszylinder beruhigt wird. Durch Schwingungsmessungen versuchten wir, die Auswirkungen der verschiedenen Aufhängevorrichtungen der Sitze objektiv zu erfassen.

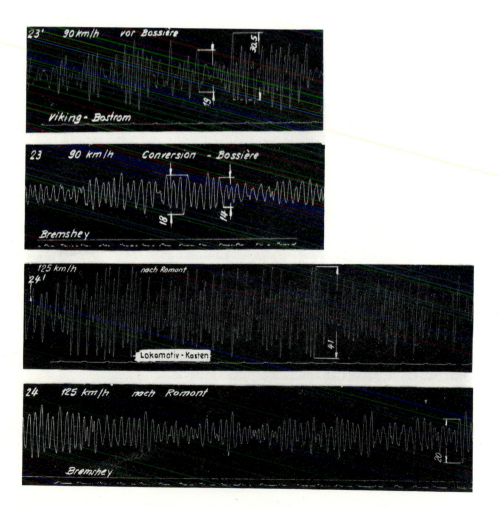

Abb. 3. Schwingungsdiagramme.

Prüfung an Gabelhubtraktoren

Wir bedienten uns eines seismischen Erschütterungsmessers vom Typ Askania. Einstellung Frequenz 11,1 Hz, Masstab der Ausschläge 8 mm (1 g), Diagramm-Vorschub 10,5 mm/sek. Versuchsstrecke: Hartbelag. Beginn und Ende der Strecke (20 m) waren durch 15 mm dicke Holzlatten markiert. Diese sowie eine in der Mitte der Versuchsstrecke angebrachte Latte (Dicke 20 mm) mussten überfahren werden.

Als Sitze wurden geprüft: Der klassische, bisher gebräuchliche SBB-Führersitz sowie Schwebesitze, deren konstruktive Merkmale ihre Anbringung auf dem Fahrzeugkasten erlaubte (Bostrom-Viking, Bostrom-Baltic, Bremshey, Grammer) (Abbildung 1).

Ergebnisse: Die maximalen Beschleunigungen des SBB-Führersitzes, des Bostrom-Viking und des Grammersitzes betrugen $\pm 1,8$ g, diejenige des Bostrom-Baltic-Sitzes $\pm 2,7$ g, des Bremshey-Sitzes $\pm 0,9$ g. Schwingungsfrequenz beim Ueberschreiten der Hindernisse circa 9-10 Hz. (Abbildung 2).

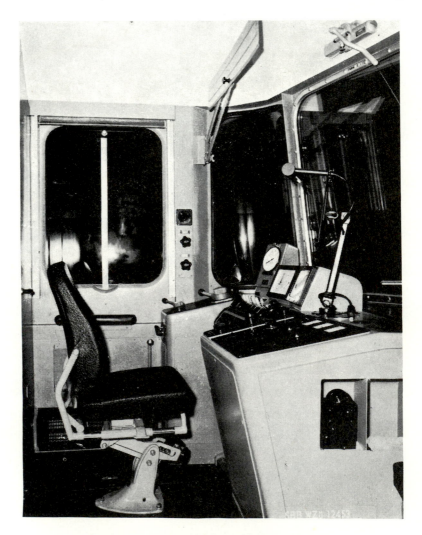

Abb. 4.   Führersitz Bremshey-SBB-Modell.

Für die *Prüfung der Sitze auf den Lokomotivführerständen* stützten wir uns weitgehend auf diese ersten Erfahrungen.

In einer ersten Versuchsserie wurde der klassische, vor kurzem verwendete Führersitz (gefederter und drehbarer mittelhart gepolsterter Bigla-Arbeitsstuhl) mit dem Bremshey-Sitz verglichen. Die maximale Beschleunigung beim Bigla-Sitz lag bei $\pm 0,5$ g, beim Bremshey-Sitz bei $\pm 0,3$. Schwingungsfrequenz bei beiden Sitzen annähernd gleich, circa 4,5–5 Hz.

In einer zweiten Versuchsreihe erfolgten vergleichende Messungen: Bremshey-Sitz/Viking-Bostrom-Sitz. Die maximalen Beschleunigungswerte des Viking-Bostrom-Sitzes betrugen $\pm 0,4$ g, des Bremshey-Sitzes $\pm 0,3$ g; diejenigi des Lokomotivkastens (bei einer Fahrgeschwindigkeit von 125 km/h) $\pm 0,6$ g. (Abbildung 3.).

Die Laufruhe der geprüften Lokomotiven entsprach noch nicht der heutigen Verbesserung (Anpassung der Sekundärfederung). Bei den neuen Ae 4/4 II-Lokomotiven liegen die Scheitelwerte des Kastens bei $\pm 0,3$ g, so dass am Führersitz künftig mit Beschleunigungen im Bereiche von höchstens $\pm 0,1$ bis $\pm 0,2$ g zu rechnen sein wird. (Abbildung 4.)

Die Erfahrungen, die wir mit den neuen Sitzen inzwischen sammeln konnten, sind durchaus positiv. Die geltend gemachten Beschwerden und Unzukömmlichkeiten sind im Abnehmen begriffen. Der erhöhte Fahrkonfort wird von den Führern der Lokomotiven sowie der Gabelhubtraktoren sehr geschätzt.

Unsere Studie stellt das glückliche Ergebnis einer engen Zusammenarbeit zwischen Ingenieur und Arzt dar, eine Voraussetzung, die vorbehaltlos erfüllt werden muss, um aktuelle arbeitsmedizinische und ergonomische Fragen lösen zu können.

Die Einführung der sitzenden Bedienung der neuen Lokomotiven und Triebfahrzeuge stellt an den Führerstuhl besondere Anforderungen, will man die Auswirkungen der Mikrotraumatisierung der Wirbelsäule einschränken und gleichzeitig die Sicherheit des Betriebes garantieren. Aehnliche Probleme liegen bei Gabelhubtraktoren vor.

Der Führersitz muss den arbeitsphysiologischen Anforderungen entsprechen und die vertikalen Schwingungen weitgehend auffangen können. Schwingungsmessungen mit einem seismischen Messapparat vom Typ Askania ergaben, dass der Bremshey-Sitz im Vergleiche zu den übrigen geprüften Schwebesitzen die niedrigsten Beschleunigungswerte aufweist. Die Einführung dieses Sitzes bei Lokomotiven und Gabelhubtraktoren der Schweiz. Bundesbahnen hat sich sehr bewährt.

The seated driving position of the new locomotives and railcars makes it necessary for the motorman's seat to meet special requirements if the micro-traumatization of the spine is to be diminished and, at the same time, the safety of operation to be guaranteed. Similar problems have been created by fork-lift trucks.

The motorman's seat must come up to the requirements of work physiology and be capable of absorbing the vertical oscillations to a considerable extent. Oscillation measurings carried out with a seismic measuring apparatus of the Askania type have shown that among the suspended seats tested the Bremshey seat exhibits the lowest acceleration values. This seat, which the Swiss Federal Railways have introduced on their locomotives and fork-lift trucks, has proved very satisfactory.

Le siège du conducteur doit être conçu de façon à réduire les effets des microtraumatismes de la colonne tout en offrant les garanties nécessaires pour la sécurité de l'exploitation. Les mêmes problèmes se posent pour les tracteurs-élévateurs. Le siège doit pouvoir satisfaire aux critères de la physiologie du travail et absorber, autant que possible, les oscillations verticales. L'étude des oscillations exécuté avec un séismographe du type Askania a démontré que, parmi les sièges suspendus expérimentés, le type Bremshey présente les plus bas coefficients d'accélération. Ce siège, qui a été introduit sur les locomotives et les tracteurs-élévateurs des CFF, a donné entière satisfaction,

Literatur

HERMANN, E., 1963, Ueber den Einfluss mechanischer Schwingungen auf den menschlichen Organismus. *Der ärztliche Dienst*, **7/8**, 196–202.

HOSHI, A., 1965, Ergonomic improvement of railcar seats. *Japanese Railway Engineering*, **3,** 22–25.

OEHMEN, H., und HERZOG, J., 1967, Entwicklung eines Fahrersitzes für schwere Kraftfahrzeuge im gleislosen Förderbetrieb. *Kali und Steinsalz*, **11,** 353–360.

SASSOR, H. J., und KRAUSE, H., 1966, Auswirkungen mechanischer Schwingungen auf den Menschen. *RKW-Reihe, Beuth-Vertrieb GmbH, Berlin 15, Köln/Frankfurt (Main)*.

WOLFF, G., 1965, Zur Praxis der Messung von Lärm und Erschütterungen in Betrieben. *Zentralblatt für Arbeitsmedizin und Arbeitsschutz*, **8,** 181–184.

The Assessment of Chair Comfort

By B. SHACKEL,* K. D. CHIDSEY† and PAT SHIPLEY

Department of Occupational Psychology, Birkbeck College, University of London, England

1. Introduction

This paper describes a series of studies made on behalf of and in conjunction with the Consumers' Association to explore the general area of seating comfort. While the primary aim was publishable comparisons between chairs, it was accepted from the start that the exploration of various testing methods would be a major aspect. Therefore the particular emphasis of our work was the study of chair comfort evaluation methods in the context of selection by or for a potential user or purchaser.

The studies have involved two experimental programmes, the first being three experiments rolled into one. In the first programme the primary aims were to explore several possible methods and to compare a homogeneous group of chairs in a range of typical user situations. From part of this a *Which?* report was published (Anon., 1966).

In the second programme, the primary aim was to study the value of individual subjective opinions or of recommended dimensions as guides for the user to select chairs. Such quicker, simpler methods would be useful. The layman is often suspicious of the expert and his experiments; are they really necessary? After all, the expert has no monopoly of the appropriate part of the human anatomy.

From these two programmes some of the results, in particular from further analyses of the data, are relevant to the general question of methodology. These analyses and results are brought together, with a discussion and some suggestions, at the end of the paper.

Thus, having found by the initial survey that theoretical formulations and prior research did not entirely answer the questions posed, an essentially empirical approach was adopted, starting with field studies and then moving to some more basic questions. Therefore, this paper is divided into three main sections dealing in turn with the field studies, the experiment on some methods, and the analysis and discussion on methodology, followed by some general conclusions.

PART I. EXPERIMENTAL COMPARISONS OF UPRIGHT CHAIRS

2. Introduction to Part I

The Consumers' Association naturally hoped that a battery of suitable measurement techniques already existed. While a number of previous investigators have measured the comfort offered by a chair (see list of references), with various methods and with some degree of success, it was evident from the

---

† Now at the Furniture Industry Research Association, Stevenage, Herts.
* Also Head of the Ergonomics Department at E.M.I. Electronics Ltd.

literature, and was confirmed positively by all the specialists who were consulted, that there was no general agreement on one or a few techniques of proven adequacy, precision and reliability.   Thus, it became a primary aim to explore as many measurement techniques as possible, of those that seemed likely to be useful, with a view to evolving a combination of suitable methods for future chair comfort studies.

From this followed the type of chair and the range of usage situations to be studied.

2.1. *Chairs and Usage Situations*

It was decided that one type of chair used for broadly similar purposes, but in a wide range of situations, would be most appropriate.   Therefore, a homogeneous group of chairs was chosen with broadly similar dimensions and structure, namely cheap wooden or plastics, upright chairs.   If significant differences could be revealed in such a group, then the measurement techniques evolved might reasonably be considered suitable also for other cases, such as a group of easy chairs which would be expected to differ much more in size and shape. This view fitted well with the wish of the Consumers' Association to study, if convenient, the polypropylene chairs recently brought onto the market. Moreover, the only British Standard recommendations available (B.S.I. 1959; B.S.I. 1965) were mainly relevant to this type of chair and would enable a rational basis for choosing wooden chairs of varying dimensions, in the hope of exploring comfort over an adequate range.

Thus ten upright chairs were chosen, the only five polypropylene designs at that time and five wooden chairs of equivalent price and type.   Some are shown in Figure 1.   Because there are hundreds of chairs which might have been chosen, and there is no way of knowing whether our sample spanned the possible range or perhaps came from the upper or lower half or some smaller segment, the Consumers' Association decided that it would be unreasonable, in this instance, to identify two particular chairs and so they are omitted from the figure. The physical dimensions of the chairs are given in Table 1, together with the relevant British Standard recommendations; some of the measurements proved difficult to make on the polypropylene chairs and the values given are therefore best estimates.

It was considered essential to sample adequately the range of possible uses for such chairs, and the following situations were chosen:

    1. long term sitting — general use
    2. sitting at a desk — office use
    3. eating a meal   — canteen use.

These reference situations were interpreted as follows.   Long term sitting is sitting as in general usage (such as in reception, meeting and reading areas) but without a desk or table for support and for periods of up to three hours or more.   Office use is sitting at a table or desk doing general desk and clerical work, but not typing; the table or desk heights should be taken as a distribution covering the B.S.I. specified range of 28 to 30 inches.   Canteen use is sitting at a table, acceptable height range as for office use, eating at least a two-course lunch in an industrial or office canteen; the sitting at the canteen table is to last for at least half an hour, even if the meal itself ends sooner.

It had been hoped, originally, also to sample a lecture room or school hall concert as a fourth situation, and thus cover the educational and related use (where the seated user is looking forward and slightly upward), but this was abandoned during planning because of obvious lack of experimental time.

Chair X   Chair A   Chair Y   Pel Fantasia Palma = F

Hille Polypropylene = N   Cox Fiona FL12 = K   Ideal 77 = L   Wilmot Breeden Polypropylene = C

Figure 1. Upright chairs used in the experiments.

It was decided to make the test situations as natural as possible, but it was also thought desirable to vary the concentration of attention upon the comfort testing aspect of the situation. Therefore, usages 2 and 3 above were tested simply and directly in suitable field trials, with the attitude indicated to the subjects that their help was most valuable to us but that they should not allow our project to interfere in any way; whereas usage 1, long term sitting, was not explored in a typical situation, such as a waiting room or lecture room, which in any case might have been rather unstructured and so possibly unfruitful. Instead, a more concentrated and controlled experiment was established with a view to ' stressing ' the testers and thus developing a measurement technique with the maximum chance of revealing significant comfort differences between chairs if any existed to be found. Therefore, a long-term sitting trial was chosen, using university students (who typically sit reading for long periods, at least before examinations) in a near-laboratory situation, so that this method could be repeated on subsequent occasions, as required for other test cases, if it proved to yield satisfactory results. Since this experiment was done in the Easter vacation, with subjects who were intending to sit and read for a given period each day and who were glad to have a regime requiring this, it is thought that the conditions in this case also were reasonably natural.

Table 1. Chair dimensions and British Standard recommendations

Chair	K	F	N	L	X	C	Z	A	T	Y	Recommended for non-adjustable office seating B.S. 3079 (1959)	B.S. 3893 (1965)
Dimension												
1. Height of seat (in.)	*18¼	*17¾	17	17	*18 (with cushion depressed)	*16¼	17½	17	17⅛	16¾	17	17
2. Effective depth (in.)	15	15	15½	14	15	16½	13½	13½	13½	*11½	not >15	14–18½
3. Minimum width (in.)	*15	*14½	16½	*14½	*15	18	*15¼	*13½	*14½	*13	16	16
4. Slope of seat (deg.)	*7	5·5	*10	*10	1 *—2 (downwards)	*—2 (downwards)	2·5	2	4	0	3–5	0–5
Backrest												
5. angle (deg.)	105	*112	105	105	*109	*116 (flexed)	102	94	105	100	95–105	95–105
6. height (in.)—lower	—†	—†	—†	9½	9½	8½	*4½	*5½	*13¼	*5	>8	>8
7. from seat—upper	*15	—†	13½	13½	13	—†	13	13	*14½	*13	<13	<13
8. horiz. curvature radius (in.)	*9½	*0	*10½	16½	*9¾	*10½	*20	11½	*20	*0	16–18 preferably and not <12	12–18
9. height of backpad (in.)	—†	—†	—†	4	3½	—†	8½ (curved)	7½ (curved)	1¼	9	about 4–5	not <4, if >5 shall be curved convexly in vert. plane
10. width between outside edges (in.)	13	15	14	14½	16	15	17	16	16	13	not >13 (max. 15 if curved in horiz. plane)	not <12
Rows 1–8 Total B.S. Failures	5	4	2	2	4	4	3	2	4	5		

Note: * Indicates that this chair dimension fails to conform to the relevant B.S. recommendation.

† There is no separate backrest or the equivalent on these chairs, which have moulded backs, but the construction of some causes a projecting area which impinges on the user's back at the height given. In one case there is no such projection and the user's back pressure would seem to be evenly distributed.

2.2. *Subjects*

The general user population assumed to be appropriate for these chairs was defined as shown in Table 2.

Table 2. User population and stature of subjects

User population
Age range   18–65 years
Equal sex distribution

Weight :	male	M.	145 pounds	S.D.	16
	female	M.	127	S.D.	18
Stature :	male	M.	67·5 inches	S.D.	2·4
	female	M.	63·5	S.D.	2·6

Subjects—stature

		Mean (inches)	Range (inches)	Percentile range
Long term :	male	68·5	66–73·5	26–99
	female	65·3	63–69	47–98
Desk :	male	68·6	64–73·5	7–99
	female	64·6	59–68·5	4–97
Meal :	male	68·5	64–74	7–99·5
	female	64·5	60–68·5	9–97

A different group of twenty subjects, ten male and ten female, was used for each experiment.   All were selected on the basis of their stature, with the aim of covering the 5 percent to 95 percent range of the normal population; for males, this range is taken as 63·5 to 71·5 inches, and for females 59 to 68 inches.   The subjects were further selected to be within ± 1 standard deviation of the mean weight for their height.   As may be seen from the stature data in Table 2, the availability and choice of subjects was satisfactory except for the long term sitting test, where the sample is deficient in shorter people of both sexes as would be expected in a student population.   This was unfortunate, but could not be overcome; in particular, short female students are extremely rare. Similarly, the age range of the student subjects was very narrow.   However, these sampling limitations were not considered to be unduly restrictive, because the aim of this long term test was to develop a measurement technique which would reveal significant differences, if any existed to be found, in the equivalent of a laboratory situation which could be repeated subsequently as required.

3.   Measurement Methods

3.1. *Alternatives*

The bases upon which comparative measurements can be and have been developed may, it is suggested, be divided into four broad groups.

1. Anatomical and physiological factors—body size, shape and structure, related orthopaedic aspects, and effects of prolonged pressure and other restrictions on physiological functions, all leading to comparisons and recommendations in terms of physical dimensions of the chairs.
2. Observations of body position and movement—closely related to the first area but essentially different in that such aspects as the number, frequency and other characteristics of movements and changes of posture are the prime variable studied, and often recorded, usually during ' natural ' and fairly lengthy sitting trials.

M

3. Observation of task performance—real, or specially devised or controlled, work tasks are measured appropriately.

4. Subjective methods—under standardised conditions the assessments and judgements are obtained by a controlled procedure.

From the literature and discussions with colleagues it was concluded that probably all these bases have some validity and that none should be considered an exclusive solution. In this context, the opinion of Burandt and Grandjean (1963) is particularly relevant: 'from this analysis we have drawn the conclusion that the exclusive application of an anatomical magnitude for the determination of chairs and tables for office use is unsatisfactory. It is only the experimental investigation that will reveal the effective interrelations between the comfort of seat and table dimensions and anatomical data '. It was therefore decided to compare the physical dimensions of the chairs with British Standard recommendations based on anatomical considerations, and to study the anatomical criterion data by comparing the chair rank orders based on the anatomical recommendations with those derived from the comfort assessments.

The observation of body movement was considered appropriate, and it was hoped to arrange a ciné-camera memomotion equipment to enable counting of numbers of posture changes, etc. This could not be done because of shortage of time and facilities. However, an attempt was made, during the long-term sitting test, to explore whether this method would be useful and would correlate well, by having two graduate psychologists observe the subjects on a time-sampling basis and record the frequency and duration of posture changes.

Any attempt to use the method of observing task performance would obviously be impossible under the field trial conditions preferred for this study.

From the evidence in the literature and in view of the time available for the study, it was decided that the primary results must be obtained from subjective measurement techniques taken during and after appropriately controlled sitting trials. A further and conclusive reason for our concentrating upon subjective measures is inherent in the context of studying chair comfort in relation to individual users choosing for themselves, namely that the ultimate criterion must be the subjective judgement of a representative sample of users. However, such subjective methods inevitably are somewhat crude measuring instruments; it was therefore decided to use a number of techniques of different types and to correlate their results, so as to improve the validity of the final evaluation.

The measurement methods finally adopted are listed below. The last two have already been described briefly. The first four are the subjective measures and are described in the next sections.

1. General Comfort Rating.
2. Body Area Comfort Ranking.
3. Chair Feature Checklist.
4. Direct Ranking.
5. Body Posture Change Frequency.
6. Chair Dimensions and Standard Recommendations.

3.2. *General Comfort Rating*

The first subjective measure was aimed to elicit from the subjects, at appropriate intervals during a trial session, a rating of their present sensation on a comfort-discomfort scale.

This rating scale was established empirically in the usual manner. Twenty statements were chosen, suggestive of a degree of comfort or discomfort. Each was printed on a separate piece of paper, and one set was distributed to each member of a group of fifty people, who were asked to rank the statements in the order from most comfortable to least comfortable. For each statement, a frequency distribution of the ranks thus given was compiled, and those statements whose frequency distribution lacked a peak, or which were not unimodal, were excluded. An 11 item, 10 interval, scale was then made up from the remaining statements. In the best tradition of intelligence test establishment the next step should have been to submit this scale to an even larger group, but lack of time prevented this.

The eleven statements of the final scale are listed below. They were printed against a vertical line 10 cm long and with a short mark at each centimetre opposite the statement. The subjects were instructed to draw a horizontal mark anywhere on the vertical scale to express their rating. Scoring was done with a 10 cm scale, by rounding the mark position to the nearest half interval (i.e. $\frac{1}{2}$ cm) and then doubling the value, so as to give a score scale of whole numbers from 0 to 20 for ease of handling.

> I feel completely relaxed
> I feel perfectly comfortable
> I feel quite comfortable
> I feel barely comfortable
> I feel uncomfortable
> I feel restless and fidgety
> I feel cramped
> I feel stiff
> I feel numb (or pins and needles)
> I feel sore and tender
> I feel unbearable pain

3.3. *Body Area Comfort Ranking*

Allen and Bennett (Bennett 1963) have described a forced-choice ranking technique for assessing the pattern of local comfort and discomfort whilst sitting, and the technique appeared appropriate for use in this context.

On the answer sheet a manikin is shown, divided into fifteen body areas each with a reference number on it. Alongside are five boxes, labelled ' 3 most comfortable ' to ' 3 least comfortable '. The subject is instructed to select the three body areas which are most comfortable for him at the moment, to write their reference numbers in the first box and delete them on the manikin; then he writes in the next box the numbers of the next three most comfortable areas, and so on until all fifteen numbers are listed in the five boxes. If he prefers or finds it easier, the subject may enter them in a different order, e.g. the three most comfortable and then the three least, and so on.

3.4. *Chair Feature Checklist*

This checklist was aimed to sample, in a different way, the same aspects as the Body Area Comfort Ranking. It was in fact a direct attempt to get people, who had been given considerable experience of a chair, to comment on those features of a chair which might produce local comfort or discomfort, instead of doing this indirectly, as in the Body Area rankings.

Using as a basis the significant parts of a chair distinguished by Floyd and Roberts (1958), various features were listed such as seat height, depth, width, etc. For each, a three part answer was provided and the subject was asked to select the one appropriate to his opinion of that feature. For the layout of this form, see Table 3.

Table 3. Layout of Chair Feature Checklist

Chair No. ....
Date ....
Tester No. ....

CHAIR FEATURE CHECKLIST

Instructions to subjects: Below is a list of all the features of a chair which contribute to the feelings of comfort produced by it. On the right hand side of the page, opposite each feature, are three brief phrases, descriptive of the feature, of the general form (too little, right, too much). Encircle the phrase which describes the opinion you have of that feature. Please examine carefully all the phrases describing a feature before choosing each response.

Section 1

Seat

1.1	Seat height above the floor	*too high* (chair presses on thighs)	*correct* height	*too low* (thigh completely clear of chair)
1.2	Seat depth (length)	*too long* (presses into hollow behind knee)	*correct* depth	*too short* (overhang seat at back)
1.3	Seat width	*too narrow* (unable to move sideways)	*correct* width	*too wide* (unable to slide out out of seat sideways)
1.4	Slope of seat	*slopes too far down towards the back* (slide down into the seat, wedged)	*correct* slope	*slopes down at front too much* (slide out of the seat)
1.5	Seat shape (if not flat)	*poor*	*adequate*	*good*

Section 2

Back support (either separate backrest, or moulded chair back)

2.1	Position of backrest, in relation to the seat	*too high* small of back unsupported	*correct*	*too low* middle of back unsupported
2.2	Moulded chair back	*fits the back* shape very well	*adequate*	*poor fit* excessive pressure at some point
2.3	Curvature of the back support (side to side)	*too curved* clamps the sides	*correct*	*too flattened*

Section 3

3.1	Clearance for feet and calves, under the chair	*too little* (difficult to get up)	*adequate*

3.5. *Direct Ranking*

For this measure, each subject was asked to sit in each chair in turn, and first to divide the chairs roughly into three groups. Then he sat on them again and ranked the chairs within each group, comparing adjacent chairs to decide a final ranking.   The subject was encouraged to take his time and to change the chairs round as much as he wished until he was entirely satisfied with the order, sitting up and down the line at least once in sequence as a final check.

3.6. *The Comfort Questionnaire*

For the convenience of subjects and experimenters, a standard form was prepared to contain the most frequent measures—the General Comfort Rating and the Body Area Comfort Ranking.   These were suitably placed side by side in the middle, with a space below designated for any comments (with another numbered manikin, similar to the Body Area one, for easy reference).   At the top there were spaces for the chair code number, the date, the subject number, and the times of departure and return for any breaks from sitting; finally, two columns of times at half-hour intervals from 0830 to 1900 enabled the time of using that sheet to be ringed.   The forms were printed on different coloured papers, which enabled a colour code also to identify the chairs.   A fresh sheet of the same form was used for each separate time of making a General Comfort Rating and Body Area Ranking; each completion of the questionnaire took about 30 to 45 seconds.

4.   Experimental Design

4.1. *Long Term Sitting*

This test was held in Birkbeck College, in March–April 1965, with ten male and ten female subjects selected from the student population.

Each subject was given standard instructions as to dress throughout the testing period, and the Comfort Questionnaire and the Chair Feature Checklist were demonstrated, with a detailed explanation of how to fill in the Questionnaire.

Testing took place in a quiet room, during the Easter vacation, and was supervised by two postgraduate psychologists.   Each student made a direct ranking of the ten chairs, before testing began, without being allowed to see them; the chairs were positioned behind him, and manoeuvred by the experimenter, to rule out the use of other cues than those derived from actually sitting.

After the direct ranking, the students each sat in one chair per day, during a three and a half hour session, only essential toilet breaks being allowed.   These averaged five minutes per subject, over each session.   Refreshments were served to the subjects during each session.   The order in which each subject sat in each chair was established in advance, using a randomized matrix design.   In practice, this was adhered to stringently, only being departed from on two occasions (out of 200), when a subject was absent.

Each subject completed the Questionnaire five minutes after the start of each session, and thereafter at half-hour intervals; at the end of his session with a particular chair, he completed the Chair Feature Checklist for that chair.

During the test session, each subject read, or made brief notes.   He was allowed a table alongside his chair, for storing books, but was not allowed to use it for support.

4.2. *Sitting at a Desk*

This test was held in the offices of the Greater London Council, during July 1965.

The ten male and ten female subjects selected for this study were engaged in clerical duties, which included filing, and the majority of their work was done whilst sitting at a desk.   The desks ranged in height from $28\frac{1}{2}$ to $30\frac{1}{2}$ inches to the top, and from $22\frac{1}{4}$ to $27\frac{1}{2}$ inches to the underside, from the floor.   The heights recommended from anthropometric data are 28 to 30 inches top) and at least 26 inches underside).   The desks themselves, for the most part, were thus not entirely matched to the recommendations, but they are typical of the range in general use.   It was therefore decided not to attempt to test the chairs at standardised desks, although this would have been desirable to show up differences between chairs more clearly.   Each subject stayed at the same desk during the entire experiment.

Since this test was taking place during the normal operation of the offices, it was not possible to control it as closely as had been achieved during the earlier test at Birkbeck.   In particular, subjects were quite frequently absent, or had to perform other duties away from their office; this affected to some extent the order in which the chairs were presented, which was planned to be randomized.

The Comfort Questionnaire was demonstrated to each subject before testing began, and each subject was asked to fill in a fresh sheet at hourly intervals during the day.   At the end of each day, he completed the Chair Feature Checklist.   After testing, it had been intended to obtain a Direct Ranking from all subjects, but due to a misunderstanding, the chairs were collected before this was complete, and insufficient results were obtained for this data to be used.

4.3. *Eating a Meal*

This test was held in the staff canteen of Penguin Books Ltd., during August–September, 1965.

The ten male and ten female subjects were selected from the population who used the staff canteen.   The tops of the canteen tables, it should be noted, were at a height of 28 inches from the floor.

At the beginning of the study, the Comfort Questionnaire was explained and demonstrated to each subject.   He was asked to eat in the canteen on each of ten successive days, and to sit in a different chair on each occasion, as directed by the experimenter.   He was asked to remain in the chair for half-an-hour, or for the duration of the meal if it exceeded half-an-hour, and at the end of the meal to complete the General Comfort Rating, the Body Area Comfort Ranking, and the Chair Feature Checklist.   When he had sat on all of the chairs, he was asked to give a Direct Ranking of the comfort they afforded, sitting in each in turn in the standardised method.

4.4. *Summary of Test Conditions*

For convenience, the various test conditions for these three experiments comparing upright chairs are summarized in Table 4.   Other conditions were

Table 4. Summary of test conditions for Part 1

Subjects	Location	Duration (per chair)	Occupation
Students	Quiet room	$3\frac{1}{2}$ hrs	Revision reading
Clerks	Own desk	All day	Normal duties
General	Canteen	$\frac{1}{2}$ hr	Lunch

unified as far as possible. Standard instructions were given to all subjects, on such aspects as asking them to concentrate upon comfort assessment, to try to ignore chair appearance and colour, and to wear similar clothing throughout the trial period (e.g. not light trousers and sweater one day and a heavy suit the next). Also, written Questionnaire instructions were given for reference if needed, in addition to the explanations and demonstration. In general, it is considered that the tests were as controlled and standardised as could be achieved under non-laboratory conditions.

5. Results

5.1. *Sex Differences*

The General Comfort Ratings were summed, for each subject separately, over all of each separate chair session in the three test situations. A statistical comparison was made, using the Mann-Whitney U Test, for each chair between the scores of the male and female subjects. No significant differences were found except for one chair, marginally at the 5 per cent level, in the desk test (chair Z), and similarly for one in the meal test (chair N). Since there are thirty separate comparisons here (ten chairs in each of three situations), between one and two differences significant at the 5 per cent level are expected by statistical definition. Moreover, the meal test data are based on one score only from each subject on each chair, and there are no obvious factors about either chair which appear relevant for sex differences such as stature to be linked with. Therefore, it is concluded that these are chance effects, and that the men and women may be regarded as part of the same homogeneous population for these sitting studies.

5.2. *Comfort Ratings over Time*

The comfort rating scores from all subjects were averaged for each chair separately, at the half-hour rating intervals of the long term test and at the hour intervals of the desk test. The results are graphed in Figures 2 and 3.

For the long term test (in Figure 2) there is a clear trend for comfort ratings to decrease with time, and for three groups of chairs to separate out. The poorer rating of the two worst chairs seems clear from the start, but the others only seem to separate clearly after 1 to $1\frac{1}{2}$ hours. There seems also a tendency for the slope of deterioration of comfort to be steeper with the worse chairs.

Note the position of chair F well up, and A and C well down, and compare this with the same graph for the desk test (Figure 3). Here, F is well down and A and C at the top. The same general trends appear, as on the previous graph, but less marked. No doubt the freedom, and perhaps necessity, to leave their desks as part of their normal duties, enabled the subjects to avoid a larger deterioration in comfort.

5.3. *Significance of Chair Differences*

The General Comfort Rating Scale is not claimed to be more than an ordinal scale. In order to test the differences between chairs, the summed scores for each subject (as in section 5.1) were compared for each pair of chairs using the Wilcoxon test, to determine which pairs of chairs were significantly more or less comfortable than each other.

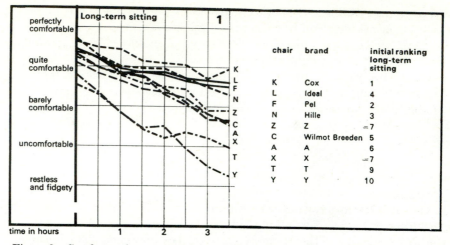

Figure 2. Comfort ratings averaged from all subjects on each chair at successive times
in the long-term test.

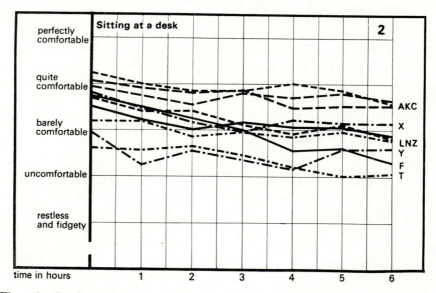

Figure 3. Comfort ratings averaged from all subjects on each chair at successive times
in the desk test.

The chairs were then ranked into groups, with differences between the groups
being significant.   The resulting rank orders are shown in Table 5 with the best
on the left.   The multiple significant differences can only be shown in a complex
table; the effective result is shown by bracketing together those chairs which
can be regarded as not different.   The coarseness of this grouping may represent the insensitivity of the measure, but it may equally reflect the insensitivity
of the sitter to the chair.   Nevertheless it is clear that there are large and reliable
differences between some chairs.

5.4. *Comparison between Comfort Ratings from the Three Tests*

Having established significant differences and thus a reliable rank ordering
(with ties), we can compare the comfort order for the different test situations.

The three Kendall *tau* correlations at the bottom left of Table 5 show that the comfort preference order for desk usage is not significantly correlated with the other two. This result is suggestive; the task is manifestly different and may well require a different posture and chair. However, we must remember that different groups of subjects were used, so the result needs confirmation to be conclusive.

Table 5. Ranks from comfort ratings

1. Long term	K	(F	N	L)	(X	C	Z	A)	(T	Y)
2. Desk	K	(A	C)	(L	X	N	F	Z)	(T	Y)
3. Meal	(N	K)	(F	Z	X	C	L	A	(T	Y)
4. Combined	K	N	(F	C	X	L	A	Z)	(T	Y)

1–2	$\tau = 0.3$	$p = 0.1$	1–4	$\tau = 0.7$	$p = 0.002$
1–3	$\tau = 0.4$	$p = 0.05$	2–4	$\tau = 0.5$	$p = 0.04$
2–3	$\tau = 0.4$	$p = 0.1$	3–4	$\tau = 0.6$	$p = 0.006$

Concordance 1, 2 and 3 $\quad W = 0.69 \quad p = 0.03$

Despite the separate correlation results, when taken together the rank orders for the three test conditions show a significant coefficient of concordance $W = 0.69$. Therefore, we may reasonably sum the ranks to give a probable preference order for all situations combined. Of course, all three rank orders correlate significantly with this final order as is shown. This result suggests that it may well be possible to design an optimum compromise to satisfy the user's needs for these different situations at least, since there appears to be a reasonable overlap in their comfort requirements.

5.5. *Comfort Ratings and Direct Rankings*

The Direct Ranking measure was taken for the Long Term Sitting situation *before* the subjects started the trials, to investigate the value of this type of short test. Of course, they were not allowed to see or touch the chairs, only to sit in them. For the other tests, the ranking was taken after all the trials, to add to the comparison data which would otherwise be limited, especially in situation 3, the meal. An organization error in the Desk test caused only half the rankings to be taken, so only the Meal data is usable.

The direct rankings from the twenty student subjects were compared for agreement, using the coefficient of concordance W, and show a high value $W = 0.5$ significant at $p = 0.001$.

It will be remembered that the decision was taken to use several subjective measures in the hope that significant inter-correlations would give reassurance of valid results. The ranks derived from the comfort ratings and the direct rankings from the long term and meal tests are shown in Table 6.

Table 6. Comfort ratings and direct rankings

1. Long term										
(*a*) Rating	K	(F	N	L)	(X	C	Z	A)	(T	Y)
(*b*) Rank-before	K	F	N	L	C	X	(Z	A)	(T	Y)
2. Meal										
(*a*) Rating	(N	K)	F	Z	X	C	L	A	(T	Y)
(*b*) Rank-after	F	K	N	(X	C)	A	L	Z	T	Y

1. *a–b* $\quad \tau = 0.87 \quad p = 0.001 \qquad$ 2. *a–b* $\quad \tau = 0.53 \quad p = 0.02$

Considering the meal data first, the direct rankings taken at the end with sight may have allowed non-comfort factors to have some effect, and the comfort ratings consist of only one reading. These may be the reasons why the agreement is less than with the long-term data; nevertheless, the correlation between the two meal test subjective measures is still considerable and significant.

The very high correlation from the long term test between the direct ranking taken beforehand by feel alone, and the subsequent comfort rating ranks from the extended sitting trials, is noteworthy and interesting. While further tests are needed to establish the repeatability of this result, it does seem to suggest the possibility that the much quicker procedure of direct ranking may prove a reliable technique for the assessment of comfort.

5.6. *Body Area Comfort Ranking*

In order to analyse the data, the number of points given to each body area was added, and the mean over the test period found. Unfortunately, the technique had two serious drawbacks for use in this context. It had previously been used to compare a small number of aircraft seats, which were essentially similar; in this study, the range of chairs was rather greater. The technique allows for the body areas to be ranked, but gives no indication of degree of comfort for an area. All of the chairs tended to score rather low for comfort afforded to area 10, the buttocks, but differences between chairs were swamped. It proved an extremely laborious technique for scoring and statistical manipulation, and the results were not promising for this particular study. However, this should not be taken as a contra-indication of the method, except for this type of study with a larger number of chairs and subjects (when computer analysis might be the solution).

5.7. *Body Posture Change Frequency*

The time-sampling count of the frequency of posture change, in the long term test, made by the psychologist observers, did not show a significant correlation with the corresponding direct ranking of the chairs. This measure is mentioned for completeness only.

It is still thought that ciné-camera memomotion would have yielded useful results, but it is doubtful whether the technique would be feasible for use with such a large number of chairs, because of the labour of film analysis.

5.8. *Chair Dimensions and Standard Recommendations*

The physical dimensions of the chairs, and the summarised recommendations from British Standards B.S. 3079 and B.S. 3893, are given in Table 1. While every attempt was made to ensure accuracy, the measurement tolerances are probably at least $\pm \frac{1}{4}$ inch and $\pm 0.5°$; with some of the plastics chairs it is not easy to be certain of some limit points to better than $\pm \frac{1}{2}$ inch. Moreover, it is unlikely that differences less than at least $\frac{1}{2}$ inch or 1 deg. have any significance.

A comparative analysis was made between chair dimensions and B.S. 3893 recommendations, ignoring differences less than $\frac{1}{2}$ inch or 1 degree. The chair dimensions which fail to conform to the B.S. recommendations are marked on Table 1, and the total number of failures for each chair, on the dimensions in rows 1 to 8, are noted in the bottom row. The chair columns in Table 1 are placed from left to right in the rank order of the comfort ratings from the long term sitting test (which is similar to the final combined comfort order from all tests, see Table 5). It is noteworthy that all chairs have at least two discrepancies from the B.S. recommendations, and that both the top two and bottom two chairs for comfort (K, F and Y, T) have the same fault totals, five and four respectively.

Allowance must be made for the fact that neither B.S. 3079 nor B.S. 3893 were prepared for application to polypropylene chairs. Nevertheless, it is felt that a justifiable conclusion from the results of this experiment would be that any use of the B.S. recommendations as a guide for chair selection should be made with due caution.

Further analyses and discussion of comfort ratings and B.S. recommendations are given in Part 3.

5.9. *Chair Feature Checklist*

The results so far have shown that there are measurable and significant differences in comfort afforded by a sample of the upright chairs available for domestic and office seating, and that a realistic rank ordering for comfort can be achieved. A second, tentative, conclusion is that, since at least one chair was considered acceptable and moderately comfortable in all three test situations for the majority of users, a general-purpose chair can be designed. However, when we seek reasons for the final rank order and for the differences between some chairs in different situations, in the hope of deriving criteria for selection and guidance for future design, the results so far help little.

It was to assist with these aims that the Chair Feature Checklist (CFC) method was tried (see Table 3 for the layout of the Checklist). In the analysis, the ratings are first summed across all 60 subjects for each chair separately on each feature to give the totals in the three possible answer cells (e.g. correct, too little, too much). To assess how the chairs compare with each other for any feature, one chair must first be chosen as the acceptable standard chair or criterion for that feature. The distribution of ratings between the answer cells for each chair can then be compared statistically with the distribution of ratings for the criterion chair. To illustrate the data and the analysis, three of the features are listed in Table 7. The chairs are placed in the serial order of the physical dimension, the value of which is given in the second row. In the third and fourth rows, the totals of the CFC ratings in the two ' unsatisfactory ' cells are given, because these show the change-over in rating (the third, ' correct ', cell contains the balance of ratings to the total of 60 subjects).

To test for its significance the difference between each chair and the criterion chair, the χ^2 one sample test was used. The criterion chair was selected to be one of those with the lowest total of ratings in the ' unsatisfactory ' cells, with the constraint that no cell must contain less than five ratings. In the example features in Table 7, the criterion chairs selected were Z, K and C respectively for height, depth and backrest curvature. When the optimum contained a cell with less than five ratings, the chair closest to it but justifying the statistical assumption was selected. This statistical requirement has of necessity led to a few minor anomalies in the significance results, which do not, however, affect the clear general trends. The significant differences on the features in Table 7 are shown in row 5.

Table 7. Chair dimensions and chair feature checklist ratings (CFC)

(Seat height, seat depth and backrest curve radius)

Chair	K	X	F	Z	A	L	T	N	Y	C
Seat height	18¼	18	17¾	17½	17	17	17	17	16¾	16¼
CFC rating :										
High	21	17	18	10	12	8	1	1	4	5
Low	2	1	6	5	16	19	20	27	40	19
Significance										

Too high   Too low

Chair	C	N	K	X	F	L	Z	T	A	Y
Seat depth	16½	15½	15	15	15	14	13½	13½	13½	11½
CFC rating :										
Long	19	7	6	5	3	1	2	1	0	0
Short	5	1	9	4	18	18	15	31	36	48
Significance										

Too long   Too short

Chair	Y	F	Z	T	L	A	N	C	X	K
Back curve radius	0	0	20	20	16½	11½	10½	10½	9¾	9½
CFC rating :										
Flat	42	24	32	25	20	10	3	6	6	0
Curved	0	2	6	6	6	5	13	10	17	27
Significance										

Too flat   Too curved

The detailed CFC results and the further analyses comparing these results in a more general way with chair dimensions and other factors, are discussed in Part 3 since they are primarily relevant to methodology.   The results from this measure are used here to identify specific design features which may be the reasons for the respective comfort ratings given in the various test situations. The chairs will be considered in the final combined comfort order (Table 5).

K. The seat was rated too high by a third of the subjects, and the backrest too curved by nearly half; of the 21 rating it too high, 15 were women; this is the only marked sex difference in all the CFC ratings.   This chair appears as best from these experiments, but probably despite rather than because of these two features, which suggests that there is still room for improved designs.

N. The seat was rated too low by nearly half and the backrest too low by a third.

F. The seat was too high and its depth too short for nearly a third; the backrest was too flat for a quarter.   The seat depth rating, despite the dimension of 15 ins., may be a design clue to the need for attention to the chair back to improve matters.

C. The seat was too low, too long and too wide for a third, and the seat shape was rated poor by nearly half the subjects.   These factors did not make the chair uncomfortable for everyone, but restricted the number who found it quite comfortable.   However, they do not seem to explain why the chair is rated better for comfort in the desk situation.

X. The seat was too high, the backrest too curved, and there was inadequate clearance for feet and calves, for nearly a third of the subjects.

L.   The seat was too low and too short for nearly a third, and the backrest too low and too flat for a third of the subjects.

A.   The seat was too short for over half, the backrest angle too upright for half and its height too low for a quarter of the subjects. These features appear to give good reason for the lower general comfort rating, but do not entirely explain why this chair is rated higher in the desk situation; however, the more upright backrest angle may be a clue in this latter case.

Z.   The backrest was too low for a third and too flat for over half the subjects. The seat height satisfied almost 80% of the subjects but the seat was too short for a quarter. This latter fact possibly accounted for the large body height range of people who found the seat height satisfactory, and perhaps implies that seat length may be traded off against seat height, thus to accommodate a greater range of people for a given seat height.

T.   The seat was too low for a third, too short for over a half, and of poor shape for a third of the subjects. Also, the backrest was rated too low by a third (but the significance of this may be a statistical anomaly), and the backrest was definitely too flat for a quarter.

Y.   The seat was too low for two-thirds, and too short and too narrow for 80% of the subjects. Also the backrest was too low for a third, and too flat for two-thirds. Further, a third rated clearance for feet and calves inadequate. As with chair T, these CFC ratings may help to explain why it is at the bottom of the comfort rank order.

5.10. *Comments Section of Questionnaire*

An appropriate place was provided, on the comfort rating questionnaire, for the testers to add their personal comments if they wished. The purpose of this was to elicit any factors which might not have been covered by the planned questions and assessment methods, and also as a form of ' safety valve ' and to help the subjects feel more directly involved and in contact with the experimenter who would score the results.

A review of the comments shows that they deal more specifically and forcefully with aches and pains in various body areas, but do not in general contradict or add anything fundamental to the Comfort Ratings and Body Area Rankings.

However, there was one exception. Chair K was reported by several subjects to have given them a shock from static electricity, usually after sitting on it for a time and when they happened to touch the metal legs of the chair, perhaps when standing up. We checked these reports and have been able to verify and experience the effect ourselves, which is not entirely pleasant. This has also occurred occasionally with some of the other plastics chairs. Clearly the assessment in favour of chair K should be slightly qualified in view of this problem.

6. Conclusions from Part I

The report to the Consumers' Association, giving the relevant results, concluded with the following opinions and suggestions on chair selection.

' The best advice we can offer is to suggest that anyone selecting a chair should first specify the primary usage and decide whether mainly taller or shorter people (or mainly men or women) will be the users.

Then the first three or four chairs in the ranking for the nearest equivalent usage should all be considered. If taller people, especially over 70 in. (178 cm), will predominate, preference should be given to, for instance, K or F; if shorter users will predominate, especially below 64 in. (163 cm), preference should be given to N or L.

Finally, we must emphasise that this study has been concerned with what chairs feel like to sit in, and has tried to minimize any influence from the visual appearance of the chairs, on the subjective assessment of comfort. But when making a choice and when living with an environment, the appearance of the chairs is also relevant. By this study we do not wish to minimize the importance of the aesthetic factor, which has tended to be a little too predominant in the past, but rather to place the ergonomics factor, the provision of comfort for the sitter, in correct perspective as at least of equal importance. But there is one other result from our study; it is highly unlikely that you will ever choose a chair about which absolutely no-one will complain.'

The experiments in this first programme seem to suggest the following conclusions on ergonomics aspects.

1. It should be possible to develop further techniques from ergonomics to assist both designers of chairs and potential users when selecting them.

2. The chair design for optimum comfort may not be the same for different uses such as general or desk sitting, but it may be possible to achieve acceptable compromise designs for general purpose use.

3. The B.S.I. dimension recommendations may be useful broad guides for designers but need further experimental checking in some aspects, particularly with regard to possible use as a selection aid.

4. It should be possible, but only by fairly extensive experiments, to isolate more clearly the dimensional factors, such as seat height, depth, slope, etc., and their range, which most affect the risk of user discomfort, so as to provide a more accurate guide for designers.

5. Some of our ergonomics principles for chair design may need re-examination. For example, one principle is that a chair should permit a variety of postures. In these tests chair K was clearly best, yet departed from B.S.I. recommendations both in $18\frac{1}{2}$ in. (47 cm) seat height and $9\frac{1}{2}$ in. (24 cm) radius backrest curvature (and in other aspects). The subjects noted both of these factors as excessive on the Chair Feature Checklist, so that they probably preferred it despite these features rather than because of them. It seems to be comfortable because it permits a very restricted range of postures and holds the sitter fairly securely. We wonder, therefore, whether the multi-posture principle, whilst perhaps holding for easy chairs and school chairs, does not hold uniformly in situations such as those for our tests.

PART II. INDIVIDUAL OPINIONS AND DIMENSION RECOMMENDATIONS AS CHAIR SELECTION METHODS

7. Introduction to Part II

7.1. *Alternative Approaches to Selection*

While the first study was in hand, other investigations revealed the size of the problem if the Consumers' Association followed its usual methods. Hundreds of chairs would have to be tested if even a fair proportion of all available was to be sampled. Therefore, emphasis grew on the study of alternative methods. Although it was not thought possible yet to study methods which might be used by the individual purchaser, it seemed that alternative methods might well be used by industry and other public bodies when selecting for bulk purchase. Opinionative assessment and interpretation of dimensions based on British Standard recommendations would seem to be the two possible alternatives to the experimental approach.

Selection and purchase of one from a group of chairs which are broadly comparable, especially in price, is usually the problem, and other criteria as well as comfort are important. Hence, consulting the subjective opinions of anyone considered relevant, from the managing director to the office boy, is often the intuitive first approach. While this method may be relevant for some criteria and sometimes a sound principle of good industrial relations, its validity for asesssing comfort has not been tested. An enlightened management might alternatively be aware of the British Standard recommendations and therefore instruct someone to measure the chairs, interpret the British Standard data, and thus choose a small group of chairs, presumed acceptable for comfort, from which to select by other criteria.

With the results of the first study providing the necessary reference data, the value of these alternative approaches was tested. It was considered that the method by which they would be most likely to succeed, and the method which intelligent managements in industry and elsewhere might well adopt, would be to have experienced research workers, qualified by having published studies of the ergonomics of chairs and sitting, to give the opinionative judgements and to interpret on the basis of B.S.I. recommendations. Therefore, the only eight ergonomists thus qualified in Britain were invited and gave their time to take part.

7.2. *Additional Rationale*

In the context of this study the logic of the selection problem led immediately and inevitably to the idea of inviting experts to act as subjects. However, a topic of some interest to the first author was served by the same solution and also enabled an alternative convincing rationale to be presented to the subjects, since it was thought preferable not to reveal the primary aims until afterwards. The rationale is explained by the following excerpt from the invitation letter sent to the experts.

I have for some time been interested in the possibility of validating the usefulness of experts' opinions; this is my aim, and the field of seating comfort happens to be the first suitable opportunity which has arisen. In ergonomics, even more than in some other subjects, a full research investigation can be very expensive. It would be of considerable value to establish the areas and the extent to which the opinions of experts can take the place of longer studies. Experts in our field are always very cautious in venturing opinions, and rightly so at present, but it may be that we could be rather more confident in our opinions than we at present believe. With regard to practical problems, ergonomists may certainly expect an increasing demand for them to give expert opinions. Therefore it seems to me important to take every opportunity which may rise to study and establish the value of the expert opinion.'

7.3. *What is Expertise?*

This more basic question will not be discussed at any length in this paper, since it is not the primary aim of this study. Nevertheless, some brief comments are offered as suggestions, particularly since the last may underlie some of the results here.

In general, the expert in a field should be distinguishable from the non-expert on the following considerations.

(*a*) He is able to estimate, from a brief experience of a situation, the further changes that are likely to occur with time.

(*b*) He is able to make use of a wider variety of cues than the non-expert.

(*c*) His opinion is more reliable (less biassed by fortuitous occurrences).

From these considerations, the expert should be able to extrapolate from a wide range of situations which he has experienced to the present situation, and thus to make a reliable assessment of a situation after a briefer exposure to it than the non-expert would require.

These distinctions between the expert and the non-expert should also apply in the situation of judgement of chair comfort, if the experts' judgements are to be a useful tool. In addition, for chair assessment a further ability is required, namely

(*d*) The expert must be able to make allowance for his own particular body size, in arriving at a judgement of the comfort which a chair would afford the 90 per cent range of sizes existing in the general population.

8. Methods

8.1. *Test Procedures*

To test the method of opinionative assessment, the experts used similar methods of making judgements as had the test panels in the sitting trials, and their performance was compared with the larger sitting trial panels under the following headings.

1. Inter-judge reliability, given only the minimum information, and then using increasing amounts of information.

2. Ability to make reliable assessment which were valid for the general population, and uninfluenced by the assessments which the expert would make for his own comfort alone.

To do these the expert must be able to make allowance for his own particular body size. If he cannot validly do this, despite his much greater experience, then clearly the approach of opinionative assessment cannot be supported for use by the layman.

To test the method of interpretation of dimensions, based on the B.S.I. recommendations, the experts were given the chair dimensions and the B.S.I.

data and their performance was to be studied under the following headings.

1. Inter-judge reliability in making the assessments.
2. Ability to make reliable assessments, on the basis of the dimensions and B.S.I. data, which were valid for the general population by correlating significantly with the comfort results from the sitting trials panels.

The same ten chairs were used as in the first study. The measurement techniques were similar:

   (a) Rating of general comfort afforded to the individual tester (chairs unseen).

   (b) Ranking in the order of comfort the chairs were assessed as affording to the 5 per cent to 95 per cent population range (chairs unseen).

   (c) Ranking similar to (b) on the basis of dimension departures from the British Standard recommendations (chairs unseen).

   (d) Ranking similar to (b) (chairs seen, handled, etc.) using all the information available and any techniques they chose to use within the context of the opinionative, as opposed to experimental, approach.

In the past we have found some laymen rather suspicious of the need for ergonomics experiments. A cynical objection might be made to the present test procedure that the opinions of the experts might be sub-consciously biassed to favour the experimental approach. We believe this is untenable, because we did not describe the whole rationale until afterwards, and in our invitation and explanation we assumed that expert opinion would be a valid method and implied, by concentration on anonymity, etc., that the individuals were, as it were, on their mettle. For obvious reasons, other information also such as data about the first study was withheld, and the entire test situation was carefully controlled, with an experimenter accompanying each subject throughout; therefore we consider that the results should carry full conviction in non-specialist circles also.

8.2. *Subjects*

Of the eight subjects, one was female and the rest male. Their height range was $62\frac{1}{2}$ to 72 inches (159 to 183 cm); the height of the female subject was 63 inches. Their weight range was 140 to 182 pounds. Their build was in all cases about medium, and their weights were within approximately ± 1 S.D. of the mean for their body height. Thus, as a group they were, apart from the sex imbalance, by chance quite a representative sample, anthropometrically, of the adult population.

8.3. *Test Sequence*

The subject was given a written copy of the instructions, the whole sequence of trials was explained and then he was taken into one of the experimental rooms containing chairs. He was blindfolded if necessary to avoid his seeing the chairs as he entered.

1. The subject was guided backwards into the chairs and was asked not to touch them with his hands; he was not told their identity code letters. The order of presentation of the chairs was randomised and different for each subject. He sat in each chair in turn for two minutes, and then marked the comfort rating scale twice:

   (a) for his personal comfort, and

   (b) as he predicted he would feel at the end of one hour.

N

2. Having completed the ratings, the subject was again asked to sit in each chair in turn, still without seeing them, and to rank them for the comfort he assessed they afforded for the general population (5 per cent to 95 per cent range) allowing for his own body size, using the Direct Ranking method.

3. Next he was taken back to the general meeting room and given a copy of the B.S.I. recommendations, a list of the relevant data from it, and the full dimensions of the chairs (as in Table 1, but without analysis and markings of B.S. faults for each chair).   Without seeing the chairs he was asked to rank them all
   (a) for comfort for long-term sitting;
   (b) for comfort for office use.

The meaning of these situations was explained by describing the equivalent test situations on the first study; ' long term sitting ' as general usage (such as reception, meeting and reading areas) for sitting without a table or desk for periods up to three hours or more; ' office use ' as sitting at a table or desk doing general desk and clerical work (but not typing), the table or desk heights to be taken as a distribution covering the B.S.I. specified range of 28 to 30 inches.

4. He was then taken again to an experimental room and allowed to see and sit on the chairs; with their dimensions and the B.S.I. recommendations, but without his previous ranking assessments, he was asked to rank them again, for the general population (5 per cent to 95 per cent)
   (a) for comfort for long term sitting;
   (b) for comfort for office use.

This was a free situation, in which he was asked to take over and do whatever he liked, using the experimenter merely as an assistant, to arrive at the rank order for comfort which he as an expert assessed the chairs would afford.

For clarity and ease of reference, a summary of the test sequence is given in Table 8.   The estimated times for these tests, and the actual range of times taken by the experts, are also noted in Table 8.   While the consideration of the dimensions and B.S.I. data took longer than expected, and while more time would always be desirable for studies of this nature, practical needs and the time which the experts could spare set an inevitable limit.   Nevertheless, it is thought that the time taken was reasonable and not unduly restrictive.

In the expectation that the general results might be positive in supporting the opinionative approach, these tests were presented in the above order with the aim of studying the relative contribution, to the experts' ultimate rankings, of what seemed likely to be the two main sources of information, his personal comfort sensations and the chair dimensions in relation to the B.S.I. recommendations.   Situation No. 1 therefore aimed to gather data on his personal assessment for his own comfort, No. 2 on his modification of this when assessing for the population range, No. 3 on his assessment using the dimension data, and No. 4 on his final opinion, as an expert, using all the information about the chairs already obtained and any more he wished to gather or ask for.   We thus hoped also to be able to determine whether the opinionative approach, based on assessment from sensation data, or interpretation, based on dimensions and B.S.I. recommendations, would be more valid and therefore more to be recommended to laymen.

It should perhaps be emphasized again that only these two approaches, opinion and dimension interpretation, were studied in this experiment, with the experts being considered the best qualified experimental subjects for evaluation of these methods. There was no question of the experts themselves using or being tested on any of their experimental methods.

Table 8. Summary of test sequence and time for Part 2

Test sequence

1. Rating of personal comfort
   (*a*) after 2 min
   (*b*) predicted for 1 hr

2. Ranking of comfort assessed for
   5%–95% population

3. Ranking based on chair dimensions and B.S.I. recommendations
   (*a*) for long-term sitting
   (*b*) for office use

4. Final ranking of comfort assessed for 5%–95% population
   (*a*) for long-term sitting
   (*b*) for office use

Task	Planned time	Actual time
1	25 min	25 to 30 min
2	10 min	10 min
3 *a, b*	30 to 45 min	45 to 90 min
4 *a, b*	15 min	15 to 20 min

9. Results

9.1. *Inter-Judge Reliability*

The coefficient of concordance (W) was computed for the rankings assigned to the chairs for each situation. The results indicate a significant level of agreement between the subjects in all the situations 1, 2 and 4.

9.2. *Interpretation from Chair Dimensions and B.S.I. Recommendations*

In situations 3 (*a*) and 3 (*b*), however, for ranking from British Standard data, there is no agreement between the subjects either for long-term sitting or for office use. This test is severe, since half the chairs were plastic, and it was difficult in these cases to apply the standard consistently. In addition, most experts found that to complete this part of the exercise satisfactorily required much longer than the planned time (see Table 8), and that even so they were not very satisfied with their conclusions. The lack of significant agreement or usable results here, together with the results from Part 1, make it clear that the B.S.I. recommendations may perhaps be used as a design guide but cannot at present be regarded as a reliable basis for selection.

9.3. *Personal Comfort Ratings and Prediction for a Population*

By converting the ratings of personal comfort after two minutes (task 1 *a*) to a rank order, this can be correlated with the ranks from the ratings for predicted comfort after one hour (1 *b*), and with the subsequent direct rankings with the chairs seen and not seen (2 and 4). The Kendall *tau* correlations are highly significant and are respectively 0·82, 0·73 and 0·69. They appear to suggest that the personal comfort assessment may have a considerable influence on the attempt to make predictive assessments of comfort for a population, and thus that personal comfort sensations may perhaps be a predominant factor in the prediction process.

This would not be unexpected, but the crucial question is whether they provide a sound or a misleading basis for the population assessment, that is whether they can be used with appropriate modification or whether they distract unavoidably. The comparison of the experts' opinionative assessments with the appropriate criterion data from Part 1 is the final test.

9.4. *Opinionative Assessments of Comfort for the General Population*

In general, the experts' pooled rankings do not correlate highly with those from the general population, as represented by the two relevant test panels from the first study.

The experts' pooled rankings (task 4 *b*) did not correlate at all with the second test panel's rankings for office use sitting at a desk.

Table 9. Experts' assessments and Part 1 test results

		Direct ranking	After 5 min	After 1 hr	Sum of $3\frac{1}{2}$ hr
	Experts	Long-term sitting test subjects			
1 *a*	Rating after 2 min		$\tau = 0.38$ $p = 0.4$	$\tau = 0.2$ $p = 0.2$	
1 *b*	Predicted for 1 hr			$\tau = 0.02$ $p = 0.4$	
4 *a*	Final direct ranking	$\tau = 0.44$ $p = 0.05$	$\tau = 0.45$ $p = 0.03$	$\tau = 0.53$ $p = 0.02$	$\tau = 0.42$ $p = 0.05$

For long-term sitting (see Table 9), the only experts' assessments which showed significant correlations were the final direct rankings (task 4 *a*). These, together with the correlations in section 9.3, might seem to suggest that the personal comfort sensations have been usefully modified, or their distraction resisted, and that the experts as a group at least may be able to assess the chair comfort rank order for the shorter periods up to one hour. However, a detailed analysis was made of the data on individual chairs from the experts' (task 4 *a*) and the long term panel's results after one hour (the best correlation). This revealed significant differences (Mann-Whitney U test, $p = 0.002$ and 0.02 respectively) on the two chairs K and Z; therefore, that correlation cannot be accepted as conclusive, and the others are of marginal significance.

The main reason for these correlations in Table 9 seems to be that the group of experts happens by chance to form a body height sample analogous to that used in the long-term sitting test representing the general population. It is therefore necessary to determine whether the experts are behaving as a homogeneous group, or as individuals, each with his own idiosyncrasies.

On calculating, for each expert, the correlation between his ranking from situation 4(*a*) and the test results, see Table 10, only one expert produces rankings which correlate significantly with the criterion. Moreover, an inspection of his actual rankings, compared with the criterion test results, see Table 11, shows considerable discrepancies with, for instance, chairs K and C and thus suggests that no predictive value can be placed even on this correlation. The differences in rankings for chairs K and Z, which were mentioned above, may also be seen in this Table 11, and the variations in the experts' ranking of chairs T and Y, clearly the two worst from the Part 1 tests, should be noted.

One may therefore deduce that, as individuals, experts differ markedly in their ability to rank chairs for affording comfort to the general population, and they cannot give accurate rankings for use as a substitute for actual sitting trials.

Table 10. Correlation of each expert's final rankings with long-term sitting test results

Expert	τ	p
a	0·16	0·3
b	0·60	0·01 sig.
c	0·29	0·1
d	0·24	0·2
e	0·07	0·4
f	0·33	0·1
g	−0·20	0·2
h	0·09	0·4

Table 11. Final rank orders from experts and long-term sitting test

Chairs	K	F	N	L	X	C	Z	A	T	Y
Test order	1	2	3	4	5	6	7	8	9	10
Experts	↕						↕			
a	9	2	1	3	4	10	5	8	7	6
b	5*	3	1	4	7	2*	6	8	9	10
c	7	5	1	3	4	2	9	8	6	10
d	5	3	6	4	2	7	10	1	8	9
e	5	8	3	6	2	4	10	1	9	7
f	6	2	4	7	5	1	8	3	10	9
g	8	7	3	6	5	2	10	1	9	4
h	10	2	1	3	9	4	8	7	6	5

* ↕

10. Conclusions from Part II

The following conclusions seem justified.

1. Experts in ergonomics research on sitting comfort do not appear able, either on an opinion basis or from chair dimensions and British Standard recommendations, to give accurate comfort assessments, e.g. to select either best or worst chairs for use by a general population.

2. This being so, and on the assumption that the experts acting as subjects in this study are the most likely to be able to give valid opinions,
   (*a*) neither assessment by opinion from one or a very few persons,
   (*b*) nor interpretation based on chair dimensions and B.S.I. recommendations,
   can be considered valid methods for assessing sitting comfort or selecting chairs for use by a general population.

3. There are some indications that expert opinion in this type of situation has to be based on personal comfort sensations; these must be dependent considerably on the individual's body size and shape, and it may be that expertise in ergonomics chair comfort research cannot compensate for the inherent subjective factors specific to such individual body differences.

4. Therefore, at the present state of knowledge, comparative evaluation trials based on research by ergonomics experts seem the only valid procedure.

PART III. METHODOLOGY—RESULTS AND SUGGESTIONS

11. Introduction to Part III

One of the primary aims of these studies was to explore as many measurement techniques as possible, of those that seemed likely to be useful, with a view to evolving a battery of suitable methods for future chair comfort selection studies.   While we would not claim that a proven battery has yet been evolved, some useful methods have been developed further, tested and show promise, and some comments and suggestions on methodology can be offered.

11.1. *Measuring Comfort*

In fact, very little has been done about measuring comfort.   Most investigators, ourselves included, have considered chair design as being concerned with the avoidance of discomfort for the majority of users.   Hence the concept of comfort is relatively unexplored; nor has the exploration of positive comfort been advanced much by this study.

Since our experiments were done, two reports of considerable interest have been written.   Kroemer and Robinette (1968) have reviewed the European literature on the ergonomics of office furniture, concentrating mainly upon the recommendations by orthopaedists, physiologists and physical anthropologists; their review is thorough and their reference list comprehensive.   Branton (1966), in an interim report, described a number of recent experiments, discussed the whole range of theoretical considerations relevant to this problem, and advanced a theory of sitting comfort in terms of ' postural homeostasis '. While we have not been able to test this concept, it is well reasoned and seems potentially fruitful.   Moreover, his excellent discussion on methods enables much to be omitted here.

The range of possible measurement methods, the reasons for our choice of certain types, and the tests chosen or developed, have been summarized in section 3 above.   The only aspect which may usefully be amplified is the choice between different subjective methods.

11.2. *Choice of Subjective Methods*

Since comfort is still an unexplored concept, and one cannot measure it in physical units, psychophysical methods which put the onus on the subject are the only ones available.   Rating methods are sometimes crude and unreliable measuring instruments, and for this study it was decided to use a number of techniques with different bases, and to achieve internal reliability or construct validity by this means, by comparing results from the different tests.

The available subjective methods are as follows.

1. Rating, on a numerical scale.
2. Rating, with verbal cues.
3. Checklist rating.
4. Absolute rating (using 0-100 scale).
5. Pair comparisons.
6. Direct ranking.
7. Forced-choice rating.

It is possible to argue a case for the use of any of them but, on the grounds of economy, it was decided to adopt a direct ranking method, and a rating method, for the measurement of overall comfort; a forced-choice rating, and a checklist rating were used for the examination of local features of the chair.

Turning now to the choice of rating scale for overall comfort-discomfort, we must first consider the range of techniques available. In summary, these are labelled numerical, graphical, standard, cumulated points or checklist, and forced-choice. The methods differ mainly in the number and kind of aids or cues given to the rater, and the fineness of discrimination called for. Previous rating scales used for overall comfort-discomfort rating have generally been of numerical form. Schlechta *et al.* (1957) used a nine-point and a 21-point scale, and Barkla (1964) used a seven-point scale, with verbal phrases only. As far as we are aware, little use has been made of absolute numerical scales, nor has anyone developed and evaluated a rating scale specifically for measuring comfort-discomfort.

It was decided that, in order to achieve a rating which used the sitters' experience directly, we would have to construct a scale starting from first principles. The form and size of the scale were fairly easily decided. The recommended number of scale points for a bi-polar scale, given in the standard work on psychometric methods (Guilford 1954) is approximately nine, but in favourable situations, with experienced raters, up to twenty-five may be used. There is certainly no logical reason for sticking to short scales if the rater can discriminate finely. In the event, it was decided to use an eleven-point scale, the two extreme positions being sufficiently distant to discourage frequent use, and to encourage testers to use the intermediate points as and when they felt that they could discriminate, giving as maximum a 20-point scale.

It is not generally sufficient to tell raters that they are to rate their own feelings of comfort or discomfort, without giving them some cues, so as to be sure that different testers give the same rating when experiencing the same amount of discomfort. Rating scale cues have the double purpose of reminding the tester constantly what is being rated, and also giving him anchors or guides to quantitative judgements. The procedure used to develop the anchor terms and the final rating scale has already been described in section 3.2. The resulting scale is assumed to be an ordered scale and no more, certainly not an equal-interval scale.

The second measure of general comfort used was the direct ranking method. This method of comparing comfort of seats has been shown by Stone (1965) to give reliable results, in comparison with the more laborious pair-comparison method. However, there are two objections to its validity. When the judgements are difficult to make, the tester will grasp at any cue which seems to be an aid, and one cannot be sure that he is, in fact, ranking the chairs for comfort, and not merely relying on a cue such as springiness, feel of the upholstery, etc., which may not be directly related to comfort in the real-life situation. The second objection is that there is always a danger, in compressing what is essentially a multi-factorial judgement into a single rank ordering, that the final order may not be meaningful.

In order to limit the irrelevant cues available to the tester, he was not allowed to see or touch the chairs directly, and could thus only use somatic cues in making his ranking.

The two methods of measuring overall comfort, direct ranking and the rating scale, were disconnected in experience and the tester had no record of the results of the one whilst using the other. In the rating, the subject is judging his own comfort, not that of the chair, and his comfort is affected by the total situation, of which the chair is only a part; it is affected by the time of day, distractions in the situation, etc. If there is a clear statistical difference between chairs despite the large amount of variance contributed by the situation, and if the order given is in agreement with that given by the ranking method, then we have grounds for inferring that there is a real difference between chairs, appreciable by the user.

An ergonomics study should be concerned with more than gross evaluation of an existing design; unless the designs tested are demonstrably the best possible, it should also be concerned to point to particular areas in which the design could be improved. In this study, the Chair Feature Checklist, and the Forced-Choice Rating of Body Area Comfort, were both intended to examine the subject's reaction to local design features of the chair. The Chair Feature Checklist was an attempt to get the subjects to comment directly on the features of a chair, after sitting in it for a considerable period. The Body Area Comfort Ranking was intended to allow them to criticise each chair in terms of its effect on their anatomy.

As has already been discussed, the Body Area Comfort Ranking did not prove very satisfactory in this study. With twenty subjects, the labour of scoring was considerable. In addition, for domestic or office seating, the critical body areas are those directly supported by the seat, and these were so frequently rated in this study as the least comfortable that, using the Allen and Bennett simple scoring technique, almost all the chairs ranked significantly poorly for the comfort afforded to these areas. This is to be expected in non-upholstered chairs and, since the technique gives no indication of the quantitative amount of discomfort, the differences between chairs are swamped. This obstacle was overcome by transforming the raw scores for each body area into z scores, using the standard deviation of scores from all ten chairs as base, but this was again a laborious procedure. It would seem to be preferable to rate the amount of discomfort in critical body areas, using the methods used by Schlechta *et al.* (*op. cit*).

The Chair Feature Checklist has provided some useful data, see section 5.9, and the results of some further analyses will be presented in a later section.

12. Further Results

12.1. *Prediction of Comfort from Brief Test*

Ergonomics sitting comfort experiments obviously take some time; therefore, any test methods are of interest which appear capable of giving equivalent results in a shorter time. The very high correlation of the long-term subjects' Direct Ranking beforehand with their Comfort Ratings, $\tau = 0.87$ (Table 6), has already been noted. To test this prediction idea further, the rank order from the long-term subjects' ratings at the five minute point were correlated with their scores after one hour, three hours, and with their averaged scores over all. The correlations are respectively: $\tau = 0.64$, $p = 0.005$; $\tau = 0.67$, $p = 0.004$;

$\tau = 0.6$ $p = 0.008$. Inspection of the rank orders confirms the good predictive value of the five minute assessment with $1\frac{1}{2}$ places being the largest change in rank order for any chair.

This result encourages further the hope that controlled subjective assessments by a test group of appropriate size may prove a reliable technique for general use, even with only the five minutes or so exposure to each chair required by either of the ranking and rating methods we have used.

12.2. *Body Height and Comfort*

Further analyses were made to see if consistent relationships between body height and comfort resulted from certain chairs or chair dimensions. The subjects for each test situation separately were divided into three groups by stature, the middle group being between $65\frac{1}{4}$ and $68\frac{3}{4}$; the nine stature groups varied in size from 4 to 9 subjects, and the size balance would not have been improved by moving the boundaries. The summed comfort ratings for each subject separately on each chair were placed in rank order for that subject, and then the rank orders were tabulated for the separate stature groups within each test situation. A rank order for the chairs was derived for each stature group from the sum of the subjects' rankings, thus representing in rather an approximate fashion the comfort preference order of the chairs which suited the different tall, medium and short stature groups.

It was hypothesised that, if these rank orders were similar, then stature could be dismissed as an important variable, at least within the range of stature, situation and chair studied here. None of the Kendall rank correlations (ranging from 0.1 to 0.3) nor the coefficients of concordance (0.4 to 0.5) are statistically significant. Therefore, we must infer that stature may have an important effect; however, it should be noted that non-significance may have arisen from other causes, particularly the small number of subjects in some of the stature groups.

To examine further the possible influence of stature, the rank orders of the groups were inspected. In each test situation there are several changes in rank order by two or three places, but only two large changes, and in each case one is explicable by stature and one is not. For example, from the long-term test chair C is ranked 1 for the short group and $8\frac{1}{2}$ for the tall, and its seat height is lowest ($16\frac{1}{4}$ in.); but chair 7 has the next lowest seat height ($16\frac{3}{4}$ ins.), and it is ranked 10 for the short group but $5\frac{1}{2}$ for the tall. Again, from the meal test, chair K is ranked 1 for the tall goup, 7 for the short group, and its seat height is highest of all the chairs ($18\frac{1}{4}$ ins.); but chair F, the next highest ($17\frac{3}{4}$ in.), is ranked 2 for the short group and $8\frac{1}{2}$ for the medium group (and $2\frac{1}{2}$ for the tall group).

Neither consideration of other dimensions, nor the Chair Feature Checklist results, suggest any further insights. It is concluded that stature may be an important factor in some cases, that other factors probably interact and confuse the picture, and that this form of analysis may be useful with larger subject groups and a more controlled sample of chair designs.

12.3. *The Chair Feature Checklist (CFC)*

It was at first thought that this measure would prove particularly useful. In section 5.9 it provided some clues about some chairs, although its results were not particularly illuminating in all cases.   The first step in further analysis was to consider its design and it's results for each feature.

For seat height (see Table 7), there is an obvious correlation between the order of ratings and the chair physical dimension.   However, other factors affect subjective impressions of seat height, for instance seat depth as in the case of chair Y, which is reported as too low by over twice as many subjects as chair C (which is lower).   A similar reason (but not seat depth and there is nothing obvious) may explain the puzzling discrepancy between the two most acceptable chairs (A and Z) and the three others with similar seat height at 17 inches, conforming to the British Standard recommendation (chairs L, T and N), which are rated significantly too low.

For seat depth, the acceptable region of 15 to 15½ inches is in the middle of the B.S. 3893 recommended range.   For seat width, the preference seems to be for 15 inches, as compared with the B.S. 3893 recommended minimum of 16 inches, but this may be a central tendency in the subjective assessment resulting from the wide range of the chairs from 11½ to 18 in. width.   For seat slope, downwards from front to back, the ratings do not show very clear trends by inspection.

For seat shape, backrest height, backrest angle and clearance for feet or calves, there is nothing obvious by inspection.   The physical dimensions of the backrest cannot be measured easily on the polypropylene chairs, and therefore comparisons may not be very meaningful; the CFC questions on the other three are badly phrased, and so the answers are only useful for one or two chairs.

For backrest curvature radius, there is a distinct trend favouring a chair with a horizontal curvature of 10½–11½ inches.   This is amongst the most clear-cut in the study, and the figure contrasts with the recommendation of 12 to 18 inches in the British Standard.

It was decided to check statistically the evident close agreement between the CFC ratings and the range of physical dimensions of the chairs.   A ' CFC vote order ' was calculated by taking the difference between the totals in the two ' unsatisfactory ' cells (e.g. CFC ratings ' high ' and ' low ' in Table 7), with regard to sign, and ranking them from 1 to 10 from the largest positive difference, with equal ranks for any where the vote difference was not at least two.   The chairs were similarly ranked for physical dimension.   The Kendall *tau* correlations are given in Table 12.

Table 12.   Chair feature checklist ratings (CFC) correlated with
physical dimensions and comfort ratings

	Feature	CFC and dimension		CFC and comfort	
1.	Seat height	$\tau = 0.6$	$p = 0.01$	$\tau = 0.1$	$p = 0.4$
2.	Seat depth	$\tau = 0.6$	$p = 0.005$	$\tau = 0.5$	$p = 0.04$
3.	Seat width	$\tau = 0.6$	$p = 0.01$	$\tau = 0.4$	$p = 0.1$
4.	Seat slope	$\tau = 0.5$	$p = 0.04$	$\tau = -0.2$	$p = 0.3$
5.	Back curve	$\tau = 0.8$	$p = 0.001$	$\tau = 0.3$	$p = 0.2$

It was decided also to compare the order of acceptability of the chairs on each feature, as indicated by the CFC ratings, with the overall ranking from the comfort ratings of programme 1 (Table 5).   To derive a ' CFC rank order ' for the chairs, based on the CFC ratings, in terms of their acceptability on the feature concerned, they were ranked in decending order of the total votes in the 'correct' cell, with equal ranks for any where the vote difference was not at least two.

This CFC rank order was then correlated with the overall ranking from the comfort ratings and the Kendall *tau* correlations are also given in Table 12.

The contrast between these two sets of correlations seems larger than might be expected if the subjects were in fact using the checklist as was hoped. A high correlation between CFC vote order and the physical dimensions must occur, if the subjects are able to discriminate adequately in relation to the range of the physical dimension covered by the chairs. However, smaller correlations would certainly be expected between CFC rank order and the comfort ratings, because the latter are made in response to the total chair and task situation; moreover, interactions also between different chair features would be expected to affect the overall comfort response. But, in view of the high correlations found on another aspect and reported in the next section, it is suspected that the subjects have responded to the CFC in a different way.

It is thought that they may not have related the answering of the CFC very directly to their recent comfort experience with the chair, but instead may have treated it as an exercise in stating how well the chair corresponds to a set of intellectual concepts. In this way, they could well, for instance, have rated the seat height of chair K as too high, even though they had just been sitting on it with no discomfort. In the end the net result might be the same, with the low correlation indicating that seat height is not a major determinant of sitting comfort within the limits studied, but that would be fortuitous.

Therefore, since this alternative must be suspected and cannot be disproved, it has to be stated that the CFC in its present form cannot be claimed to have achieved the aim of revealing which aspects of a chair are the predominant features for determining the response of comfort. Nevertheless, the results here are by no means entirely negative. The high correlations of CFC vote order with chair physical dimensions certainly indicate good discrimination ability, as has also been shown by Kirk *et al.* (1967). This suggests that the intended purpose of the CFC could be achieved. Therefore, it is suggested that the principle of this method merits adoption and further research.

12.4. *Comfort and Dimension Recommendations*

This section also is concerned with the question of whether the comfort results can be shown to depend upon particular features of the chairs. However, the analysis here is directly in terms of the comfort responses and the chair dimensions. The combined ranking based on the comfort ratings from Part 1 (Table 5) sets the chairs in an order for optimum comfort from best to worst. To arrange the chairs in a similar order, for comparison, based on their dimensions, they were ranked for each feature, in sequence of the amount by which they departed from the optimum value recommended in British Standard 3893 (1965). If an optimum range was recommended, the centre value was usually adopted. Other values were also tried, to find a better fit to the data. Each time, the chairs were set in the order of the differences, regardless of sign, of their dimensions from the chosen optimum, with rank 1 being given to the smallest difference. The chair dimension rank orders thus obtained were then correlated with the combined comfort rating order from Part 1. The resulting Kendall *tau* correlations are shown in Table 13.

Here again, seat height does not seem to be a significant factor. However, chairs were from $16\frac{1}{4}$ to $18\frac{1}{2}$ in. (41 to 47 cm) which is about the recommended range for adjustable seat heights; our subjects covered the full spread of population body height, but did not represent in numbers the majority of people in the middle, and this probably prevented a clear trend in favour of an optimum.

Table 13. Comfort ratings correlated with chair dimensions (chair dimensions ranked from various possible 'optimum' values)

	Feature	τ	p	Optimum used
1.	Seat height	-0.3	0.1	17 in (43 cm)
	Seat height	0.4	0.1	18 in (46 cm)
2.	Seat depth	0.6	0.01	15 in (38 cm)
3.	Seat width	0.6	0.01	16 in min
4.	Seat slope	-0.6	0.01	$2\frac{1}{2}°$
	Seat slope	-0.2	0.2	$5°$
	Seat slope	0.6	0.01	$7\frac{1}{2}°$
5.	Back curve	0.2	0.2	15 in (38 cm rad)
	Back curve	00.5	0.03	12 in ($30\frac{1}{2}$ cm rad)
	Back curve	0.6	0.01	10 in ($25\frac{1}{2}$ cm rad)

The significant correlations of the comfort results with depth and width dimensions seem to support the British Standard recommendations. The four least comfortable chairs were the four shortest, and were all below the bottom of the B.S. recommended range of 14 to $18\frac{1}{2}$ in. ($35\frac{1}{2}$ to 47 cm); however, no comment can be made on the top limit, since the deepest chair was $16\frac{1}{2}$ in. (42 cm). Again, we cannot comment on the validity of the B.S. recommendation for width to be 16 in. ($40\frac{1}{2}$ cm) minimum, since only two chairs exceeded this size.

The results on seat slope and backrest curvature are most interesting. In both cases the chairs covered a wide range (slope down from front to back from $-2°$ to $10°$, and curvature radius from $9\frac{1}{2}$ in., 24 cm, to flat). The B.S. recommendation for slope is a range from $0°$ to $5°$, but with these chairs an optimum of $2\frac{1}{2}°$ from which to rank them gives a significant negative correlation, whereas an optimum of $7\frac{1}{2}°$ gives a significant positive correlation with comfort ratings. Again, with backrest curvature the recommendation is a range from 12 to 18 in. ($30\frac{1}{2}$ to 46 cm) radius, but taking an optimum of 15 in. (38 cm) gives no correlation, whereas an optimum of 12 or 10 in. ($30\frac{1}{2}$ or $25\frac{1}{2}$ cm) gives a significant result.

Since the exact position of the backrest could not be measured on several of the plastics chairs, no statistical analysis is possible on this feature. However, it should be noted, from Table 1, that three of the four lowest rated chairs (Z, A and Y), broke the B.S. recommendation for the height of the bottom of the backrest by a considerable margin, and the fourth (T) had a peculiar backrest also not conforming to the B.S. figure. The general tendency of these data suggests that the backrest position is an important factor.

Since the B.S. recommendations were made for wooden and not polypropylene chairs, we must be cautious in our interpretation. However, it does seem equally, that one should be cautious in using the B.S. recommendations as a design guide. Moreover, these results and the resulting suggestion are directly confirmed by the purely dimensional analysis which was made in section 5.8.

More important, these experiments suggest a method for establishing better dimension limits to improve chair comfort or at least to avoid discomfort. The above analysis raises some crucial questions in an unavoidable way, but no changes can be made on the basis of these specific chair studies alone. Clearly a more comprehensive programme of research is needed, which should concentrate upon this type of approach. Any recommendations must be firmly based on extensive validation studies, using full ranges of chair dimensions, and relying primarily on the comfort criterion as measured by controlled subjective tests with large groups of subjects.

12.5. *Discussion of These Results*

It should be emphasized that, in this section 12, the analysis has been aimed to develop fresh insights to the general problem. The statistical calculations have all been cautious, because of groupings, tied ranks, and possible distribution anomalies; the values least favourable have been taken to avoid inflating the results, particularly the correlations. However, it could be argued that the raw data have been stretched and possible interactions ignored, since the chairs sampled were probably all suboptimal on one dimension or another, and these variations along different dimensions in many cases may overshadow significant preferences along other dimensions. In addition, some chairs have upholstered seats and some were flexible, making it difficult to take measurements which would be representative for when someone is sitting on the chair. For example, chair L has a very flexible backrest, which deforms under the weight of the sitter, the angle of the backrest to the horizontal being then a function of the height and weight of the sitter. Further, the seats had a wide range of surface textures, which altered the position in which the sitter was supported. The physical measurements are, therefore, in many cases only best estimates. It is not possible, in a practical study such as this, to control these various factors. It must be noted that they were probably affecting the data in varying degrees, and therefore the results must be taken as, it is believed, strongly suggestive but not necessarily conclusive.

A number of further specific developments may perhaps be suggested. First, a test-retest validation is needed with another group of subjects, to establish firmly the main findings from the long-term sitting experiment, especially the predictive ability of the direct ranking method, and to check that comfort preferences are reasonably stable over a sizeable time gap, for instance 6 months.

Secondly, the Comfort Rating scale could well be improved, by refinement at the comfort end. It will be remembered from Figure 2 that the top group of chairs were rated well into the comfortable region, and in fact very few individual ratings from the subjects put them in the discomfort region, even after $3\frac{1}{2}$ hours of sitting. One may question whether, for the majority of chairs which we are likely to come across, real discomfort is unlikely to be a major factor, and that we might profitably begin to look at the minimal unholstery and flexibility characteristics which produce actual comfort, i.e. to concentrate upon the comfort of chairs, rather than the discomfort.

Thirdly, the Chair Feature Checklist could be thoroughly revised and retested, and perhaps joined with the type of approach in section 12.4 linking directly comfort ratings and dimension recommendations. These and the CFC results here demonstrate, it is believed, that this kind of study is an essential follow-up to anthropometry, for establishing design criteria for chairs, and for confirming that these criteria hold in practice.

13.   General Discussion and Suggestions

13.1. *Chair Selection*

It is thought that the most important result of this study is that it has highlighted a whole area for which, despite the extensive ergonomics research on sitting comfort, there is as yet no answer and little knowledge: how can we help the individual or institutional purchaser of chairs to select those comfortable for his own or his organization's use? It is suggested that research in the following areas might be appropriate.

1. The Direct Ranking method found to give a good prediction with the Birkbeck student subjects should be tested for validity with other similar groups and types of people.   If the general result is confirmed, then the size and selection criteria for the test group and a standardized test method should be explored and established.

2. As an extension from No. 1, can a layman population predict its comfort results from a similar brief controlled test?   If so, are there any limitations on the type of people who can do this?

3. If, as seems likely, the group prediction by Direct Ranking is confirmed, then a relatively fast and straightforward method will have been established for selecting chairs for groups of people (i.e. the institution selection problem), but this does not necessarily select the best chair for a particular individual. Therefore, there would still remain the untouched area of what, if any, method can be recommended to the prospective domestic purchaser who cannot carry out any sophisticated tests.   Shall we eventually be able to recommend a validated ' do-it-yourself ' chair selection method?

4. Equal study and development is needed on observational and objective methods to supplement and, where appropriate, to replace the subjective methods.

It should be noted that these suggestions are not meant in any way to minimize the importance of the ultimate goal, and the research leading to it from which advice on good design will come, that no bad or even suboptimum chairs should be produced.   However, the practical fact remains that, until this goal has been reached, many poor chairs will be produced and selection will still be needed on comfort criteria.

13.2. *Chair Design*

It is suggested that research related to ergonomics advice for chair design might be directed, among other things, to the following areas.

1. What deviations from the B.S. recommendations, or any other design guide optima, are tolerable and what are serious?   These would have to be studied along each chair feature dimension separately and also in all the interacting combinations.   In other words, the ultimate desideratum is a range of three dimensional limit profiles (external and internal limits) between which a chair design will be optimal, and some indication of the percentage degradation in comfort in each main direction.   Of course, account would have to be taken of construction materials, compressibility and resilience allowed,

surface texture, etc. This is a mammoth task, but at least the goal must be acknowledged.

2. Methods must be explored, developed and standardized for testing the comfort-chair dimension relationship. This and much other work, of course, has to come first, before the ' grand scheme ' of No. 1 above.

3. On the subject of methods, perhaps the biggest need is for research with the aim of developing and validating observational methods particularly, and objective methods in general, which correlate well with controlled subjective sitting trial results.

4. As a final suggestion, which might equally be placed under the heading of Chair Selection, there is the question of appearance. How far do the appearance, colour and related factors enter into or influence a person's choice, and his subjective responses when acting as a subject, and how far do they affect his ultimate acceptance and regular usage (or tolerance) of a chair? As part of this area of investigation, it would be desirable also to include factors such as the influence of fashion and status.

13.3. *The Criterion for Seating Comfort*

From these two studies, we wish to mention briefly one more general thought for further discussion. The basic question of what is the ultimate criterion for seating comfort, and indeed what is comfort, still seems unanswered. The three main experimental approaches at present used study

(a) the fit of the chairs to the user's anatomy,

(b) the user's performance and/or behaviour, and

(c) the user's subjective assessment, as measured by controlled methods in a controlled situation.

Have the methods at times obscured the relevant criteria? In so far as the term and concept ' comfort ' is being studied, it would seem self-evident, since this is an abstract noun for a personal sensation, that the ultimate criterion must be what a range of human subjects, using suitably defined and proved measurement methods during a properly controlled sitting trial, assess to be comfortable in the situation given.

It may be objected that this is inadequate because it takes no account of the reason for sitting and the effect of sitting on the task being done; hence the performance should be the criterion. This, we suggest, is wrong or at least debatable; the performance measure does not answer ' what is comfort in this situation ' but answers ' what level of comfort is appropriate to the task and the performance required in this situation '. On the other hand, the observation of behaviour and performance as part of the experimental method is a different matter. What we wish to suggest is the importance of the user's subjective assessment, and its essential primacy as the ultimate criterion of comfort against which other more convenient and perhaps more objective methods may be validated.

We believe our studies have shown that it will be possible, after further work, to establish and accept certain methods as reliable measures of this ultimate criterion. This at least would be a significant step forward, but there will remain much more work yet on other criteria and other simpler and quicker methods.

14. General Conclusions

Turning finally to our general conclusions from these two programmes, and the methodology discussion, three important factors seem to have emerged.

First, the different orientation shows a major gap in our knowledge. Previous research seems mainly, and rightly, to have aimed for methods and results to aid the designers. But there is no data on whether, and if so how, the individual prospective user could validly compare and choose to suit his own needs. We must at once say that we have not yet tackled this gap, but it seems the most significant one.

Second, when we started, the experts with whom we discussed the aims could not recommend a proved or generally accepted set of comparison tests, and were unanimous in only one view, that experiments were necessary. Our results show that evaluation trials can be run fairly easily on quite a large scale and with meaningful results. Since there have been very few similar studies, we venture to suggest that more are needed and that our methods may be of interest.

Third, the experts' recommendations on the need for more experiments, in the present state of knowledge, are fully substantiated. The use of individual subjective opinion, or of interpretation of chair dimensions based upon size recommendations, must be considered of doubtful value for comparative assessment.

The final conclusion seems clear, that seating comfort is still a very complex problem and the only valid approach is the experimental method. We hope that the methods tried out and the results found in these studies, may add to the growth of knowledge and assist further work in this important area of ergonomics.

We wish to express our thanks to the Greater London Council and to the Directors of Penguin Books Ltd., for permission to use their facilities, and our gratitude to the Birkbeck students and the G.L.C. and Penguin staff who were so diligent and helpful as subjects. Similarly, we wish to express our very sincere appreciation of the assistance and time given to us by our colleagues who took part in the second programme, without them the study would literally have been impossible: Mrs. Joan Ward, Mr. D. Barkla, Mr. P. Branton, Prof. W. F. Floyd, Mr. J. C. Jones, Dr. D. F. Roberts, Mr. P. T. Stone, and Dr. R. J. Whitney. Finally, we would acknowledge the assistance given by the Consumers' Association in providing the initial stimulus, and our thanks to the Editor and the Project Officers for their great help and for permission to publish the results.

The purpose of the two experimental programmes described in this paper was the study of chair comfort evaluation methods in the context of selection for a potential user.

Programme 1 aims were :
1. to explore methods ;
2. to compare a group of chairs in typical usage tasks.

Programme 2 aim was :
    to study the value of (1) individual opinions, and (2) B.S.I. dimension recommendations as methods for users to select chairs.

In programme 1, three separate experiments were made, with the same chairs and three panels of 20 subjects, under conditions of long-term sitting, sitting at a desk, and eating a meal. The results were :
1. hardly any significant differences between male and female subjects ;
2. significant decrease in comfort ratings with time ;
3. significant differences in comfort ratings between chairs ;
4. only small correlation between chair comfort preference order for different tasks, suggesting the need for different optimum designs, yet significant concordance between tasks to suggest that an acceptable compromise can be achieved in one design ;

5. significant correlation between rankings before trials and comfort test results, suggesting a possibly useful technique.

In programme 2, experienced research workers in the ergonomics of chairs and sitting agreed to be the test subjects, since if these methods are valid then positive results would be most likely to come from such experts. However, the results did not support the validity of assessment based on subjective opinions or B.S.I. recommendations ; with the methods and time used in this study, it is clear that opinionative assessments cannot be given as valid predictors for the general population. The opinionative methods cannot therefore be recommended for use by laymen either.

In the third part of thispaper, methodological aspects are considered more fully. Further analyses of some results yield, in particular, significant correlations and differences between comfort test results and chairs ranked by size from B.S.I. recommended dimensions, suggesting the need for further work to improve recommendations and a useful technique for such studies. Problems of comfort criteria, chair selection, chair design and test methods are discussed.

The general conclusion seems evident that seating comfort is still a very complex problem, and that the only valid approach is the experimental method.

L'objectif des deux programmes d'expérimentation décrits dans cet article était de mettre au point des méthodes d'évaluation du confort d'une chaise, dans le contexte d'une application par les utilisateurs.

Les objectifs du 1er programme étaient:

a. de définir des méthodes,

b. de comparer un groupe de chaises utilisées dans l'accomplissement de tâches typiques.

L'objectif du 2e programme était d'étudier la valeur (*a*) des opinions personnelles et (*b*) des normes recommandées par le B.S.I., en tant que méthode de sélection des chaises et qui pourrait être pratiquée par les utilisateurs.

Dans le programme 1, trois expériences différentes ont été exécutées avec les mêmes chaises et avec trois groupes de 20 sujets, dans des conditions de position assise prolongée, de position assise devant un bureau et durant un repas.

Les résultats ont été les suivants:

1. Il y a, à peine, une différence significative entre hommes et femmes.
2. Les notes attribuées au confort décroissent significativement avec le temps.
3. Les chaises diffèrent significativement entre elles du point de vue de l'évaluation du confort.
4. Les ordres de préférence pour le confort d'une chaise ne donnent lieu qu'à des correlations faibles, en ce qui concerne les diverses tâches; ce qui suggère la nécessité de conceptions optimales différentes. Cependant, il existe une concordance significative entre les tâches, ce qui indiquerait qu'un compromis acceptable puisse être réalisé dans une conception unique.
5. L'existence de corrélations significatives entre les rangs avant les essais et les épreuves d'évaluation du confort suggère l'applicabilité d'une technique déterminée.

Pour la réalisation du programme 2, des chercheurs compétents dans la conception ergonomique des chaises ont accepté d'être sujets, puisque, si ces méthodes sont valides, on devra s'attendre à ce que des résultats positifs puissent être obtenus avec de tels experts.

Cependant, les résultats ne confirment pas la validité des jugements basés sur l'appréciation subjective ou celle des recommandations du B.S.I. Compte-tenu des méthodes utilisées et du temps dont on dispose, il est évident qu'une appréciation par l'opinion ne peut être considérée comme un prédicteur valide pour la population globale. Les méthodes basées sur l'opinion ne peuvent donc pas être retenues non plus pour l'usage des non-spécialistes.

Dans la troisième partie de cet article, on examine plus en détail certains aspects méthodologiques. L'analyse plus détaillée de quelques résultats montre qu'il existe des corrélations significatives et des différences entre les résultats des épreuves d'appréciation du confort et les rangs attribués aux chaises d'après le classement des dimensions préconisées par le B.S.I., ce qui suggère la nécessité d'autres recherches pour améliorer les normes du B.S.I. et la possibilité d'une technique utilisable pour de telles études.

La discussion porte sur les problèmes des critères du confort, le choix et la conception des chaises, ainsi que sur les méthodes d'évaluation.

La conclusion générale est que le problème du confort de la position assise est vraiment très complexe et que la seule voie d'approche valable est la méthode expérimentale.

Die beiden hier beschriebenen Programme beabsichtigten, Methoden der Bewertung des Sitzkomforts in Hinblick auf ihre praktische Anwendung zu studieren. Programm 1 versucht (1.) die Methoden zu analysieren, (2) eine Gruppe von Stühlen im Gebrauch zu vergleichen. Programm 2 versucht, den Wert individueller Meinung und die Empfehlungen der British Standard Institution als Methoden für die Auswahl von Stühlen für den praktischen Gebrauch zu studieren.

Im Programm 1 wurden mit denselben Stühlen und 3 Gruppen von 20 Versuchspersonen 3 verschiedene Versuche gemacht, unter Bedingungen langdauernden Sitzens, Sitzens an einem Tisch und Essens einer Mahlzeit. Die Resultate waren:

1. Kaum ein Unterschied zwischen männlichen und weiblichen Personen;
2. Eine signifikante Abnahme der Komfortbeurteilung mit der Zeit;
3. Signifikante Unterschiede der Komfortbeurteilung verschiedene Stühle;
4. Nur geringe Korrelation swischen Stuhl-Komfort-Deurteilungswert für verschiedene Aufgaben lässt vermuten, dass verschiedene optimale Entwürfe notwendig sind, aber ein annehmbarer Kompromiss in einem Entwurf verwirklicht werden kann;
5. Signifikante Korrelation zwischen Rangfolge im Versuch und Komfort-Test- Resultat lässt eine möglicherweise nützliche Technik vermuten. Im Programm 2 stellten sich erfahrene Experimentatoren auf dem Gebiete der Stuhl—und Sitz—Ergonomie als Versuchspersonen zu Verfügung, in der Hoffnung, von diesen die besten Resultate bei subjektiver Beurteilung zu bekommen, wenn diese Methode gültig ist. Die Resultate bestätigten aber nicht den Wert persönlicher Beurteilung oder Der British Standard Institution—Empfehlungen. Es ist klar, dass die persönlichen Meinungen keine gültige Vorhersagen für das allgemeine Publikum liefern können. Sie können auch nicht dem Laien zum Gebrauch empholen werden.

Im 3. Teil dieser Arbeit werden die methodischen Aspekte ausführlich besprochen. Weitere Analysen einiger Ergebnisse signifikante ergeben Korrelationen und Unterschiede zwischen Komfort-Beurteilungs-Werten und Stühlen in der Rangfolge der British Standard Institution— Masse und-Empfehlungen, welche dazu führen, weitere Untersuchungen auf diesem Gebiet zu fordern. Probleme der Komfort-Kriterien, der Stuhl-Wahl, des Stuhl-Entwurfs und der Test-Methoden werden diskutiert.

Allgemein ist zu schliessen, dass Sitzkomfort immer noch ein sehr komplexes Problem ist, und dass die einige brauchbare Methode die experimentelle ist.

References

AKERBLOM, B., 1954, Chair and sitting. In *Symposium on Human Factors in Equipment Design* (Edited by W. F. FLOYD and A. T. WELFORD) (London : H. K. LEWIS) pp. 29–35.

ANON., 1966, Which chair for comfort ? *Consumers' Association magazine WHICH?*, April, pp. 108–112.

BARKLA, D., 1961, The estimation of body measurements of British Population in relation to seat design. *Ergonomics*, **4**, 123–132.

BARKLA, D., 1964, Chair angles, duration of sitting and comfort ratings. *Ergonomics*, **7**, 297–304.

BENNETT, E. M., 1963, Product and design evaluation, Study 3 : The feeling of comfort. In *Human Factors in Technology* (Edited by E. M. BENNETT, J. DEGAN and J. SPIEGEL.) (New York : McGRAW-HILL) pp. 543–552.

BRANTON, P., 1966, The comfort of easy chairs ; an interim report on the present state of knowledge. *FIRA Tech. Rept.*, 22 ; *unpublished restricted circulation report from Furniture Industry Research Association, Stevenage, Herts.*

BRITISH STANDARDS INSTITUTION, 1959, Anthropometric recommendations for dimensions of non-adjustable office chairs, desks and tables. B.S. 3079 : 1959.

BRITISH STANDARDS INSTITUTION, 1965, Specification for office desks, tables and seating. B.S. 3893.

BURANDT, U., and GRANDJEAN, E., 1963, Sitting habits of office employees. *Ergonomics*, **6**, 217–228.

FLOYD, W. F., and ROBERTS, D. F., 1958, Anatomical and physiological principles in chair and table design. *Ergonomics*, **2**, 1–16.

GUILFORD, J. P., 1954, *Psychometric Methods* (2nd edition) (New York : McGRAW-HILL).

KARVONEN, M. J., KOSKELA, A., and NORO, L., 1962, Preliminary report on the sitting postures of school children. *Ergonomics*, **5**, 471–477.

KIRK, N. S., WARD, J. S., ASPREY, E., BAKER, E., and PEACOCK, B., 1967, Discrimination of chair seat heights. *Paper to Third International Congress on Ergonomics Birmingham, England.* (To be published.)

KROEMER, K. H. E., and ROBINETTE, JOAN C., 1968, Ergonomics in the design of office furniture ; a review of European literature. *Tech. Rept. AMRL–TR–68–80* ; *unpublised report from USAF Aerosp. Med. Res. Labs., Wright-Patterson A.F.B., Ohio.*

SLECHTA, R. F., WADE, E. A., CARTER, W. K., and FORREST, J., 1957, Comparative evaluation of aircraft seating accommodation. *Tech. Rept. No. 57–136. WADC, Wright-Patterson Air Force Base, Ohio.*

STONE, P. T., 1965, An approach to the assessment of the comfort of foam cushioning. *Automotive Body Engng*, **135**, 28–30.

WACHSLER, R. A., and LEARNER, D. B., 1960, An analysis of some factors influencing seat comfort. *Ergonomics*, **3**, 316–320.

The Development of a Rest Chair Profile for Healthy and Notalgic People

By E. Grandjean, A. Boni and H. Kretzschmar

Department of Hygiene and Applied Physiology of the Swiss Federal Institute of Technology, Zurich, and Department of Physical Therapy of the Cantonal Hospital of Zurich, Switzerland

1. Introduction

The number of people afflicted with periodic or permanent notalgia is great: statistics indicate that over 50 per cent of adults suffer from backache at least once in their lives.

Starting from the conception that it is important for persons experiencing backache to find the most relaxed and pain-free posture possible, the intention of the present study was to develop an appropriate seat profile for a rest chair realizing that even a suitable seat profile would not result in a lasting correction of spinal deformity. If the pain commonly caused by abnormal positions of the vertebrae and discs or by myospasms could be relieved, this might provide a first prerequisite for an improvement in the condition.

2. Method

2.1. General Procedure

With the aid of a ' seating machine ' enabling any seat profile to be formed the development of a seat profile for notalgic persons was attempted by systematic adjustment of the seat elements. There were three stages:

    development of a seat profile for normal persons;
    testing the normal seat profile on notalgic persons;
    development of a rest chair causing the minimum discomfort to notalgic persons.

The method applied at the outset involved recording the movements of subjects and the pressure distribution on the seat surface and back-rest during the seating tests. At the beginning and end of the tests the subjects were given a questionnaire in which they were asked to note the subjective sensations in various parts of their bodies and their evaluation of individual units of construction.

The recording of movement and pressure distribution gave no results for the different seat profiles tested either for normal or for notalgic subjects. These two measuring methods were therefore not applied in the subsequent tests and no results are tabulated.

2.2. The Seating Machine

Figure 1 shows the seating machine. It enables the back-rest and seat to be given any inclination; the arm-rests and seat any height. The seat surface and back-rest consist of frame members into which adjustable wooden slats are clamped. These members make it possible to adopt a variety of profiles for the seat and back-rest surfaces. A foam rubber sheet of 6 cm thickness was placed on the entire seat.

Figure 1.   The ' seating machine '.

3.   Seat Profiles for Normal Subjects

3.1. *Experimental Procedure*

In the first series, five known and frequently recommended profiles were selected and tested with 10 men in seating tests of 150 min each.

In the second series, 36 men and 16 women tested for 8 min each of the profiles which had been developed on the basis of the results obtained in the first series.   Using the same subjects (Vp), adjustments were subsequently modified until the optimum position was found for each Vp.   All tests were performed twice, once for reading and once for rest.

3.2. *Results*

All individual results and a detailed discussion of the literature have already been published by Grandjean and Burandt (1964).   Only the most significant results are summarized here.

Table 1 gives the individual data about the most comfortable inclination angles of the seat surface and back-rest and the preferred seat heights.

Table 1. The distribution of seat dimensions considered to be the optimum by 52 persons

Seat inclination angle (°)	21 & 22	23 & 24	25 & 26	27 & 28	29 & 30	31 & 32
Number considering inclination optimum at rest	0	3	29	14	6	0
Number considering inclination optimum for reading	14	30	7	0	1	0
Back-rest inclination angle (°)	99 & 100	101 & 102	103 & 104	105 & 106	107 & 108	109 & 110
Number considering inclination optimum at rest	0	0	2	15	28	7
Number considering inclination optimum for reading	2	21	24	4	1	0
Seat height (cm)	<37	37 & 38	39 & 40	41 & 42	43 & 44	>44
Number considering height optimum at rest	7	20	15	8	1	1
Number considering height optimum for reading	0	6	31	8	5	2

Seat inclination angle is the angle between the seat surface (chord) and the horizontal plane. Back-rest inclination angle is the angle between the seat surface and the prolonged tangent to reversing point of the back-rest.

The table shows that the following angles and dimensions were most frequently the most comfortable:

Reading	Seat inclination	23–24°
	Back-rest inclination	101–104°
	Height	39–40 cm
Rest	Seat inclination	25–26°
	Back-rest inclination	105–108°
	Height	37–38 cm.

The striking element in these results is the comparatively large angles of the seat surfaces. We presume that this is due to their preventing the slipping forward of the buttocks and thus improving the support of the back-rest. On the other hand it must be noted that getting up from such seat inclinations is inconvenient, and this is why we have used lesser seat inclinations in later tests and suggested them in our recommendations.

All seating tests clearly revealed that every Vp gave preference to a back profile which offered satisfactory support of the loins. This resulted in a profile of the back-rest characterized by a loin welt of which the most pronounced bulge is located 11 to 14 cm vertically above the effective seat surface (upholstery compressed). The shape of this back-rest profile corresponds to the recommendations of Åkerblom (1948) who however provided the loin welt at a higher level (about 18 cm).

The two seat profiles evolved in our experiments made on normal persons are shown in Figure 2.

Proposal for reading   Proposal for rest

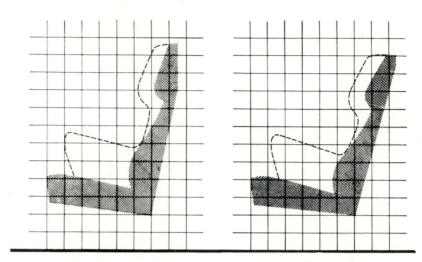

Figure 2.   Seat profiles for healthy persons.   Module: 10×10 cm.   The broken line corresponds
to the arm-rests and a possible outer contour.   The dots show the surface of the seat
profile including upholstery 6 cm thick.

4.   Seat Profiles for Notalgic Persons

4.1.   *Testing the Developed Seat Profile on Notalgic Persons*

4.1.1.   *Experimental procedure*

In the first experiment we used 17 women and 21 men who had previously
been treated for lumbar complaints.   All patients had roentgenological and
clinical disc trouble between the 5th lumbar and the 1st sacral vertebrae.
At the time of the seating experiments the acute stage had ended but all Vps
complained of backache.

Their ages ranged between 30 and 75 years for the women (average 59) and
between 33 and 82 for the men (average 55).

According to the clinical findings, the Vps were divided into a hyperlordosis
group and an alordosis group, and we had 20 cases in the former and 18 in the
latter.

It is known that women more frequently have hyperlordoses and men more
frequently alordoses.   This was also the case with our patients: hyperlordoses
13 women, 7 men; alordoses 4 women, 14 men.

The seating experiments lasted one hour each using the profile 'for rest'
shown in Figure 2.   The questionnaire was completed twice in every experiment: after 5 min and after 60 min of the sitting posture.

4.1.2.   *Results*

We checked whether the assessments were different after 60 min from those
after 5 min.   A comparison of the replies after the two periods showed differences which were in no case statistically significant.

Table 2 shows the evaluation of the feeling of comfort in the back and loins
for the two periods.   It shows that *the subjective assessments after 5 min were
not substantially different from those made after 60 min.*

Table 2. The effects of sitting time on the subjective assessments of 38 persons

	After 5 min sitting			After 60 min sitting		
	uncomfortable	medium	comfortable	uncomfortable	medium	comfortable
In back (above lumbar spine)	1	3	34	1	6	31
In loins (lumbar spine)	4	4	30	2	4	32

Table 3. Assessment of seat profiles for ' rest ' by notalgic and healthy persons

	Notalgic persons ($n=38=100\%$)			Healthy persons ($n=52=100\%$)		
	(%) uncomfortable	(%) medium	(%) comfortable	(%) uncomfortable	(%) medium	(%) comfortable
Head and back of neck	21	52	27	35	29	36
Shoulders	3	8	89	0	25	75
Back	3	8	89	6	8	86
Loins	11	11	78	8	17	75
Buttocks	0	3	97*	0	21	79*
Thighs	13	32	55	13	35	52
Arms	8	50	42	38	21	41
	too much	good	too little	too much	good	too little
Seat height (42 cm)	13	52	35†	23	67	10†
Arm-rest height (25 cm)	11	34	55	4	38	58
Seat depth (48 cm)	11	50	39*	8	75	17*
Seat inclination (20°)	11	78	11	12	77	11
Back-rest inclination (108°)	11	76	13	6	86	8
Head-rest inclination (156°)	11	26	63	10	44	46

Duration of test: 5 and 8 min. respectively. $* p<0.025$. $† p<0.01$.

Table 3 compares the evaluation of the notalgic Vp with the healthy Vp. Only the values after 5 and 8 min sitting time in the profile according to Figure 2 are considered here.

Applying the χ^2 test, we examined which differences between the evaluations were significant. These calculations revealed that notalgic Vps more frequently considered the sensation in the buttocks to be comfortable. On the other hand, they more often found the seat height to be too low and the seat depth too short.

A comparison of all assessments shows that *those considering the feeling of comfort of the notalgic Vps did not substantially differ from those of healthy Vps.*

A comparison of the subjective judgments of the hyperlordosis group with those of the alordosis group showed no significant difference for any question. This also appears from the results shown in Table 4.

Table 4. Assessment of general seating comfort of seat profile for rest after 60 min sitting

Group	Number of Persons	Very uncomfortable	Uncomfortable	Medium	Comfortable	Very comfortable
Hyperlordosis	20	0	1	4	15	0
Alordosis	18	0	1	3	12	2

4.2. *The Individual Optimum Adjustment of the Seat Profile for Notalgic Persons*

4.2.1. *Experimental procedure*

For these experiments we had a total of 68 persons (33 women and 35 men) available. According to the clinical findings they were divided into the following groups: hyperlordosis 29 Vps; alordosis 30 Vps; normal lordosis 9 Vps.

The average age was 52 (extremes 27 and 82) and the average stature 166 cm (extremes 147 and 186 cm).

With all Vps the treatment for disc and spinal trouble in the lower lumbar region in the clinic had been completed, but they all had occasional backache.

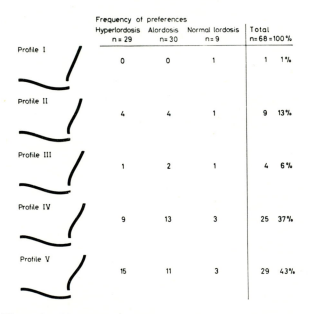

| | Frequency of preferences | | | | |
	Hyperlordosis n = 29	Alordosis n = 30	Normal lordosis n = 9	Total n = 68 = 100 %	
Profile I	0	0	1	1	1 %
Profile II	4	4	1	9	13 %
Profile III	1	2	1	4	6 %
Profile IV	9	13	3	25	37 %
Profile V	15	11	3	29	43 %

Figure 3.   The frequency of the preferred back-rest profiles.

Each of the 68 Vps examined the 5 different back-rest profiles shown in Figure 3. For each Vp's preferred back-rest profile all adjustable elements of the seating machine were varied until the individual optimum adjustment was obtained. Subsequently, the Vps sat on the seat for 45 min, then rose and sat down again. An individual optimum adjustment was then again carried out. Finally the Vps were asked for their judgments as to the convenience of the seat for rising.

4.2.2. *Results*

The frequency of preferences for the 5 back-rest profiles tested is shown in Figure 3.

The majority (81 per cent) of Vps preferred profiles IV and V. This indicates that *notalgic persons prefer a back-rest profile which has a slightly concave shape to the front above the lumbar spine and a pronounced frontal convex shape on the lumbar spine.* On the other hand, a straight back-rest (profiles I and III) is rejected by almost all Vps (63 of 68 Vps).

Checks using the χ^2 test revealed no significant differences between the hyperlordosis and the alordosis groups. Consequently, the preference for profile IV by alordosis and of profile V by hyperlordosis is not statistically significant.

Later experiments designed to find the individual seat profile which was as comfortable and convenient as possible showed that no substantial differences were noted between the first adjustment and that after 45 min sitting time.

The average values of the most comfortable individual adjustments of profiles IV and V are shown in Table 5.

Table 5. Average values of the most comfortable individual settings of profiles IV and V.
  Forty-eight subjects. Mean values are given together with standard deviations. The diagram indicates the angles and linear dimensions given in the table. The loin height is the distance between the lowest point of depressed seat surface and the main rest point on the loin welt

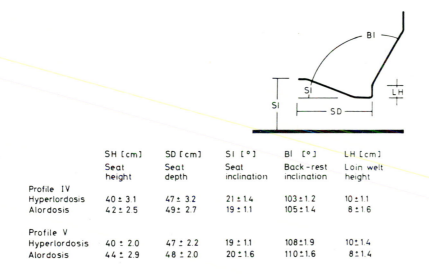

	SH [cm] Seat height	SD [cm] Seat depth	SI [°] Seat inclination	BI [°] Back-rest inclination	LH [cm] Loin welt height
Profile IV					
Hyperlordosis	40 ± 3.1	47 ± 3.2	21 ± 1.4	103 ± 1.2	10 ± 1.1
Alordosis	42 ± 2.5	49 ± 2.7	19 ± 1.1	105 ± 1.4	8 ± 1.6
Profile V					
Hyperlordosis	40 ± 2.0	47 ± 2.2	19 ± 1.1	108 ± 1.9	10 ± 1.4
Alordosis	44 ± 2.9	48 ± 2.0	20 ± 1.6	110 ± 1.6	8 ± 1.4

With both profiles the Vps preferred almost the same angles and length dimensions. Only in the case of the back-rest angle they preferred a rearward inclination greater by 5° with the seat profile V. This is likely to be due to the more pronounced loin welt of profile V. Furthermore, it is striking that all notalgic persons desired a comparatively low-positioned main rest point in the loins; 8 to 10 cm above the depressed seat surface.

By virtue of the results shown in Figure 3 and Table 5 we evolved a seat profile for notalgic persons which is represented in Figure 4. This proposal is based on profiles IV and V which yielded no particular differences in individual adjustment. As regards the back-rest angle, we compromised at 106 to 107°. The other dimensions arrived at were:

seat inclination	19–21°;
seat depth (depressed)	47–51 cm;
seat height (depressed)	38–42 cm.

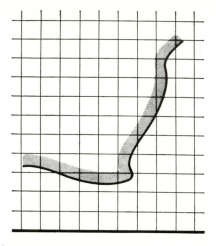

Figure 4. The seat profile for notalgic persons evolved from the experiments. Module 10 × 10 cm. The full line represents the solid base, the shaded area a uniformly thick upholstery of latex ('hard' quality) 6 cm thick. This design produces the desired seat profile when depressed.

Of all Vps, 8 stated that getting up was uncomfortable; the others experienced no difficulties. With a seat angle of 16° rising was comfortable for all.

None the less a seat angle of 20 to 21° is to be preferred because this effectively eliminates slipping frontwards. If the buttocks remain on the rear portion of the seat area, the back will rest on the back-rest throughout its length, which may be highly conducive to relaxation and elimination of possible backache.

Our warm thanks are expressed to the Fritz Hoffmann–La Roche Foundation for the Promotion of Scientific Work Panels for research grants. We would also thank the Swiss Industrie-Gesellschaft, Neuhausen a. Rh. and Giroflex Entwicklungs-AG, Koblenz, for the assistance given and the test seats they made available.

With the aid of a 'seating machine' which makes it possible to give the seat surface and back-rest of a test chair any profile, we tested various seat profiles with healthy and notalgic persons. After the experiments these persons were given a questionnaire in which they had to report their subjective sensations in the various parts of the body and their evaluation of various angles and dimensions.

A first experiment conducted with 52 healthy persons indicated the following optimum values:

    Reading: seat inclination 23–24°; back-rest inclination 101–104°,
    Rest: seat inclination 25–26°; back-rest inclination 105–108°.

A back-rest having a frontal convex loin welt and a frontally slightly concave back contour proved to be the most comfortable configuration.

In two further series of tests with 38 and 68 notalgic persons who had all been treated for disc derangements in the region of the lower lumbar spine the seat profiles were evolved which were as comfortable and free from inconveniences as possible during a sitting period of 45 min. From the results of these experiments, the following recommendations for a rest seat for notalgic persons were arrived at. The back-rest must be provided with a frontal convex loin welt and, above the lumbar spine, a frontally slightly concave contour.

The main rest point of the loin welt must be vertically spaced from the depressed seat surface (lowest point under the tubers of the ischium) by 7 to 12 cm.

The most significant data are:

back-rest inclination	105–108°,
seat inclination	19–21°,
seat depth (depressed)	47–51 cm,
seat height (depressed)	38–42 cm.

Les auteurs utilisent un siège réglable dont la surface d'assise et le dossier peuvent présenter des proportions ou un profil quelconques. A l'aide de cette " machine à s'asseoir ", différents profils d'assise ont été testés sur des personnes en bonne santé ou souffrant de dorsalgies. Après l'expérimentation, les sujets remplissaient un questionnaire dans lequel ils mentionnaient leur sensation subjective de confort pour les différents angles et dimensions proposés.

Une première expérimentation, effectuée sur 52 sujets en bonne santé, a montré les valeurs optimales suivantes:

pour la lecture:	inclinaison de l'assise:	23°–24°
	inclinaison du dossier:	101°–104°
pour le repos:	inclinaison de l'assise:	25°–26°,
	inclinaison du dossier:	105°–108°.

La configuration la plus confortable du dossier est celle présentant une courbure convexe vers l'avant au niveau des reins et une courbure légèrement concave dans un plan frontal pour le dos.

Dans deux autres séries d'expériences furent testées 38 puis 62 personnes souffrant de dorsalgies, qui avaient toutes été traitées pour des lésions discales de la colonne lombaire basse. Le profil de siège le plus confortable et présentant le moins d'inconvénients a été recherché au cours de périodes de position assise d'une durée de 45 minutes. Ces expériences ont montré qu'un siège pour dorsalgique doit présenter les caractéristiques suivant:

le dossier doit comporter une courbure convexe vers l'avant au niveau des reins et une forme légèrement concave dans le plan frontal;

le point d'appui principal de la région lombaire doit être distant verticalement de 7 à 12 cm de la partie la pluse basse de la surface d'assise (point le plus bas sous les tubérosités ischiatiques);

les dimensions les plus importantes sont:

inclinaison du dossier:	105°–108°;
inclinaison de l'assise:	19°–21°;
profondeur du siège:	47–51 cm;
hauteur du siège:	38–42 cm.

Mit Hilfe einer " Sitzmaschine ", die es ermöglicht, der Sitzfläche unde dr Rückenlehne eines Teststuhls jedes beliebige Profil zu geben, prüften wir verschiedene Sitzprofile an gesunden Personen und an Personen mit Rückenschmerzen. Nach den Versuchen hatten diese Personen ihre subjektiven Empfindungen in den verschiedenen Körperpartien auf einem Fragebogen anzugeben und die verschiedenen Winkel und Dimensionen zu bewerten. Eine erste Versuchsreihe an 52 gesunden Personen ergab folgende Optimalwerte:

beim Lesen: Sitzneigung 23–24°;	Neigung der Rückenlehne 101–104°;
beim Ruhen: Sitzneigung 25–26°;	Neigung der Rückenlehne 105–108°.

Eine Rückenlehne mit einer frontal konvexen Lendenabstützung und einer frontal leicht konkaven Rückenkontur gab die besten Werte.

In zwei weiteren Testserien mit 38 und 68 Personen mit Rücken—schmerzen, die alle wegen Bandscheiben-Störungen im Bereich der unteren Lumbalwirbel behandelt worden waren, wurden Sitzprofile entwickelt, welche während einer Sitzperiode von 45 Minuten möglichst bequem und frei von Störungen waren. Aus den Ergebnissen dieser Versuche ergaben sich folgende Empfehlungen für einen Ruhesitz für Personen mit Rückenschmerzen.

Die Rückenlehne muss mit einer frontal konvexen Lendenabstützung versehen sein. Sie soll oberhalf der Lumbalgegend eine frontal leicht konkave Kontur besitzen. Der Hauptstützpunkt der Rückenlehne soll vertikal 7–12 cm über der zusammengedrückten Sitzfläche liegen (tiefster Punkt unter dem tuber ischii). Die wichtigsten Daten sind: Neigung der Rückenlehne 105–108°, Sitzneigung 19–21°, Sitztiefe (nicht zusammengedrückt) 47–51 cm, Sitzhöhe (nicht zusammengedrückt) 38–42 cm.

References

AKERBLOM, B., 1948, *Standing and Sitting Posture* (Stockholm: NORDISKA BOKHANDELN).
GRANDJEAN, E., and BURANDT, U., 1964, Die physiologische Gestaltung von Ruhesesseln. *Bauen und Wohnen*, **6**, 233–236.

Behaviour, Body Mechanics and Discomfort

By P. BRANTON

Medical Research Council, London, England*

1. Introduction

Surveying the 20 years since the publication of Åkerblom's (1948) study, one cannot but be struck by the fascination the subjects of sitting and seating exert on ergonomists, designers and the general public. Yet this interest does not appear to have led to much improvement in the general quality of seats, nor has it stemmed the tide of clinical complaints about backpains. The assertion of a causal connection between seats and body malfunctions would therefore not be altogether unjustified, however difficult to prove such a connection may be. Even so, general principles for the improvement of seats still remain to be formulated.

With technological advances and the continuing increase in the time for which people sit, the problem is unlikely to diminish in importance and, even if some may harbour the suspicion that there are no ideal solutions, the search must continue. Perhaps this is the moment to pause and reconsider the conceptual framework of research on sitting and seats. This paper will try to show that behavioural study reveals some gaps to exist in this framework and will suggest ways of bridging them. In our view the gaps result from the fact that the problems are of an interdisciplinary nature and there are three areas for further research in which our understanding could be considerably advanced by joint effort.

The first is the area between body mechanics and behaviour, insofar as it lies between the biological and the behavioural approaches to human action. The second is the area between behaviour and subjective feelings, where we might look for a theoretical basis for research into comfort—and its measurement. In the third area, the technological, demands raised in the other two are to be translated into hardware and then tested systematically to secure validity as a precondition for acceptance by the public at large.

Before discussing each of these research areas, it is necessary to make clear that we shall confine ourselves here to discussing situations of sitting in which comfort and relaxation would normally be the major consideration. Thus we may include not only easy chairs, but also other seats without tables or desks within reach, such as are found in lounges, lecture and concert halls, as well as passenger seating in buses, trains and aircraft. In fact the limiting conditions are that the sitter's limbs perform little or no overt work and that his pelvic complex and spine rest on seat pan and backrest.

2. Behaviour and Body Mechanics

2.1. *Postural Variety*

Naturalistic studies of sitting (e.g. Branton and Grayson 1967) have shown that spontaneous behaviour regularly produces a variety of postures with highly significant differences in frequency and duration. These results have

* Now Ergonomist, British Railways Board, London, N.W.1, England

since been substantially confirmed by other researchers using both instantaneous observation and time-lapse films and nearly 45,000 observations are now available for analysis. That sitting postures will, in fact, vary all the time is therefore a basic concept which must be incorporated into our research framework. How can this variety be accounted for? And could we predict which postures will be taken up in a particular seat?

At first sight it might be thought that this variety represented merely random changes of position, an 'urge to move' ('Bewegungsdrang') caused by ischemia as postulated explicitly or implicitly in previous studies of 'fidgeting' (Grandjean *et al.* 1960, Coermann and Rieck 1964). Indeed, it is highly likely that ischemia is a contributory factor, in that it creates bodily states which make changes of position desirable. Another likely contributory factor, supporting the idea of such an urge is the considerable pressure on the skin and tissues under the ischial tuberosities (Herzberg 1955). Ischemia may thus be a necessary condition for that urge to become manifest. But it does not seem to be a sufficient one because it does not explain why on observation certain postures were found to be taken up more frequently than others and held for longer. Neither does 'urge to move' explain why the same person's behaviour should differ so greatly as between seats. Statistical treatment of the behavioural data leaves no doubt that what was observed could not have been postural variation for its own sake and a more complex and specific explanation will be required of how and where ischemia is generated or how it influences spontaneous behaviour.

Another factor usually thought likely to account for the observed postural variations is the possible discrepancy between linear anthropometric and chair dimensions. Here again the relationship may turn out to be more complex than would appear at first sight. It may be remembered that, in the above-cited study of behaviour in train seats, postures differed significantly between tall and short subjects, as well as between types of seat. As the linear dimensions of these two types of seat were almost identical, the differences in behaviour cannot easily be related to misfit between the body dimensions and those of the seats. In this connection the statement by Burandt and Grandjean (1963)—in a somewhat different context—is recalled: ' the exclusive application of an anatomical magnitude for the determination of chairs and tables for office use is unsatisfactory '. Since neither of the usually accepted factors, ischemia and dimensional misfit would account directly for the differences in behaviour, we feel that our observations warrant so far only the limited conclusion that seats do something to the body, or that some kind of interaction takes place between seat and sitter. The nature of this interaction needs to be explored.

2.2. *Dynamics of Sitting Postures*

The time scale of overt events is very slow and it is therefore not surprising that the mechanics of so-called resting postures have so far attracted little attention. We were first brought to realise the importance of the time factor on viewing the time-lapse films when run 160 times faster than real time. The description of one film in particular will be recalled, as it revealed gradual changes in sequences each of which took 10 to 20 minutes or even longer. These sequences recurred at least 12 to 16 times during a 5-hour period of observation. In each case the sitter slid into a backward slumped posture, propped himself up first with his arms, crossed his knees and then stretched his legs forward, only to end up in a nearly horizontal position. Speaking in terms of interaction, the seat slowly and repeatedly ejected the sitter.

On close analysis the events can be reconstructed as follows. Initially, and because the seat depth was too great for this subject, we assume he had no support in the sacro-lumbar region. Thus, even when sitting upright, his pelvis would rotate backwards over the tuberosities. The upholstery both under and behind him gave little resistance to this movement. If the armrests are used, the trunk is easily raised and suspended by the shoulders rather than resting on the seat. Then, under the backward pressure of the shoulders and with diminished pressure (friction) underneath him, the sitter's pelvis slides forward until a point is reached when the arms can no longer counteract the horizontal component of forces. At this point the angle formed by trunk and thighs is so large that these two levers are driven apart like the unfolding of a claspknife.

To regard trunk and thighs as mechanical levers is, of course, justifiable and not altogether a new concept. Dempster (1955) in a seminal paper refers to the work of Fischer in 1907. Describing the human body as predominantly an open-chain system of links, and applying the kinematic model to sitting, Dempster proceeds as follows.

> ' The fingers of the hand may be interlocked . . .; the legs may be crossed for seated stability; the arms may be crossed or placed on the hips. In such actions as these, temporary approximations to closed chains are effected. . . . Link chains may be cross-connected, as in crossing the knees (viz., pelvis and right and left thighs). . . . To the extent that these temporary closed chains approximate a closed triangular, or pyramidal pattern, the less muscles are called upon for stabilizing action at the joints. One may recognize many rest positions involving this principle: crossed arms, hands in pockets, or such sitting positions as crossed knees, ankle on opposite knee, or head in hand. . . . As additional joints come into the linkage, the accessory tensions of muscles become more and more important for stability or for directing forces in specific ways. Temporary closed chains that involve extrinsic environmental objects also may be recognized.'

Our problem is now to apply such bio-mechanical model to the behaviourally found variety of postures. For if we could in some way measure what a posture does to the body, and what a seat does to a posture, then we would begin to gain control over the design of seats.

The mechanical characteristics of the spine above the waist are known and need no elaboration now. Instead, we wish to consider the structure from the waist down. A schema of this structure, and of its mobility can be constructed graphically (Figure 1). The point to note in this model is that there are at least four ' degrees of freedom to move ', even if the feet are considered to be firmly planted on the floor:

> the pelvis can rock over the tuberosities;
> the thigh can rotate in relation to the pelvis;
> the lower leg can rotate at the knee; and
> the lower leg can rotate in relation to the feet.

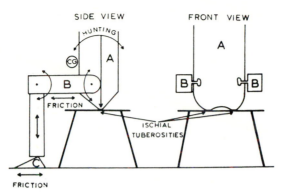

Figure 1. The mechanical characteristics of the human body below the waist.

The notion of freedom to move rests on the consideration that, in the sitting position, the angles at the hip, knee and ankle joints are at about the mid-points of their range of movement and hence in a state of maximum mobility. Any apparent rigidity in this part of the system would therefore depend on muscle action or on external contact with the backrest. These restraints will be discussed later.

If this model reflects the true state of affairs, it draws attention to the mobility of the pelvis and to the flexibility of the link with the lumbar spine at the sacro-lumbar joint. The freedom of the pelvis to move implies not only the possibility of occasional rotation, hinted at when reconstructing the sliding to the slumped posture of the small man in the film. It further necessitates envisaging the possibility of continuous hunting, or relatively fast oscillatory movements of the pelvis rocking over the tuberosities. This possibility arises in all sitting postures in which the top of the sacrum is not resting against a back rest. Indeed, if the lumbar spine, say above L3, were supported while the pelvis remained free to move, such movements would set up a kind of sheering action at the sacro-lumbar joint. This consideration lends weight to the importance Keegan (1953) gives to this joint and the stresses and strains to which it is subjected.

Åkerblom (*op. cit.*), Keegan (*op. cit.*), and Schoberth (1962) have already shown that the angulation of the sacral plate varies with different individuals and Schoberth argues that sitters may rotate their pelvis in an endeavour to bring the plane of the joint to the horizontal ' in order to reduce vertical stress.' This would explain, at least partly, why backward-slumped postures and/or large seat-to-back angles (105° or more) are so often preferred. Such subjective preferences were reported in a number of studies of normal persons (e.g. Barkla 1964) and of patients with backache complaints (Kretzschmar and Grandjean 1967).

We arrive thus tentatively at reasons for some of the behavioural observations, in particular of slumped postures. The role of the legs in sitting, and in particular the crossing of knees could also be explained in terms of this model. It will be remembered that the small man at some point in his sliding almost always crossed his knees, and that crossed knees occurred in approximately 30 per cent of all our observations. We believe a fitting explanation for this phenomenon is as follows. It reduces the tendency of the pelvis to rotate. Indeed, when the thighs are adducted and one knee is imposed on the other, a lock is effected across the pubis and rocking is likely to be prevented altogether. Or, if the legs are stretched forward and crossed at the ankles so that the knee joints are locked, the thighs are again adducted and the effect on the pelvis is similar to that of crossed knees. These two ways of counteracting the hypothesized rocking of the pelvis are examples of attempts at stabilization by internal rigidification of the body structure. In accord with Dempster's view, little muscle action would be involved in holding these postures since they approximate triangular patterns. Further considerations in the assessment of these postures are that stretching the legs forward must shift the overall centre of gravity in the same direction, and that at the same time the legs can be used as stanchions to the floor against the forward slide of the body.

2.3. *Muscle Action*

Our model leads to the assertion that even in the maintenance of so-called resting postures muscles are involved to a greater extent and degree than seems commonly appreciated. This activity is of the tonic type and its relation to relaxation as an elementary factor in 'comfort' is obvious. It would therefore be of critical importance to measure muscle involvement in various resting postures. This, however, seems to be very difficult. Most published electro-myographic studies of sitting have been concerned with work seats and muscle action potentials were mostly demonstrated rather than measured. Moreover, some workers (Lundervold 1951, Floyd and Silver 1955, Basmajian 1962) claim to have detected periods of 'electrical silence' in the sacrospinalis during upright sitting. Yet, in a recent study, Nachemson (1966) by recording simultaneously from the sacrospinalis and the psoas, demonstrated the functional relationship between anterior and posterior spinal muscle groups. In upright sitting, free from backrest, when the sacrospinalis was silent, the psoas was very active, while in leaning forward from the hip the reverse was the case. Evidently one of these muscle groups does the anti-gravity work; which of them it is at a given moment depends upon the position of the centre of gravity at the time.

In the slumped posture, as observed, the spine can be regarded as a flexible open chain of links and, with the expected rocking of the pelvis, if back support is given only under the shoulders, the activity of the antagonistic sets of muscles must be considerable. Since the psoas is accessible only by deep needle electrodes, demonstration of this antagonism is extremely hazardous and it would be fair to say that EMG alone could reveal only a partial picture of the complexity of resting postures.

Attempts to obtain an objective measure of relaxation by more indirect means have been made by the present author. The assumption was that the expected muscle tremor would be shown in fine and fast movements of the centre of gravity and that these might be measured on a force platform. These studies have, however, so far not shown conclusive results. Nevertheless, muscular involvement is a problem still to be faced, since, as Nachemson (*op. cit.*) reminds us, 'The lack of inherent and intrinsic stability of the vertebral column and the importance of trunk muscles are clearly demonstrated if one tries to hold an unconscious person upright.'

2.4. *Seat Features as Supports and Stabilizers*

If it is accepted that the spine and the pelvic complex are loose chain links, it is necessary to consider what support the body structure could derive from normal resting seats. Following on Hertzberg's study (*op. cit.*) Swearingen *et al.* (1962) in an analysis of sitting areas and pressures provide some evidence on this point. They found in 104 subjects that one-half of the body weight is supported by 8 per cent of the seat area (under the ischiae). About one-third (35·2 per cent) of bodyweight is borne by the combination of footrests (18·4 per cent), arm rests (12·4 per cent) and a backrest slightly sloping at 15° rake (4·4 per cent). In effect, the seat pan carries 65 per cent of the weight. Schoberth (*op. cit.*) arrived at similar values by calculations based on Braune and Fischer's work.

To consider the relationship of the seat pan to the structure first. It is the area of contact with the two ischiae which bears over half the body weight, which means that the sitting body is unstably suspended in respect to the seat pan. This would appear to be the case whether the person is seated on a bare plank or on an upholstered chair, because a soft cushion under the ischiae would impart as little stability and hence rigidity to the system as flabby muscle or adipose tissue.

This is not to say that, to the sitter, the difference between a hard and a cushioned seat is immaterial. Cushioning does affect the relative distribution of skin pressure over the seat, such that the pressure of body weight would be spread over more than 8 per cent of area. The onset of ischemia is thus likely to be delayed on cushions. If, however, such a cushion were compressed by the ischiae to the point of becoming solid (i.e. if 'bottoming' occurs), that part of the cushion would have the same effect on the system as a hard seat.

As regards the contact of the body with backrests, unless they are raked more than 30° their contribution to support of weight—only about 4 per cent—is astonishingly small. It becomes apparent then that, contrary to expectation and common belief, the effective function of backrests is not to support the body in the sense of bearing weight, but rather that they act as stabilizers.

The extent to which sitters in the behaviour study actually utilized these stabilizers can be gauged from frequency of occurrence of four main groups of postures, and from mean durations as shown in Table 1.

Table 1. Frequency of occurrence of four groups of postures

Code	Posture	Frequency observed (%) ($N = 1644$)	Mean duration (min) ($N = 10$)
1111	Minimal support derived from seat	3·3	2·3
1221, 1222	Full back support from seat and armrests	49·5	15·1
1321, 1322	Slumped, some support from seat and armrests	23·4	11·4
	Other postures	23·8	5·5

This comparison brings out the considerable difference between least supported postures and well supported ones, which tends to strengthen the argument for the operation of stabilization.

P

To sum up the consideration of body mechanics in relation to behaviour, the sitter is now seen as a dynamically balanced system of open links. Most observed sitting postures are maintained by participation of muscles and the relation a specific posture has to ' comfort ' depends, at least partly, on the degree of muscular relaxation it permits. Apart from the possibility of measuring fine and fast oscillatory (or ondulatory) movements of the trunk, the various postures may be assessed for their intrinsic stability or rigidity. *In both respects it can be said that at some points in time the seat imparts motions to the body and these tend to be similar for many sitters.*

3. Behaviour and Subjective Sitting Comfort

3.1. *Concepts of Subjective Comfort*

The second main area in which conceptual orientation is required if we are to progress towards generalisable statements in sitting and seat research concerns the relation of behaviour to subjective comfort. In the many studies in which subjective sitting comfort has been investigated, very few attempts have been made to define the concept of comfort accurately or to make inter-experiment comparison possible. Some little consolation may be derived from the fact that a similar situation exists in other fields of comfort research. For instance, Teichner (1967) cogently shows this in a discussion of subjective thermal comfort. He concludes that

' All things considered the problem of assessing human subjective thermal responses appears to have suffered from

(1) Lack of use of psychophysical technique, that is methodology which relates behaviour to physical or bodily conditions;

(2) Lack of a systematic, theoretical approach to guide research and define validity, and finality, to be discussed below;

(3) Failure to consider motivation as a factor in the human response.'

Our general problem is similarly twofold. What is to be measured? and How is this to be done?

At this point, it seems to us essential to bear in mind that the aim of our research is to find out about the comfort of seats rather than the comfort feelings of a person and that the experimenter uses persons as channels of information about seats. Experimentation must therefore not merely aim at obtaining statements about general comfort. It must attempt to point to specific design features and seat characteristics. Moreover, we must consider the possibility that comfort as such is not amenable to measurement and that working definitions of it may have to be modified.

Most investigators assume that the term comfort denotes a feeling or an affective state, which varies subjectively along a continuum from a state of extreme comfort through indifference to a state of extreme discomfort. We find it most difficult to envisage deriving extreme feelings of well-being merely from sitting however good the chair may be. In our view, therefore, the possible continuum extends only from indifference to extreme discomfort. The absence of discomfort denotes a state of no awareness at all of a feeling and does not necessarily entail a positive affect. Similarly, the absence of pain does not necessarily entail the presence of pleasure. This consideration does not merely affect the construction of comfort scales for the eliciting of subjective judgments, to be discussed later. It leads further to the search for criteria in the sitter's behaviour which validly express comfort. It may be that motivation, or the purpose of sitting, provides a criterion. After all, seats are used for a variety of purposes and it is perhaps by their efficiency for these purposes that we may judge them.

3.2. *Motivation to Sit*

In the previous section attention was drawn to the need for understanding the mechanical characteristics of given postures and we arrived at the view that postural behaviour may consist of attempts at attaining relative stability of body structure in a physical sense. This implies that in observing postural changes we observe the operation of an underlying purpose and, indeed, our major premise here is that behaviour is not random but purposeful, motivated or expressing needs. Welford (1966) has said ' If we know how motives operate, what they are becomes of secondary importance.' Nevertheless to speculate in psychological terms, the purpose of sitting might seem to be the achievement of maximal comfort. But this could easily overstate the matter, for maximal comfort would be found in sleep. In that case we should more appropriately investigate the efficacy of beds than that of seats. But the human is not to be regarded only as a comfort seeker. He may have a motivation to rest, but if too much rest is given he also seems to seek stimulation from the environment. We would thus not equate maximum bodily comfort with an optimum state of *sitting* comfort. Our case is that we normally sit to some purpose quite unrelated to the shape and properties of the seat and that sitting, like all postural activity, is only a means to another end. In rest seats we sit for a compound of primarily social and personal reasons with the secondary purpose of ' taking the weight off our feet ' while listening, conversing, looking at television, or just daydreaming while being transported from A to B. We do not seek comfort for its own sake but rather seek to attain a state which is optimal for the pursuit of these other purposes. Observations of verbal and postural behaviour are then to be interpreted in terms, not of the experience of comfort, but of motivation to avoid interference with primary activities, or of avoiding *dis*comfort.

Comfort achievement is now assigned to a secondary, as it were auxiliary role, which by no means diminishes the importance of research into it, but rather raises it to a psycho-physical problem of attention and discrimination.

This conception may bring sitting comfort research into a field of current psychological experimentation in which performance of two concurrent tasks is measured. A seat may be measurably ' inefficient ' to the degree to which it interferes with the primary activity. It also makes experimentation amenable to treatment in terms of signal detection theory (Tanner and Swets 1954), in that the seat characteristics can be regarded as a source of ' noise ' interfering with the ' signals ' emanating from other sources competing for the subject's attention. For example, during a lecture we may become aware of disturbing features of our seat to the detriment of our intake of the speaker's words. What we would be trying to measure would be fluctuations in the person's *tolerance of discomfort*.

3.3. *Problems of Measurement*

Some problems of measurement of subjective responses remain to be solved. Two of these, set by the limitations in human functions, can be raised here: they concern verbal and memory capacity.

The eliciting of verbal judgments in seat assessment presupposes that something can be verbalized. The verbalization of feelings presents, of course, a longstanding problem in experimental psychology. Moreover, to gain general statements that, say, one seat is uncomfortable or even comparative statements that one seat is less uncomfortable than another are insufficient guidance for the improvement of seating. Judgments should at least be such that practically useful information is gained about either a distinct feature of the seat or a body region, as was done for instance by Grandjean and Burandt (1964).

But, unlike in the areas of environmental comfort research, in seat assessment the subject's attention cannot always be readily focussed and related to physical variables as yet not clearly defined. Also, if the subject knew what was wrong with a seat, our work would soon be done. The analysis of postural behaviour presented in the previous section suggests that the difficulty lies in the fact that posture maintenance is a very primitive and deeply ingrained skill, acquired in early childhood. In the brain, this skill, like other motor skills, has very likely an 'enactive representation' rather than a 'symbolic' one (Bruner 1964). That is to say, like walking or bicycle riding, sitting behaviour is demonstrable but not readily accessible to introspection and verbalization. It may thus be that observation of behaviour is at least as good a guide to comfort as verbal judgment.

It is, of course, not advocated here that we do away with subjective judgements in comfort research. In order to avoid the problem of verbalization, the present author has experimented with the use of a hand dynamometer for the expression of feelings of bodily tension, albeit without success in relating such responses to given seat features. This would nevertheless be one way in which the subject's task of evaluation could be facilitated.

The other measurement problem, concerning memory capacity, relates to experimental method and the formulation of questions to the subject. It seems that in experiments in which two or more seats are compared and ranked in any order the fact is overlooked that reliance is placed on kinesthetic memory. The use of rankings or of pair-comparisons of seats to determine their 'comfort-retaining quality' (Slechta *et al.* 1957) requires subjects to remember what the previously presented seat 'felt like'. Any interval of more than a few minutes between presentations must make these comparisons between feeling states very volatile. With very careful experimental arrangements these methods might produce stable information about a factor of short-term comfort, but judgments of long-term comfort would be very unstable and unreliable.

Of the alternative methods, employing absolute ratings of one seat at a time, an extensive survey of past work found that the most consistent results were obtained from simple, unstructured rating scales (Slechta *et. al op. cit.*). Two further aids have been used with some success to focus the subject's attention. One is the use of a body-part manikin (Bennett 1963) or a body-map (Shackel and Chidsey 1966). These greatly facilitate subjective reports by avoiding verbal description of the locus of felt discomfort. The other device for eliciting simple but reliable judgment seems to be the use of checklists, e.g. by forced-choice questions like ' The seat is too high/just right/too low '. Though some of them are very laborious, these methods have the double advantage that they may lead to practically useful information about specific seat features as well as allowing more extensive statistical treatment and hence greater refinement of theoretical results. Whatever judgments are elicited, the basic scientific requirement remains to be satisfied, namely that such judgments be set against some objective, physical or behavioural measure. Perhaps here we must take up the challenge Teichner (*op. cit.*) has stated, to devise a ' behavioural preference ' method.

4. Some Technological Problems of Rest Seat Design

From the foregoing considerations some demands of a technical nature arise for seat design and construction. They concern dimensions as well as other physical properties of the materials to be used in rest seats. The first demand relates to *linear dimensions*. It is unlikely that discomfort can be avoided by simply matching each body dimension with the equivalent seat dimension, as if the interface were static. What seems to be critical is the relation of anyone dimension with the others and with expected sitting behaviour. For instance, the relation of seat depth to seat height depends on the purposes of the seat. If a domestic chair is intended for resting postures, a relatively deep seat of 43 cm (17 in.) may be satisfactory for women of the 5th percentile, provided that seat was low enough. In that case, legs could be stretched forward without dangling or causing excessive pressure under the popliteal area. A mass-produced rest chair of the same depth but 41 cm (16 in.) high would be inefficient for the purpose of stretching the legs forward—a posture found frequently. Short persons stretching their legs forward would either lose back stabilization at the sacrum or lose support under the feet. Another instance of an anthropometric demand, mentioned in a previous study, was found in the position of wings in an otherwise very much preferred train seat. These wings were either too low, or too close together, or protruded too far to allow the tall, broad-shouldered person full use of the lower backrest and they thus encouraged slumping.

The second demand presents a mixture of linear and angular dimensional and upholstery problems. It is that the seat's *support and stabilizing functions* are brought within the reach of all likely users. Essentially, the supports/stabilizers are required under or immediately in front of the points on the seat pan where the tuberosities rest. In the back, at least the area in contact with the sacrum is critical. The load of the sitter is bound to distort all cushioning and the location of the critical body areas on the seat profile has, to our knowledge, not been systematically investigated. The specification of unloaded profiles may therefore have to be modified, or at least augmented by specific instructions about the mechanical characteristics of underlying cushioning and springing.

The technical difficulty seems to lie in two demands which appear to be mutually exclusive. On the one hand, cushioning should relieve pressure points and spread the sitter's load over wider areas of the seat pan; on the other, it should provide such support that the sitter's sliding into the slumped posture is counteracted. This counter-action is however to be more a restraining or channelling action than total prevention of posture change. Behaviour seems to show that exertion of the combined strength of hamstring and back muscles is irrepressible. Again, if the seat pan is tilted back so that the included angle becomes less than 95°–100°, while cushioning is very firm, complaints of restriction to body movements and digestive functions may arise. If seat pan and back cushions are soft and thick, support and stabilizing becomes almost non-existent. An effective trunk-thigh angle of about 105° should cater for the observed propensity of sitters to lean back and this can be achieved by shaping and by varying thickness and hardness of cushioning over different parts of the interface with the body. Ideally, the mechanical properties of the seat pan upholstery would therefore be such that it reacts to lateral distortion while being vertically compressible. In modern foam cushioning this is perhaps a technical impossibility.

As regards the demands from the combination of cushioning and seat covering, the provision of a moderate degree of surface roughness would add usefully to the resistance against sliding movements of the body. This matter, mentioned long ago by Åkerblom, has frequently been ignored.

Lastly, another required characteristic of coverings and upholstery is that they should encourage and not restrict the dissipation of perspiration moisture and heat at the interface. Very little research on this topic appears to have been done, but a recent study by Garrow and Wooller (1968) demonstrates the importance of this factor in car driving seats.

In application to particular designs, the solutions adopted to fulfil the above demands would, of course, require experimental confirmation in which both technical experts and ergonomists would need to be involved. Validation and fitting trials (e.g. Jones 1963) can then be carried out with well established procedures.

This survey of problems raised by behavioural observation in relation to sitting comfort research attempted to show their possible complexity. It suggested that sitters can provide information about seats not only by verbal judgments but also through other channels, such as performance on concurrent tasks. Observation of sitters may be used in addition to indicate the limits of tolerance to discomfort. Sitting behaviour could be regarded as the operation of a balance between needs for physical stability and for environmental and intrinsic stimulation.

The relationship between a seat and the sitter's discomfort is not thought to be direct because his tolerance may be influenced by an overriding but fluctuating motivation to sit. A more direct relationship appears to exist between a seat and the sitting posture taken up in it. Thus seats may cause postures and a closer analysis of postural mechanics may add to understanding about the quality of seats.

La confrontation des données comportementales et des sensations de confort lors de la position assise, soulève maint problème complexe dont cette revue de travaux tente de rendre compte.

Il apparaît que le sujet assis peut apporter des éléments d'évaluation des sièges, non seulement au moyen d'un jugement verbal, mais également à travers d'autres méthodes d'évaluation telles que la prise en compte des performances dans les tâches doubles. L'observation du sujet assis peut servir, en outre, à objectiver les limites de tolérance de l'inconfort. Le comportement de posture assise peut être considéré comme une recherche d'optimisation entre le besoin de stabilité physique et le besoin de stimulations intrinsèques fournies par l'environnement.

Le relation entre la conception de siège et l'inconfort qu'il procure n'est sans doute pas directe, puisqu'on supporte l'inconfort d'un siège moyennant une motivation qui, bien que fluctuante, est surtout faite du désir de se trouver assis. Une relation plus directe semble exister entre le type de siège et la posture assise qu'il permet d'adopter. La posture est alors directement dépendante du siège, ce qui permettrait, à partir d'une analyse plus approfondie de la mécanique posturale, de mieux définir les caractéristiques qualitatives d'un siège.

Beobachtungen über das Sitzverhalten werfen Probleme auf, die für Untersuchungen der Sitzbequemlichkeit von Bedeutung sind. Es wird darauf hingewiesen, dass Auskunft über Komfortgefühle des Sitzenden nicht nur aus wörtlichen Urteilen sondern auch auf anderen Wegen bezogen werden kann, z.B. durch Ausübung gleichlaufender Tätigkeiten. Beobachtung Sitzender führt zur Bestimmung der Grenzen der Erträglichkeit von unbequemen Sitzen. Das Sitzverhalten wird als Wechselwirkung zwischen einem Bedürfnis für Stabilität und einem Bedürfnis für Stimulierung betrachtet.

Es scheint keine direkte Beziehung zwischen dem Sitz und dem Komfort des Sitzenden zu bestehen, nachdem seine Toleranz von den Zwecken überwiegend beeinflusst ist, wegen denen er sitzt. Die Beziehung erscheint direkt zwischen einem Sitz und den Sitzhaltungen, die darin eingenommen werden. Sitze bewirken Sitzhaltungen, und Untersuchungen der Haltungsmechanik können zu besserem Verständnis der Sitzqualität verhelfen.

References

ÅKERBLOM, B., 1948, *Standing and Sitting Posture* (Stockholm : A. B. NORDISKA BOKHANDELN).

BARKLA, D. M., 1964, Chair angles, duration of sitting and comfort ratings. *Ergonomics*, **7**, 297–304.

BASMAJIAN, J. V., 1962, *Muscles Alive* (London : BALLIERE, TINDALL & COX).

BENNETT, E., 1963, Product and design evaluation through the multiple forced-choice ranking of subjective feelings In BENNETT, E., DEGAN, J., and SPIEGEL, J., *Human Factors in Technology* (New York : McGRAW-HILL).

BRANTON, P., and GRAYSON, G., 1967, An evaluation of train seats by observation of sitting behaviour. *Ergonomics*, **10**, 35–51.

BRUNER, J. S., 1964, The course of cognitive growth. *Amer. Psychologist*, **19**, 1–15.

BURANDT, U., and GRANDJEAN, E., 1963, Sitting habits of office employees. *Ergonomics*, **6**, 217–228.

COERMANN, R., and RIECK, A., 1964, An improved method for determining the degree of sitting comfort (in German). *Int. Z. angew. Physiol.*, **20**, 376–397.

DEMPSTER, W. T., 1955, The anthropometry of body action. *Ann. N.Y. Acad. Sci.*, **63**, 559–585.

FLOYD, W. F., and SILVER, P. H. S., 1955, The function of the erectores spinae muscles in certain movements and postures in man. *J. Physiol.*, **129**, 184.

GARROW, C., and WOOLLER, J., 1968, The use of sheepskin covers on vehicle seats. *Ergonomics* (in press).

GRANDJEAN, E., and BURANDT, U., 1964, Die physiologische Gestaltung von Ruhesesseln. *Bauen und Wohnen*, pp. 233–236.

GRANDJEAN, E., JENNI, M., and RHINER, A., 1960, Eine indirekte Methode zur Erfassung des Komfortgefühles beim Sitzen. *Int. Z. angew. Physiol.*, **18**, 101–106.

HERTZBERG, H. T. E., 1955, Some contributions of applied physical anthropology to human engineering. *Ann. N.Y. Acad. Sci.*, **63**, 616–629.

JONES, J. C., 1963, Anthropometric data : limitations in use. *Archit. J.*, **137**, 317–325.

KEEGAN, J. J., 1953, Alterations of the lumbar curve related to posture and seating. *J. Bone and Joint Surg.* **35-A**, 589–603.

KRETZSCHMAR, H., and GRANDJEAN, E., 1967, Comfort requirements of patients with back pain complaints. *Paper given at 3rd I.E.A. Congress, Birmingham.*

LUNDERVOLD, A., 1951, Electromyographic investigations during sedentary work, especially typewriting. *Brit. J. phys. Med.*, 32–36.

NACHEMSON, A., 1966, Electromyographic studies on the vertebral portion of the psoas muscle, with special reference to its stabilizing function of the lumbar spine. *Acta Orthop. Scandinav.*, **37**, 177–190.

SCHOBERT, H., 1962, *Sitzhaltung, Sitzschaden, Sitzmöbel* (Berlin : SPRINGER).

SHACKEL, B., and CHIDSEY, K. D., 1966, The need for the experimental approach in the assessment of chair comfort (Abstract). *Ergonomics*, **9**, 340.

SLECHTA, R. F., WADE, E. A., CARTER, W. K., and FORREST, J., 1957, Comparative evaluation of aircraft seating accommodation. *USAF WADC Tech. Rep.*, No. 57-136.

SWEARINGEN, J. J., WHEELWRIGHT, C. D., and GARNER, J. D., 1962, An analysis of sitting areas and pressures of man. *US Civil Aero-Medical Res. Inst., Oklahoma City*, Rep. No. 62-1.

TANNER, W. P., and SWETS, J. A., 1954, A decision-making theory of visual detection. *Psychol. Rev.*, **61**, 401–409.

TEICHNER, W. H., 1967, The subjective response to the thermal environment. *Human Factors*, **9**, 497–510.

WELFORD, A. T., 1966, The ergonomic approach to social behaviour. *Ergonomics*, **9**, 357–369.

Easy Chair Dimensions for Comfort — A Subjective Approach

By E. F. Le Carpentier

Furniture Industry Research Association, Stevenage, England

1. Introduction

The comfort afforded by an easy chair depends upon the shape of the surface between the sitter and the chair, and the pressure distribution upon this interface. These factors are determined by the various ' give ' properties of the surfaces and by the physical dimensions of the chair. The present study is concerned with the physical dimensions of the chair only. The dimensions investigated were the seat height, the seat depth, the angle of tilt of the seat, the seat to backrest angle, and the angle of rake of the backrest, the armrest height above the seat and the position of the headrest. The objectives of the study were

1. to determine the optimum values for the dimensions of easy chairs for the user population, and any outstanding differences in preference due to the duration of sitting, activity performed whilst sitting, sitter's sex, age, and stature;

2. to determine the magnitudes of the departures from the preferred dimensions which a sitter considers:
   just noticeably less comfortable than the preferred level,
   decidedly less comfortable than the preferred level;

3. to examine whether and to what extent the preferred seat heights and depths are related to each other and to the sitter's lower and upper leg measurements;

4. to arrive at figures which will help designers of easy chairs to achieve a better fit for the majority of users.

These problems can be approached using either of two basic measuring techniques. The first is to present each of a group of experimental sitters with either a number of dimensionally differing chairs in turn, or with an easy chair adjusted by a discrete amount before each sitting, and then to assess the reaction of sitters by interview or questionnaire. The second approach is to allow each sitter to set an adjustable easy chair to the settings which he considers conform to a given criterion such as ' most comfortable ' or ' a little too high '.

In deciding which approach to adopt it had to be taken into account that 4 dimensions, each taken at 3 levels, can be combined in 81 ways, and that some important effects of chair dimensions upon comfort might appear only after several hours sitting. It follows that a properly balanced experiment, using previously set chairs, would have been difficult to design and lengthy in execution.

A second point is that people are more accurate and reliable in making comparative than in making absolute judgments. A sitter using a number of set chairs for a set period on subsequent days would have to depend upon memory to make a comparative judgment. In contrast, the method of self adjustment allows the sitter to make use of an important cue—the levels for each dimension at which he feels decidedly uncomfortable. The subject can adjust a dimension forward and backward until his preferred setting is reached by successive approximations. For the foregoing reasons the method of adjustment was considered best suited to the purpose of the study. The adjustment of the armrests and headrests was carried out by the experimenter on the subject's instructions, as mechanization of these parts would have been cumbersome and it was desirable to preserve as much as possible the normal appearance of the chair.

2. Subjects and Apparatus

2.1. *Subjects*

Twenty subjects, ten men and ten women, were selected on the basis of stature and age, so that the average stature of the group was close to that of the British adult population, and half the men and women were younger than 40. A sub-group of eight subjects was selected from the group and used in the first part of the experiment. The remaining twelve subjects were used in the final part of the experiment. The anthropometric measurements of the groups and of the British adult population are given in Table 1.

Table 1. Group measurements (in.)

Group	Statistic	Sample categories						All categories combined. Stature	Popliteal height	Knee to buttock length	Seated elbow height
		Sex		Age		Stature					
		men	women	younger than 40	older than 40	below 66″	above 66″				
Initial sub-group (8)	Mean	67·5	63·8	28·0	51·7	63·2	69·7	65·6	16·2	18·4	
	S.D.	3·6	2·9	4·9	5·0	1·7	2·0	3·6	1·3	2·5	
	Range	64–72	60–67	23–36	45–58	60–65	67–70	60–72	13·5–17·8	16·5–20·5	
Total group (20)	Mean	67·7	64·7	25·6	46·1	63·4	69·0	66·2	16·3	18·4	8·7
	S.D.	3·0	3·1	4·7	4·9	1·8	1·9	3·4	1·0	1·4	1·1
	Range	64–72	60–70·5	18–36	41–58	60–65	67–72	60–72	13·5–19·0	16·0–20·5	7·3–9·5
British adult population	Mean	67·3	63·3					65·3	16·0	18·5	8·5
	S.D.	2·6	2·6					3·7	1·1	1·4	1·5
	90% Range	63–71·5	59–67·5					59·3–71·3	14·3–17·8	16·3–20·8	6·0–11·0

2.2. *Apparatus*

This consisted of a specially constructed easy chair. The four chair dimensions, the height of the seat above the floor, the seat depth, the seat tilt, and the seat to backrest angle, could all be independently adjusted by the subject operating switches built into the arm of the chair, which actuated four reversible electric motors built into the chair. The armrests and headrest could be adjusted manually by the experimenter on the subject's instructions. The seat pan and backrest consisted of rubber webbing straps on wooden frames overlain with unprofiled polyether foam cushions 3 in. thick for the seat and 2 in. for the backrest, the seat cushion resting on the wooden frame at the front of the seat. The armrests were foam padded. The headrest consisted of a rectangular plywood strip measuring 15 in. × 8 in., padded with soft grade synthetic foam 2 in. thick. Other fixed dimensions of the chair were the seat and backrest width of 21 in., armrest separation of 22 in., height of top of the backrest above the unloaded seat 21 in. The chair was upholstered in moquette. The chair was placed in a warm well-lit room with a television set and magazines provided for use at the appropriate points in the experiment.

The other part of the apparatus was a display panel fed by cables from the chair and from which the experimenter, who sat outside the room, could take recordings of the values of the dimensions of the chair at any given time. The experimenter used a system of switches and lights to flash instructions to the subject. The layout of the subjects' display is shown in Table 2.

Table 2. Subjects' signal box

Seat height O	Seat depth O	Seat tilt O	Seat to backrest angle O
O Decidedly less comfortable		(+2)	(switch up)
O Just noticeably less comfortable		(+1)	(switch up)
O Most comfortable		O	
O Just noticeably less comfortable		(−1)	(switch down)
O Decidedly less comfortable		(−2)	(switch down)

2.3. *Experimental Procedure*

The selection of the 20 subjects was based upon the anthropometric measurements and other factors shown in Table 1. Initially a balanced sub-group of 8 subjects was selected from the 20 subjects. Each of the 8 subjects was asked to sit in the chair for a practice period of one hour to get used to the operating of the chair and to become familiar with the signals which the experimenter, who in the experimental sessions was to sit outside the room, used to indicate which adjustments were to be made. The instructions as to when to start making adjustments, which dimension to adjust, and which criterion to adjust the dimension to suit, were communicated to the subject as follows. Table 2 shows the layout signal box display. A light in the top row indicated the dimension to be adjusted. The 4 control switches in the chair arm were of the 3-way centre-off variety. If one of the upper two lights in the left-hand column showed, then the subject switched the control switch up from its centre-off position until that dimension of the chair moved to a position corresponding to the caption opposite the light. If one of the lower 2 lights showed, the subject adopted the same procedure, but switching the control switch down. The control switches were wired so that switching in the upward direction increased the linear or angular values of the dimension concerned. The central light in the left-hand column indicated that the dimension was to be adjusted until the subject felt it to be at the most comfortable value.

The armrests were set at $8\frac{1}{2}$ in. above the loaded seat. No headrest was used.

In a typical experimental session the following procedure was used.

1. The subject was told he was to sit for 3 hours watching a television programme and to make adjustments to the chair as and when signalled to do so.
2. The subject sat for 5 minutes adjusting all the dimensions to his preferred values.
3. The experimenter signalled for the seat height to be adjusted to the value $+2$, then 0, then -1, then $+1$, 0, -2, 0 again, noting each value from the display panel after the adjustment was made.
4. The experimenter signalled for the seat tilt angle to be adjusted in the same way but to the values -2, 0, $+1$, 0, -1, 0, $+2$, 0. Adjustments were made to the seat depth and the seat to backrest angle in turn. All results were noted.
5. The experimenter signalled for the subject to alter any of the settings he felt necessary to achieve the most comfortable optimum. The results were noted.
6. The subject sat for 40 minutes watching television, then the adjustments were repeated. The dimensions were adjusted in a different order from before.
7. This was repeated through 4×40 minute intervals, which together with the time for adjustments made up the 3 hours of the sitting.
8. Within a few days the same subject repeated the sitting, but instead of watching television, read the magazines provided or a book of his own choice.

Analysis of the results from these 8 subjects showed that no differences in settings existed as a result of reading or television viewing, nor as a result of the duration of sitting. For this reason the remaining 12 subjects were treated in the following way. They were trained to make preference settings only and each sitting lasted for 20 minutes only. All the 12 subjects read books or magazines on both sittings. At each sitting each subject made 2 sets of preference settings, one after 5 minutes, the other after 10 minutes. At the conclusion of the final sitting the armrests and headrest were adjusted by the experimenter on the subjects' instructions, to the positions which were considered most comfortable for use for reading. The way in which the experiment was balanced throughout the group is shown in Table 3.

3. Results

The preferred settings and tolerance limits of the initial group of 8 subjects were analysed. The result showed that there were no significant differences between the settings made when reading than when television viewing. No trend among the settings made at 45 minute intervals existed, showing that neither preferences nor tolerance were related to duration of sitting.

Table 3.   Balance of the experimental sittings among subjects

Subject	Sex	Age category	Stature category	Sitting 1	Sitting 2	Used in initial group of 8
1	M	O	T	R	TV	✓
2	M	O	T	R	R	
3	M	Y	T	TV	R	✓
4	M	Y	T	R	R	
5	M	Y	T	R	R	
6	M	O	S	TV	R	✓
7	M	O	S	R	R	
8	M	O	S	R	R	
9	M	Y	S	R	TV	✓
10	M	Y	S	R	R	
11	W	Y	S	R	TV	✓
12	W	O	T	R	R	
13	W	O	T	R	R	
14	W	Y	T	R	TV	✓
15	W	Y	S	R	R	
16	W	O	T	R	R	
17	W	O	S	TV	R	✓
18	W	O	S	R	R	
19	W	Y	T	TV	R	✓
20	W	Y	S	R	R	

Key: R   =   Reading
        TV   =   Television viewing

The results of the remaining 12 subjects were pooled with those of the initial 8. The average values of the preferred dimensions with S.D.s and 90 per cent ranges are given in Table 4. The average tolerance limits and their S.D.s are given in Table 5. The population variables, sex, age and stature, were tested by splitting the group results for each factor and applying the Mann–Whitney *U* Test. The significant results are given in Table 6.

The subjects on average preferred the seat 0·5 in. deeper than their ' back of knee–buttock ' length. The range was from 2·0 in. shorter to 3·5 in. longer than the ' back of knee to buttock ' length. The subjects on average preferred the front of the seat (unloaded) 1·5 in. lower than their ' floor to underside of thigh ' length. The range was from 6 in. below to 1·5 in. above their ' floor to underside of thigh ' length.

A higher seat front was preferred by the women than by the men, the values being 16·3 in. and 15·1 in. The difference is not included in the table as the significance on the Mann–Whitney *U* Test was only $p = 0.1$.

The product moment correlation coefficient between preferred seat height and the corresponding preferred seat depth measured over all subjects was -0.38 in., significant at the $p = 0.05$ level. The preferred seat heights and depths were tested for correlation using Kendal's *tau* with the corresponding leg dimensions. The correlation values, -0.03 and $+0.005$ were found to be insignificant. The relationship between preferred seat height and popliteal heights are plotted in Figure 1.

The values of the height of the armrests given in Table 4 are from the loaded seat. Two and a half inches were allowed for depression at the centre of the cushion. The average preferred headrest position is given in Table 4.

A reliability estimate based on the first preferred settings made at each sitting is given in Table 7. The values given are the average values taken

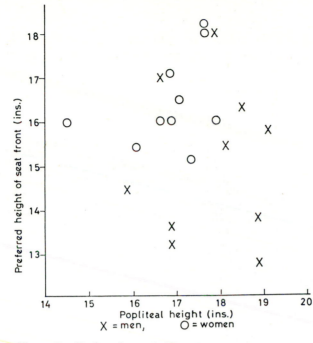

Figure 1. Preferred seat heights and popliteal heights.

over all subjects, of the differences between the settings made on each of the two occasions.

4. Discussion

As in all studies of comfort, the validity of the results rests upon a subjective link in the argument. The link in this study was the sitter's own judgment of his setting of the chair. If he considered that a setting met the criterion which the caption provided, he left the setting as it was; if not, he altered the setting until he felt that the criterion was met. The exact wording of the captions was critical. A number of criteria were tried. The criteria which were finally used are shown in Table 2. The other criteria which were tried, e.g. barely comfortable, unsatisfactory, just comfortable, uncomfortable, led the subjects to doubt their own judgment by relying upon their personal interpretation of the captions.

Table 4. Preferred dimensional values

Dimension	Preferred values (inches and degrees)			Number of subjects
	Mean	S.D.	90% Range	
Height of front of seat (unloaded)	15·7	1·5	13·2– 18·3	20
Seat depth	18·9	1·3	16·8– 21·0	20
Seat tilt angle	10·3	3·4	4·7– 15·9	20
Seat to backrest angle	109·1	5·5	100·0–118·2	20
Backrest rake angle	119·2	4·6	111·6–126·8	20
Height of armrest above loaded seat	6·3	0·6	5·3– 7·3	12
Displacement of centre of headrest, forward from plane of backrest	1·5	2·2	0·5– 3·8	12
Angle of headrest to plane of backrest	7·9	3·3	2·5– 12·5	12

4.1. Preferences

The most noteworthy result was that the men preferred a horizontal or lounging posture while the women preferred to sit more upright and with a steeper seat tilt angle.

Table 5.   Average comfort tolerance limits (8 subjects)

Tolerance limits (inches and degrees)

Dimension	Decidedly less comfortable (+2)		Just noticeably less comfortable (+1)		Just noticeably less comfortable (−1)		Decidedly less comfortable (−2)	
	Mean	S.D.	Mean	S.D.	Mean	S.D.	Mean	S.D.
Height of front of seat (unloaded)	1·4	0·8	0·8	0·8	—	—	—	—
Seat depth	1·5	0·6	1·0	1·0	1·3	0·7	2·1	1·0
Seat tilt angle	6·0	2·0	3·1	1·7	2·9	1·6	5·5	2·3
Seat to backrest angle	8·6	4·1	4·0	3·3	3·9	2·9	7·3	3·5
Backrest rake angle	10·1	9·9	4·6	3·6	3·7	3·1	7·0	3·7

Table 6.   Significant difference in preference among group factors

Dimension	Factor	Average values	Significance level
Seat tilt angle	Sitter's sex	8·8° m, 11·7° w	$p = 0·06$
Seat to backrest angle	Sitter's sex	113·0° m, 105·2° w	$p = 0·02$
Backrest rake angle	Sitter's sex	121·5° m, 116·9° w	$p = 0·04$

Table 7.   Reliability estimate of preferred settings

Dimension	Mean difference between settings on 2 occasions
Seat height	1·1 in.
Seat depth	0·8 in.
Seat tilt	1·7°
Seat to backrest angle	2·5°
Backrest and rake angle	2·1°

The absence of a simple relationship between subject's stature and the preferred linear dimensions was confirmed by the relationship between popliteal height and the preferred unloaded seat front height (Figure 1).   The absence of correlation between the preferred dimensions and the corresponding anthropometric measurements for seat height and seat depth suggests that the significant negative correlation between preferred seat height and preferred seat depth was due to personal choice factors other than the simple anthropometric ones.   The negative correlation between preferred seat height and preferred seat depth suggests that the subjects chose a fairly constant degree of ‘looseness of fit’.   The average value of the sum of the preferred seat height and preferred seat depth was 34·6 in. with a standard deviation of only 1·4 in. The limits to which the seat height and depth can compensate for each other is not known, so the 90 per cent range 31·5–37·7 in. can serve only as a general guide to the suitable sum value of the seat depth and the seat height for comfort.

4.2. *Tolerance Limits*

The inner tolerance limits ± 1 were the levels above and below the preferred setting at which the subject began to have doubts about the suitability of the setting. The outer limits ± 2 were the levels at which the subject had no doubt whatever about his dissatisfaction with the setting.

During everyday sitting, sitters adopt postural strategies to reduce discomfort as much as possible. These strategies if over-employed are themselves a source of ill-ease. The inner tolerance limits represent the limit at which the sitter can continue to sit without becoming conscious of either pure discomfort or the ill-ease resulting from the effort to avoid discomfort. For maximum comfort to be approached the dimensions of the chair should fall within the inner tolerance limits of the sitter.

The mean dimensional values for chairs made in one size only were arrived at as follows. The 90 per cent range about the mean value for each dimension was divided in the ratio of the upper and lower tolerance limits. The recommended single value thus departs from the mean value in the direction which sitters on the whole find most easy to tolerate.

In the case of the seat height the 5th percentile sitter has been considered since it is probable that where this dimension is much too large, then physical pressure under the sitter's thigh will become an important factor. The value of 15 in. was arrived at by adding the outer tolerance limit of 1·4 in. and 1 in. for shoes, to the 5th percentile popliteal height for the British adult population, which is 14·3 in., and subtracting 1·5 in. which was the average space between leg and seat preferred by the subjects in this experiment.

Ideally both the men's and women's chairs should be available in the two seat heights so that the differences in preference among the general population would be catered for. Where chairs are made in two models only the angles should differ, but the seat height should be the same for men as for women. The optimum value is 15 in. The other dimensions given for two chairs are for men and women and are marked m and w in Table 8.

These values apply to easy chairs which are similar in their general shape and upholstery to the chair used in this study.

Table 8. Dimensional values recommended for easy chairs

	Height of seat front (un-loaded)	Seat Depth	Seat Tilt Angle	Seat to Backrest Angle	Backrest Rake Angle	Armrest height above seat	Headrest centre in front of backrest plane	Headrest angle forward from backrest plane
	(in.)	(in.)	(°)	(°)	(°)	(in.)	(in.)	(°)
Single chair	15·0	18·5	10·5	110·0	120·5	6·5	2·5	8
Two chairs	15·0	18·5m	9·0m	113·0m	121·5m	6·5m	2·5m	8m
	16·5	18·5w	12·0w	105·0w	117·0w	6·5w	2·5w	8w

5. Conclusions

For easy chair sitting there were shown to be no significant differences in preferred dimensions for reading as opposed to T.V. viewing, nor any significant changes in preferences with duration of sitting.   No differences in preference were shown between the taller and the shorter, or between the older and the younger sitters.   Significant differences in the preferred seat and backrest angles were shown to exist between men and women.   The results suggest that two postural strategies exist.   In one the legs are outstretched for much of the time and the angle of recline of the back is comparatively large.   This was the posture mainly adopted by the men.   In the other posture the lower legs are kept closer to the vertical and the trunk more erect.   This posture was mainly adopted by the women.

Since the differences in preferred angles between men and women was greater than the estimated range of comfort tolerance of the average sitter, it follows that the adult population can only be adequately catered for by two models of chair.   The optimum values for a single chair and for two chairs are given in Table 8.

Eight easy chair dimensions were examined to determine for each the value which subjects preferred for comfort in two situations, reading and watching television.   The dimensions were the height of the seat front from the floor, the seat depth, the angle of tilt of the seat, the seat to backrest angle, the rake of the backrest, the height of the armrests above the seat, and the position of a flat padded rectangular headrest.   In the case of the first five of the above named dimensions, a pair of additional values were determined representing the borderlines of comfort, or tolerance limits, either side of the preferred values.   The experiments were carried out using a power driven adjustable chair controlled by the subject.   Twenty subjects each sat in the chair for periods of up to three hours, and at intervals adjusted each dimension to the level which by his judgment matched a written criterion of comfort supplied by the experimenter.

The experimental design allowed trends in preference to be tested across subjects' sex, age and stature.   Tests of correlation were carried out between the subjects' anthropometric measurements and their preference settings, and between the preference settings in different dimensions. The other results consist of distributions of the preferred settings and tolerance ranges for each dimension.   These results collectively suggest that easy chairs for use by the general population should be available in at least two models, differing in shape as well as in size.

Huit dimensions d'une chaise ont été étudiées afin de déterminer lesquelles paraissent préférables dans deux situations, la lecture et la vue de la télévision.   Ces dimensions étaient: la hauteur du siège à partir du plancher, la profondeur de l'assise, l'angle d'inclinaison de l'assise l'angle entre le siège et le dossier, l'inclinaison du dossier, la hauteur des accoudoirs à partir su siège, et la position d'un repose-tête rectangulaire rembourré.   Pour les cinq premières de ces dimensions, deux autres valeurs ont été déterminées qui représentent les valeurs extrêmes du confort ou les limites de tolérance de part et d'autre des valeurs moyennes.   Les essais ont été réalisés à l'aide d'un siège ajustable contrôlé par le sujet.   Vingt sujets ont testé la chaise durant des périodes de une à trois heures; ils ajustaient de temps en temps chacune des dimensions d'après des critères de confort fournis par l'expérimentateur.

Le mode expérimental permet de voir des tendances de préférence selon le sexe, l'âge et la stature des sujets.   Des tests de corrélation entre les dimensions anthropométriques des sujets et les positions adoptées, de même qu'entre les différentes dimensions choisies, ont été effectués. Les autres résultats consistent dans la distribution des positions recherchées et des limites de tolérance pour chaque dimension étudiée.

Ces résultats suggèrent que les sièges à utiliser par l'ensemble de la population soient de deux modèles au moins, différents de forme et de dimensions.

Acht Sessel-Dimensionen wurden untersucht, um für jeden den Wert zu bestimmen, den Versuchspersonen als komfortabel in zwei Situationen bevorzugten, beim Lesen und beim Fernsehen.   Die Dimensionen waren die Höhe der Sitzfront vom Fussboden, die Sitztiefe, der Sitzwinkel, der Winkel zwischen Sitz und Rückenlehne, die Form der Rückenlehne, die Höhe der Armlehnen über dem Sitz und die Stellung einer flach gepolsterten rechtwinklichen Kopflehne.

Im Fall der ersten fünf genannten Dimensionen wurde ein Paar zusätzlicher Werte bestimmt, welche die Grenzlinien oder Toleranzgrenzen beiderseits des bevorzugten Komfortwertes darstellten. Die Experimente wurden an Sesseln ausgeführt, die von der Versuchsperson kraftgesteuert verstellt werden konnten. Zwanzig Versuchspersonen saßen bis zu drei Stunden in dem Stuhl und paßten in Zwischenzeiten jede Dimension der Höhe an, die nach eigenem Urteil einem vom Versuchsleiter vorgeschriebenem Kriterium entsprach.

Dieses experimentelle Verfahren erlaubte es, bevorzugte Trends über Alter, Geschlecht und Grösse der Versuchspersonen zu testen. Korrelationstests wurden durchgeführt zwischen den anthropometrischen Daten der Versuchspersonen und ihren bevorzugten Sitzeinstellungen, und zwischen den bevorzugten Einstellungen in verschiedenen Dimensionen. Die anderen Resultate geben die Verteilung der bevorzugten Sitzeinstellungen und die Töleranzgrenzen für jede Dimension. Diese Resultate zeigen insgesamt, dass Sessel für den Allgemeingebrauch weningstens in zwei Modellen zur Verfügung stehen sollten, die sich sowohl in der Form wie in der Grösse unterscheiden.

Untersuchungen über das Sitzverhalten von Büroangestellten und über die Auswirkungen verschiedenartiger Sitzprofile

Von U. Burandt und E. Grandjean

Institut für Hygiene und Arbeitsphysiologie, Zurich, Schweiz

1. Fragestellungen

1. Wie und aus welchen Gründen weicht das Sitzverhalten von Büroangestellten auch beim Sitzen auf Stühlen, die als " gut " empfohlen werden, von den medizinischen Erwartungen ab?

2. Kann das Sitzverhalten durch eine zweckgerichtete Gestaltung der Sitzflächen positiv beeinflußt werden?

2. Untersuchungen über das Sitzverhalten

2.1. *Methode*

In zwei Banken und einem Industrieunternehmen der Schweiz wurden insgesamt 380 Arbeitsplätze untersucht.

Es wurden systematisch notiert:

1. die vorgefundenen Abmessungen und konstruktiven Besonderheiten eines jeden Tisches, Stuhles und jeder gegebenenfalls vorhandenen Fußstütze;

2. die wichtigsten Körpermaße des zum Arbeitsplatz gehörenden Angestellten;

3. Beanstandungen des Angestellten am Arbeitsplatz;

4. Beschwerden, die der Angestellte auf die Arbeitsbedingungen bezog.

Darüber hinaus wurden 246 dieser Angestellten je 20-mal im Verfahren einer modifizierten Multimomentaufnahme beobachtet, wobei Ereignisse über die augenblickliche Tätigkeit und augenblickliche Körperhaltung in Strichlisten notiert wurden.

Die Ergebnisse wurden ausgezählt, korreliert und mit dem *Chi²*-Test auf Unterschiede geprüft.

2.2. *Ergebnisse*

2.2.1. *Die Höhendifferenz*

Aus der Analyse der Häufigkeitsverteilungen der gemessenen Werte erhielten wir qualitative Hinweise für die Abmessungen von Arbeitsplätzen und Baudetails an Tischen und Stühlen. Auf alles kann hier nicht eingegangen werden. Als wichtigstes Maß erwies sich die Höhendifferenz zwischen der Sitzfläche und der Tischfläche. Sie wird durch die Einstellung der Sitzhöhe optimiert und ist ausschlaggebend für die Bequemlichkeit im Rumpf.

Die Verteilungen der frei gewählten, also von den Angestellten selber eingestellten Sitzhöhen sind getrennt nach den Tischhöhen, an denen sie gemessen wurden, in Bild 1 dargestellt.

Die schiefe Verteilung der Sitzhöhen bei den 72/68 cm hohen Tischen wurde durch die Stühle verursacht, die sich nicht tiefer als 42 cm einstellen ließen, die schiefe Verteilung bei den 78/63 cm Tischen (rechts) geht zu Lasten der Schublade unter der Tischplatte, die eine geringere Differenzhöhe nicht zuläßt, weil ein Raum von mindestens 13 cm für die Oberschenkel erforderlich ist. Nur die Verteilung in der Mitte am Tisch von 78 cm Höhe mit 66 cm Beinraumhöhe ist symmetrisch. Wir schließen aus diesen Ergebnissen, daß das häufigste Differenzmaß bei Büroarbeiten von 27 bis 30 cm um einen Durchschnitt von etwa 28,5 cm liegt.

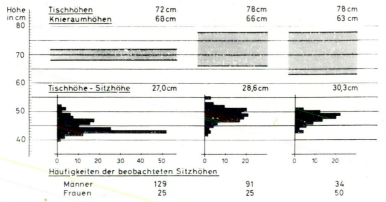

Bild 1. Häufigkeitsverteilungen von Sitzhöhen und durchschnittliche Differenzen Tischhöhe - Sitzhöhe an drei Bürotischausführungen.

Beim Lesen und ähnlichen Arbeiten, bei denen man die Arme aufstützt, wird die Differenz größer, bei Arbeiten mit frei bewegten Armen, wie z.B. beim Maschinenschreiben, wird sie kleiner gewählt. Diese Häufigkeitsverteilungen deuten darauf hin, daß die Sitzhöhe in Abhängigkeit zur Tätigkeitsart und im Interesse größter Bequemlichkeit im Rumpf eingestellt wird.

Welche Bedeutung aber hat die Unterschenkellänge, die so gerne als für die Sitzhöhe maßgebend genannt wird?

Die statistische Berechnung der Korrelation von Körperlänge und Sitzhöhe zeigte keine Abhängigkeit dieser beiden Variablen. Und überraschend zeigte die Regression eine Tendenz, nach der kleine Personen auf höher eingestellten Stühlen sitzen als große.

Diese Tendenz bestätigt unsere Feststellung: Der Büroangestellte sorgt primär dafür, daß sein Rumpf optimal und der Arbeit bequem zugeordnet ist. Die Situation der Beine ist ihm weniger wichtig. Kleine Personen sitzen darum häufiger auf der Stuhlvorderkante als große Personen.

Diese Erkenntnis unterstreicht die Forderung, daß die Verfügbarkeit des Raumes zwischen der Tischplatte und der Sitzfläche nicht eingeschränkt werden darf. Zargen und Schubkästen unter der Tischplatte über den Knien, oder zu hohe Aufbauten über der Tischplatte dürfen nicht geduldet werden.

2.2.2. *Das tätigkeitsbedingte Sitzverhalten*

Von den Ergebnissen der Multimomentaufnahmen kann auch nur selektiv berichtet werden. Da später auf Sitzflächen eingegangen wird, soll nur das gezeigt werden, was unsere Sitzflächenuntersuchungen motivierte.

Die Protokolle der Multimomentaufnahme wurden folgendermaßen ausgewertet: Für jede beobachtete Person wurden 20 Randlochkarten angelegt, in die die unveränderlichen Ereignisse zur Person (Geschlecht, Größe, Beschwerden usw.) und zum Arbeitsplatz (Tisch- und Stuhlkennzeichen) übereinstimmend eingeschnitten wurden. Die veränderlichen Ereignisse jedoch, die die Tätigkeit und Körperhaltung betrafen, wechselten entsprechend der bei jeder Einzelaufnahme jeweils anders vorgefundenen Verhaltenssituation.

Für die statistische Prüfung der Ereignisverteilungen mit dem *Chi*²-Test wurden immer zwei Kartensortimente miteinander verglichen. Die Karten des einen Sortiments trugen das Referenzereignis, die des anderen trugen es nicht. Geprüft wurde die 0-Hypothese, daß die Ereignisse in beiden Sortimenten relativ gleich häufig auftreten.

Es zeigte sich jedoch, daß in allen Fällen einige Ereignisse signifikant unterschiedlich häufig waren.

Als konkretes Beispiel sei das Beziehungsnetz des Referenzereignisses " Maschinenschreiben " in Bild 2 gezeigt.

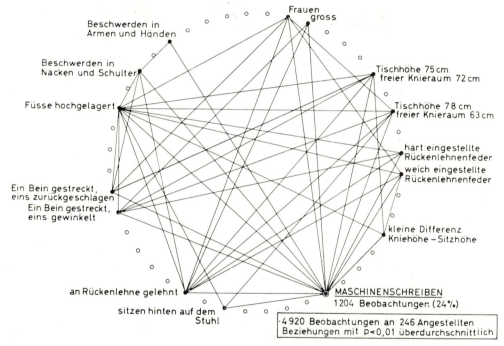

Bild 2. Beziehungen des Referezereignisses " Maschinenschreiben ". Die Linien verbinden Ereignisse, deren Beziehungen durch große Häufigkeit mit einer Wahrscheinlichkeit von $p < 0,01$ gesichert sind.

Die Darstellung ist z.B. folgendermaßen zu lesen: An Schreibmaschinen arbeiteten signifikant häufig Frauen. Die maschinenschreibenden Personen klagten signifikant häufiger über Beschwerden in Nacken und Schultern sowie Armen und Händen als die mit anderen Arbeiten beschäftigten Personen.

Die stärksten Beziehungen zum Maschinenschreiben haben die Ereignisse ' Füße hochgelagert ' und ' an Rückenlehne gelehnt '.

Noch aussagekräftiger wird das Bild durch die Ereignisse, die signifikant selten beim Maschinenschreiben auftreten.

Die wichtige Erkenntnis war die, daß die von uns unterschiedenen Tätigkeiten Maschinenschreiben, Lesen, Akten bearbeiten, Reden und Telefonieren charakteristische Verhaltensbilder zeigten. Eine Ausnahme machte das Maschinenrechnen.

Es lag nahe, dem 'Warum' dieser Verhaltensbilder nachzugehen. Wir konnten im Verlauf der Nachforschungen einerseits Mängel an den Arbeitsplätzen aufdecken, andererseits mußten wir unabänderliche Bedingungen der Tätigkeitsanforderungen akzeptieren und uns Gedanken darüber machen, wie man jede tätigkeitsbedingte Körperhaltung besser als mit den konventionellen Allgebrauchsstühlen unterstützen könnte.

Wir fragten nach den Funktionen der Rückenlehne und der Sitzfläche.

2.2.3. *Die Bedeutung der Rückenlehne*

Die Auszählung der Ereignisse 'an Rückenlehne gelehnt' zeigte folgende Ergebnisse: In der gesamten Aufnahme trat das Ereignis zu 42% auf. Es stand signifikant häufig in Beziehung zu den Ereignissen

Frauen
große und mittelgroße Personen
78/63 cm und 75/72 cm hohe Schreibtische
Stühle mit einstellbaren Rückenlehnen
relativ geringe oder durchschnittliche Sitzhöhe
Maschinenschreiben
auf der Mitte des Stuhles sitzend
beide Knie um ca. 90° gewinkelt
Beine gestreckt und gewinkelt
Füße überkreuzt, Füße hochgelagert
Beschwerden im Rücken, im Nacken und Schultern

Es überraschte, daß die Techniker und Kaufleute des Industriebetriebes, die an den 72 cm hohen Tischen und auf einstellbaren Drehstühlen saßen, die Rückenlehnen nicht signifikant häufig benutzten.

Sie taten es nicht wegen der Besonderheiten ihrer Arbeitsanforderungen.

Bemerkenswert ist ferner, daß die Rückenlehne zum Maschinenschreiben wichtig ist, und daß man sie dann benutzt, wenn man auf der Mitte der Sitzfläche sitzt.

Am häufigsten, zu 83% benutzten maschinenschreibende Angestellte mit einstellbaren Bürodrehstühlen die Rückenlehnen. Beim Akten bearbeiten, Reden und Telefonieren wurde sie signifikant selten benutzt.

Entsprechen die Stühle den Anforderungen?

Der heute handelsübliche und medizinisch empfohlene Bürodrehstuhl wurde mit folgender Logik entwickelt:

1. Man muß sich bei der Arbeit anlehnen können;
2. Wenn man sich anlehnt, neigt das Gesäß dazu, zur Vorderkante der Sitzfläche zu rutschen;
3. Um das Vorrutschen des Gesäßes zu verhindern, wird die Sitzfläche um 3 bis 5 Grad nach hinten geneigt.

Wie die Ergebnisse der Multimomentaufnahme zeigen, gilt diese Logik streng genommen nur für Schreibmaschinenstühle.

Warum benutzt man zum Bearbeiten von Akten, zum Lesen und Dis-
kutieren den gleichen Stuhl? 45% der Tätigkeitsereignisse entfallen auf
' Akten bearbeiten '; kann man auf den Stühlen auch ohne Rückenlehne gut
sitzen?

2.3. *Zusammenfassung dieser Erkenntnisse*

Die Sitzhaltung wird primär durch die Anforderungen der Tätigkeit bestimmt.
Die heute handelsüblichen Bürostühle sind nach einer Logik gestaltet, die
das zurückgelehnte Sitzen voraussetzen. Tatsächlich üblich ist diese Sitzhaltung nur beim Maschinenschreiben.

Es bleibt fraglich, ob die handelsüblichen Stühle auch für die anderen
Tätigkeiten bestmögliche Sitzbedingungen bieten.

Diese Frage kann nur an der Sitzfläche untersucht werden. Die Rückenlehne wird bei diesen Arbeiten zu selten benutzt.

3.   Untersuchungen mit verschiedenen Sitzflächen

3.1. *Vorbemerkungen*

Als Alternativen zu den heute allgemein gebräuchlichen Sitzflächen finden
sich in der Literatur zwei bemerkenswerte Empfehlungen: Staffel (1884) und
Schlegel (1956) befürworten eine nach vorne hin abfallende Sitzfläche für nach
vorn orientierte Arbeiten; Schneider (1961) schlägt einen unter das Steißbein
geschobenen Sitzkeil vor, der unter der Bezeichnung ' actilord ' (aktive
Lordosierung) propagiert wird.

Die Unterschiede in den Wirkungen dieser drei Sitzflächentypen ist am
Bewegungsverhalten des Beckens erklärbar.

Sobald das Becken beim Hinsetzen auf die Tubera ischiadica gelagert und
durch den Oberkörper belastet wird, kippt es unter der Last und bei knickender
Lendenwirbelsäule nach hinten ab, bis es von den supra- und interspinalen
Bändern und der Rückenmuskulatur aufgehalten wird.

Dieser Vorgang wird durch eine nach hinten abfallende Sitzfläche bestärkt,
weil das Becken wie eine Walze bergabrollen kann.

Die nach vorne geneigte Sitzfläche wirkt diesem Vorgang entgegen, weil das
Becken bergaufrollen muß.

Der Sitzkeil hält den Vorgang auf wie ein Bremsklotz das rollende Rad.

3.2. *Voraussetzungen*

Das Ziel unserer Untersuchungen verfolgten wir mit folgenden hypothetischen
Voraussetzungen.

1. Langdauerndes Sitzen ohne Rückenlehne mit extrem kyphotischer
   Lendenwirbelsäule soll vermieden werden, weil
   a) die Rückenmuskulatur und deren Sehnenansätze sowie die
      gespannten inter- und supraspinalen Bänder Spannungen ausge-
      setzt sind, die sie auf die Dauer nicht schadlos ertragen können;
   b) weil die Beweglichkeit der Wirbelgelenke in dieser extremen
      Haltung stark beeinträchtigt ist und nur ungenügend zum
      Wechsel der Haltung anreizt.
2. Langdauerndes Sitzen ohne Rückenlehne sollte in einer aufgerichteten
   Sitzhaltung geschehen mit möglichst häufigem Spannungswechsel in
   der Muskulatur und guten Bewegungsvoraussetzungen für die Wirbel-
   geienkkette.

Diese Voraussetzungen dürften durch die konventionellen, nach hinten geneigten Sitzflächen am wenigsten erfüllt sein; durch den Sitzkeil nur dann, wenn man sich fest gegen ihn setzt; durch die nach vorne geneigte Sitzfläche am ehesten.

3.3. *Methoden und Ergebnisse*

3.3.1. *Untersuchung*

Die Bequemlichkeit des Sitzkeils.

36 Männer und 16 Frauen verglichen die Bequemlichkeit des ' actilord ' mit einer gewöhnlichen Sitzfläche, während sie 15 min Lochkarten zählten. Die Rückenlehnen durften nicht benutzt werden.

25 Vpn bevorzugten den ' actilord '-Sitzkeil.

27 Vpn bevorzugten den handelsüblichen Sitz mit 3° nach hinten geneigter Sitzfläche.

3.3.2. *Untersuchung*

Beziehungen zwischen Sitzbrettneigung und Bequemlichkeit.

36 Männer und 16 Frauen nannten die ihnen bequemste Neigung der hinteren Hälfte einer geteilten Sitzfläche, auf der die Sitzhöcker aufsaßen Die vordere Hälfte blieb unverändert horizontal, eine Rückenlehne wurde nicht benutzt.

Keine Vp entschied sich für eine nach hinten geneigte Sitzfläche. Der Entscheidungsbereich reichte von 5° bis 22° Vorneigung, der Durchschnitt betrug 12,1°.

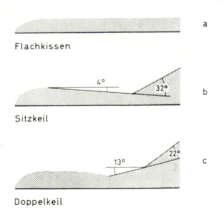

Bild 3. Die drei im Laboratorium untersuchten Sitzflächen: (a) flache Einstellung, (b) Sitzkeil, (c) Doppelkeil. In Querrichtung sind die Sitze unprofiliert flach.

3.3.3. *Untersuchung*

Der Einfluß der Sitzflächen auf die Sitzhaltung unter Laboratoriumsbedingungen.

6 Frauen und 6 Männer saßen je 40 min lang auf jedem der drei in Bild 3 gezeigten Sitzflächen. Sie mußten mit unaufgestützten Armen an einem hohen Schreibpult einen Zahlentest lösen. Über die Dornfortsätze waren

Markierungsplatten geklebt, die den Profilverlauf des Rückens auf Foto-
aufnahmen erkennen ließen, die in Abständen von 4 min gemacht wurden
(Bild 4).

Die Auswertung der Fotoprotokolle zeigte für diesen Versuch keine
signifikanten Unterschiede im Profilverlauf der Wirbelsäulen.

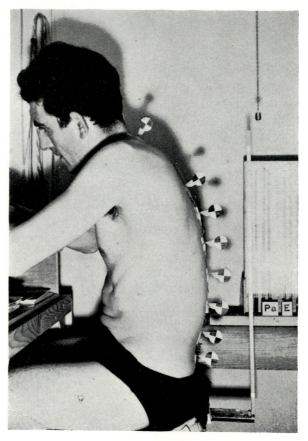

Bild 4. Versuchsanordnung zur Bestimmung des Rückenprofilverlaufs beim Sitzen auf drei
verschiedenen Sitzflächen.

An jede der eben beschriebenen Sitzungen angeschlossen war ein weiterer
Versuch. Die Vpn mußten sich mit maximaler Kyphose so hinsetzen, daß
die Gehörgänge genau senkrecht über den Sitzhöckern sich befanden.

Die Auswertung dieser Fotoprotokolle zeigte:

    1. der Doppelkeil verursachte eine signifikant stärkere Thoraxkyphose als
       die glatte Sitzfläche;

    2. der Doppelkeil verursachte eine signifikant stärkere Lendenlordose als
       die glatte Sitzfläche und der Sitzkeil;

    3. der Sitzkeil verursachte eine signifikant stärkere Lendenlordose als die
       glatte Sitzfläche.

Diese Resultate bestätigten unsere Vermutung, daß der Doppelkeil, bei
dem die Sitzhöcker auf einem nach vorne hin abfallenden Flächenabschnitt
aufsitzen, die stärkste aufrichtende Wirkung auf die Wirbelsäule ausübt.

Die Unterschiede zeigten sich besonders deutlich bei der Vp, deren Rücken-
konturen in Bild 5 wiedergegeben sind.

Neben den Unterschieden in den Rückenprofilen ist der kompensatorische Verlauf der Wirbelsäulenabschnitte bemerkenswert: bei flacher Lende stellt sich auch der Thorax flach, bei der lordotischen Lende ist auch die Thoraxkyphose verstärkt.

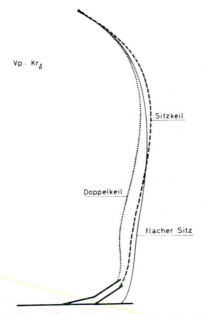

Bild 5. Die Wirkung der drei Sitzprofile auf den Profilverlauf des Rückens einer Versuchsperson.

3.3.4. *Untersuchung*

Die Wirkung profilierter Sitzflächen auf die Bequemlichkeit von Angestellten in Bürobetrieben.

Um die Wirkung der Sitzflächen auf die Bequemlichkeit bei der täglichen Büroarbeit zu prüfen, wurden je 20 Stück von den in Bild 3 gezeigten Sitzflächenprofile aus filzüberzogenem Gummihaar gefertigt.

Die Sitzkissen wurden an 60 Bankangestellte verteilt. Jede Vp sollte auf jedem der Kissen eine Woche lang sitzen. Nach jeder Woche war von jeder Vp für das gerade besessene Kissen ein Fragebogen mit Angaben über Tätigkeit, Beschwerden und Bequemlichkeit auszufüllen.

43 Vpn führten die Versuche bis zum Ende durch.

Einen wichtigen Teil der Ergebnisse zeigt Tabelle 1.

Tabelle 1. Häufigkeiten der Qualifikationen und Angaben über Beschwerden für drei Sitzkissen von 43 Büroangestellten

Qualifikation	Sitzkeil	Doppelkeil	Flachkissen
sehr gut / gut	20	14	24
befriedigend / ausreichend	16	12	15
schlecht / sehr schlecht	7	17	4
Körperteil mit Beschwerden			
Nacken	3	4	0
Rücken	13	13	5
Gesäß	5	13	1
Oberschenkel	13	16	6
Summe	34	46	12

Das Flachkissen wurde am besten beurteilt und verursachte am wenigsten Beschwerden. Es änderte wenig an der gewohnten Sitzweise. 20 der 43 Vpn bevorzugten es. Kaum schlechter wurde der Sitzkeil beurteilt. Er verursachte jedoch starke Beschwerden im Rücken und an den Oberschenkeln und wurde darum von nur 11 Vpn bevorzugt. Am schlechtesten war der Doppelkeil; er wurde aber dennoch überraschend von 12 Vpn bevorzugt. Unter diesen Vpn waren 7, deren Stühle unverstellbar feste Rückenlehnen hatten, während der Sitzkeil von 7 Vpn bevorzugt wurde, die sich die Rückenlehne einstellen konnten. Es scheint, daß der Doppelkeil die Rückenlehne eher entbehren läßt.

Die Wirkung des Doppelkeils unterschied sich noch in anderer Weise von der der anderen Kissen: Sie wurde *am entschiedensten* befürwortet oder abgelehnt!

Aus den Angaben der Vpn konnten wir entnehmen, daß der Doppelkeil die Körperhaltung tatsächlich am stärksten veränderte, so stark sogar, daß vielen Vpn latente Schäden im Rücken schmerzhaft spürbar und unerträglich wurden. Wir wissen nicht, ob die Beschwerden im Laufe der Zeit durch Adaptation an die neue Situation verschwunden wären.

3.4 *Besprechung der Ergebnisse der Sitzflächenuntersuchung*

Die Ergebnisse dieser Felduntersuchung dürfen meines Erachtens nicht als Beweis dafür angesehen werden, daß der Sitzkeil und die vorgeneigte Sitzfläche —die hier durch den Doppelkeil repräsentiert war—ungeeignet sind. Ich habe vielmehr den Eindruck, daß die Profilierungen zu stark waren und noch wesentlich verbessert werden können.

Der Sitzkeil ist unter der Bezeichnung ' actilord ' im Handel und wird trotz seiner Nachteile, die im Gebrauch bekannt wurden, von vielen Personen gerne benutzt.

Auch mit vorgeneigten Sitzflächen wurden inzwischen Erfahrungen in der Praxis gemacht. Ein großes deutsches Werk der Radioindustrie hat mehrere Montageabteilungen mit solchen Stühlen ausgerüstet und kam mit ihren Versuchen auf eine optimale Neigung von 4 Grad. Die Sitzflächen sind die gleichen, die auch an den sonst üblichen Stühlen Verwendung finden. Die Einführung der ungewohnten Stühle stieß anfänglich auf steife Ablehnung der Arbeiter und mußte vom Betriebsarzt energisch durchgesetzt werden, doch nach etwa 2 Monaten legte sich der Widerstand. Die Arbeiter äußerten sich zufrieden. Zweifelsfrei nachteilig zeigte sich dieses Sitzprinzip bisher nur für schwangere Frauen.

Doch aus einem Grunde sollte man vielleicht noch zurückhaltend mit der Einführung solcher Stühle sein.

Die Aufrichtung des Rumpfes durch die Sitzfläche scheint zwar zu einer Verbesserung der Haltung des Beckens, der Lende und des Thorax zu führen, sie verschlechtert aber sehr wahrscheinlich das Belastungsbild im cervikalen Bereich. Um die Arbeiten der Hände bei locker gehaltenen Armen im natürlichen Bereich vor der Brust beobachten zu können, muß beim Sitzen auf diesen Stühlen der Kopf stärker geneigt werden. Man muß mit Krankheitsbildern rechnen, die von vergleichbaren stehenden Beschäftigungen bekannt sind.

Dieser Fragestellung und der Feingestaltung der Sitzflächen müssen weitere Untersuchungen nachgehen.

Es werden Untersuchungen referiert, die in den Jahren 1961 bis 1963 im Institut für Hygiene und Arbeitsphysiologie in Zürich unter der Leitung von Herrn Prof. Dr. med. E. Grandjean durchgeführt wurden.

Entgegen der Gewohnheit, Stühle mit orthopädischen Überlegungen und Untersuchungen in Laboratorien zu entwickeln, wurde in diesen Untersuchungen zunächst das Sitzverhalten von Angestellten bei ihren alltäglichen Büroarbeiten festgestellt. Es zeigte sich eine starke Abhängigkeit zwischen Sitzhaltung und Tätigkeitsanforderungen. Aus den Erkenntnissen wurden Thesen für die Sitzflächengestaltung abgeleitet und in der Praxis geprüft. Die Ergebnisse der Untersuchungen bestätigten die Annahme, daß sich die arbeitsbedingten Körperhaltungen durch speziell angepaßte Sitzflächen wirksam unterstützen lassen.

This paper discusses research work done in the Institut für Hygiene und Arbeitsphysiologie in Zurich between 1961 and 1963 under the leadership of Prof. Grandjean.

In contrast to the usual theoretical and experimental approach in orthopaedy—the problem of seating, this research looked into the sitting habits of employees during their actual everyday work in the office. It was shown that there is a definite connection between sitting postures and the demands of the particular work being done. Ideas for the design of seat surfaces were formulated from these observations and put into practice. The results of this research work confirmed the view that the positions of an individual's body during his work can be effectively supported by specially adapted seat surfaces.

Cet article passe en revue les travaux accomplis à l'Institut d'Hygiène et de Physiologie du Travail de Zürich entre 1961 et 1963, sous la direction du Professeur Grandjean.

A l'opposé de l'approche classique, aussi bien théorique qu'expérimentale, de l'orthopédie de la position assise, cette recherche est consacrée à l'étude des postures assises d'employés de bureau, au cours de leur journée de travail. Il apparaît qu'il existe une relation bien définie entre ces postures et les exigences particulières du travail à accomplir. A partir de ces observations, on a pu faire un certain nombre de recommandations pour la conception et la réalisation d'un nouveau type de siège. Les résultats atteints dans cette étude démontrent qu'il est effectivement possible d'influer sur la posture du corps au cours du travail en adaptant spécialement la surface du siège.

Literatur

BURANDT, U., und GRANDJEAN, E., 1964, Die Wirkungen verschiedenartig profilierter Sitzflächen von Bürostühlen auf die Sitzhaltung. *Int. Z. angew. Physiol. einschl. Arbeitsphysiol.*, **20**, 441–452.

GRANDJEAN, E., und BURANDT, U., 1962, Das Sitzverhalten von Büroangestellten. *Industr. Organisation*, **31**, 242–250.

SCHLEGEL, K. F., 1956, Sitzschäden und deren Vermeidung durch eine neuartige Sitzkonstruktion. *Med. Klinik.*, **51**, 1940–1942.

SCHNEIDER, H., und LIPPERT, H., 1961, Das Sitzproblem in funktionell-anatomischer Sicht. *Med. Klnik.*, **56**, 1164–1168.

STAFFEL, F., 1884, Zur Hygiene des Sitzens. *Zbl. allg. Gesundh.-Pfl.*, **3**, 403–421.

Der sitzende Angestellte und der stehende Kunde

Von J. R. de Jong und C. Roggeveen

Raadgevend Bureau Ir. B.W. Berenschot, Amsterdam

Beim Auskunft geben und Informationen empfangen, bei der Arbeit an Schaltern, Ladenkassen u.ä. können die betreffenden Angestellten ihre Tätigkeiten oft am besten *im Sitzen* ausüben. Die Kunden dagegen können wegen der kurzen Dauer der von Ihnen zu leistenden Dienste dabei häufig am besten *stehen*.

In solchen Fällen scheint die Kombination der sitzenden Bedienungsweise und des Stehens des Kunden also die naheliegende Lösung zu sein. Trotzdem wird die stehende Bedienungsweise noch manchmal vorgezogen. Gründe können sein:

- . Man hat keine Lösung gefunden, die befriedigendes Sitzen ermöglicht— zum Beispiel weil sich bei sitzender Maschinenbedienung eine schlechte Körperhaltung nicht vermeiden lässt, da ungenügend Platz für die Knie vorgesehen ist. Dies kommt u.a. vor bei Registrierkassen und Fahrkartenmaschinen.

- . Es wird angenommen, dass sich bei der Kombination Sitzen-Stehen einen für unannehmbar geachteten Unterschied der Augenhöhen der beiden Beteiligten nicht umgehen lässt.

- . Man ist der konventionellen Auffassung, dass es unhöflich sei, stehenden Kunden im Sitzen Rede und Antwort zu stehen.

Auf das erstgenannte Bedenken soll hier nicht eingegangen werden. Es wird nur bemerkt, dass manchmal durch eine unherkömmliche Aufstellung der Maschine (z.B. schräg vor dem Mitarbeiter) eine gute Lösung ermöglicht werden kann.

Die Augenhöhen und die Bodenhöhen

Unerwünschte Unterschiede zwischen den Augen-, Schulter- und Ellbogenhöhen der beiden Beteiligten (man denke z.B. an das Überreichen von Dokumenten) lassen sich dadurch vermeiden, dass der Stuhl des Angestellten auf einem erhöhten Boden steht.

Aus der Figur 1 geht hervor, wie gross diese Erhöhung sein soll, wenn möglichst gleiche Augenhöhen des Angestellten und der Kunden bezweckt werden.

Augenhöhendifferenzen lassen sich durch die interindividuellen Streuung der menschlichen Körpermasse natürlich nie vollkommen umgehen.

Die Differenzen können dadurch noch vergrössert werden, dass für Apparate, Geld, Formblätter u.ä. Raum zwischen dem Angestellten und dem Kunden gebraucht wird. Es kann hinzukommen, dass solche Hilfsmittel für die Kunden unerreichbar oder sogar unsichtbar sein sollen.

Welche Höhenunterschiede sind nun noch annehmbar? Wie erleben die Beteiligten die Kombination eines sitzenden und eines stehenden Gesprächspartners?

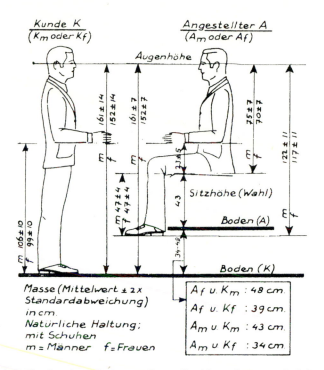

Abb. 1.   Die Bodenhöhe für sitzende Angestellten, die sich ergibt, wenn bei den angenommenen Massen die Augenhöhe der stehenden Kunden möglichst mit der der Angestellten übereinstimmen soll.

Eine Bewertungsstudie

Anlässlich der eben gestellten Fragen soll kurz berichtet werden über Daten, die gesammelt werden konnten hinsichtlich der durch die Figur 2 dargestellten Situation.

Die betreffenden Arbeitsplätze hatten vor 15 Monaten Arbeitsstellen ersetzt, in denen die Angestellten zwar über einen Stuhl verfügten, bei der Arbeit— d.h. bei der Abfertigung von abreisenden Fluggästen—jedoch standen.

Der Stuhl stand hier auf einem Boden, der nur 9 cm höher war als der Boden auf dem die Fluggäste standen.   Das hing zusammen mit der erwünschten Höhe des Transportbandes, auf dem die Passagiere ihre Koffer zu stellen hatten.   Die Angestellte konnte dieses Band mittels eines Pedals in Bewegung setzen oder stoppen.   Sie hatte u.a. die Aufgabe an jeden Koffer den Bestimmungsort anzubringen.   Die Sitzhöhe der Stühle war einstellbar.

Daten wurden gesammelt mittels:

1. Beobachtung.
   Es wurden u.a. 43 Sitzhöhen ermittelt.

2. Befragung.
   Es wurden 65 Passagiere befragt.   Die Interviews dauerten zwischen 1 und 2 Minuten.

3. Fragebogen.
   22 Fragen wurden von 65 Angestellten schriftlich beantwortet.   Zwei Angestellte zogen aus unbekannten Gründen vor, den Fragebogen nicht auszufüllen.

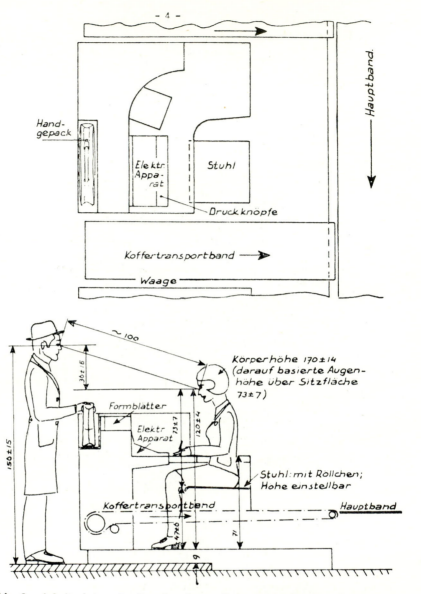

Abb. 2.   Arbeitsplatz mit sitzender Angestellten und stehendem Passagier.   Masse
(gegebenenfalls Mittelwert ± 2 × Standardabweichung) in cm.

Mit Ausnahme von den letzten zwei Fragen (nach dem Monat des
Dienstantritts und der Körperhöhe) sollte jede Frage beantwortet
werden durch das Anbringen eines Kreuzchens auf einer Skala.   Es
folgen drei Beispiele solcher Skalen:

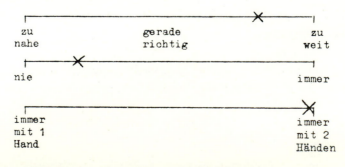

Die Sitzhöhe

Die durchschittliche Sitzhöhe betrug 47 cm. Wie aus der Figur 3 hervorgeht, sass man im allgemeinen höher, je nachdem man kleiner war ($r = -0{,}51$, die Wahrscheinlichkeit p, dass der Korrelationskoeffizient nicht wesentlich von Null abweicht, ist $< 10\%$).

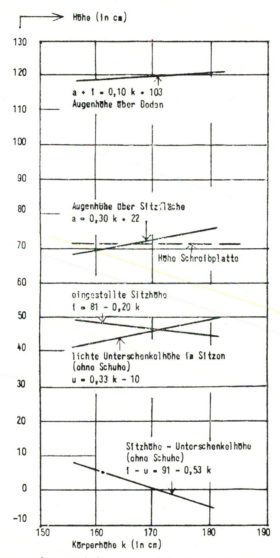

Abb. 3. Der Zusammenhang zwischen Körperhöhe k und eingestellter Sitzhöhe, Augenhöhe (im Sitzen) und Unterschenkelhöhe (im Sitzen). Die Formel für die Sitzhöhe basiert auf unmittelbare Messungen. Die Formeln für die Augenhöhe a und die lichte Unterschenkelhöhe u beruhen auf dem Schrifttum entnommenen Zusammenhängen mit der Körperhöhe k.

Die in der Figur 3 vermeldeten Zusammenhänge zwischen Körperhöhe und Augenhöhe über *Sitzfläche* und lichte Unterschenkelhöhe basieren auf Daten aus dem Schrifttum (1). Der Zusammenhang zwischen Körperhöhe und Augenhöhe über dem *Boden* beruht auf den beiden bereits erwähnten Beziehungen.

Die Einstellbarkeit der Höhe der Stuhlsitze wird hier offenbar dazu benutzt, die interindividuellen Körpermassunterschiede grossenteils auszugleichen.

Die Meinungen der Passagiere

Passagiere wurden befragt, nachdem alle Förmlichkeiten (einschliesslich der Passkontrolle) erledigt waren.

Auf die erste Frage, nach dem Verlauf der Abhandlung des Gepäcks und des Flugscheins, wurde immer positiv geantwortet.

Auch auf die zweite Frage, die die sitzende Arbeitsweise der Angestellten betraf, wurde mit einer Ausnahme immer sofort günstig reagiert: " stört gar nicht ", " normal ", " nicht unhöflich ", " möchte selber auch nicht stehen ", " sehr vernünftig ", " ist mir gar nicht aufgefallen ", " rationell ", usw. Mehrere Fluggäste fanden es offenbar fast sonderbar, dass die sitzende Abhandlungsweise zur Sprache gebracht wurde. Ein älterer Passagier war der Meinung, dass Abhandlung im Stehen vielleicht höflicher sei; er habe jedoch sicher keine Bedenken gegen die sitzende Methode.

Ein Einfluss des Alters, der Körperhöhe oder des Geschlechts der Passagiere auf ihre Meinungen konnte nicht festgestellt werden.

Die Meinungen der Angestellten

Es wurde bereits erwähnt, dass die Angestellten mittels eines Fragebogens gebeten wurden 22 Fragen zu beantworten.

Die Fragebogen boten die Gelegenheit nach der Beantwortung einer Gruppe zusammenhängender Fragen (d.h. nach der 4., 14., 17., 20. und 22. Frage) Bemerkungen zu machen, Wünsche zu äussern und Vorschläge zu formulieren. Diese Möglichkeit wurde von 82% der Teilnehmer benutzt; die mittlere Zahl ihrer Bemerkungen belief sich auf etwa vier. Hier sollen nur die Meinungen, Wünsche u.ä. besprochen werden, welche die Höhe der Schreibplatte, den Kontakt der Angestellten mit den Passagieren, das Sitzen und die Erreichbarkeit der benötigten Formblätter betrafen.

Wir werden je Frage die Häufigkeiten der Meinungen (Scores) vermelden; die verwendeten Skalen wurden dazu in 11 Klassen eingeteilt. Weiter wird jeweils der mittlere Score angegeben.

Frage 1: Wie finden Sie die Höhe Ihrer Schreibplatte?
Die Häufigkeiten der Antworten je Klasse und der mittlere Score waren:

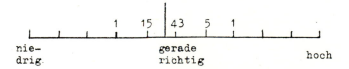

Im allgemeinen ist man zufrieden. Es besteht eine schwache Korrelation zwischen der Körperhöhe und dem Urteil über die Höhe der Schreibplatte: kleine Personen haben Neigung die Platte hoch, grosse Personen dagegen sie niedrig zu finden ($r = -0,19$; $p = 0,1$).

Frage 3: Wie sitzen Sie hinsichtlich der Passagiere, was das Annehmen und Herreichen der Flugscheine betrifft?

Häufigkeiten der Antworten:

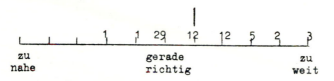

Man findet den Abstand zum Passagier offenbar öfters zu gross, aber meistens nicht viel zu gross.

Kleine Personen meinen häufiger, dass der Abstand für sie gross sei, als grosse Personen (Korrelation Körperhöhe—Score: $r = -0,44$; $p < 0,001$).

Frage 4: Wie sitzen Sie hinsichtlich der Passagiere, was den weiteren Kontakt anbelangt (Fragen stellen, Aufschluss geben)?

Häufigkeiten der Antworten:

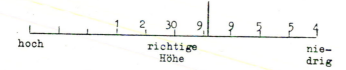

Hier wird, wie zu erwarten war, ähnlich reagiert, wie auf die vorige Frage. Korrelation Körperhöhe—Score: $r = -0,23$; $P < 0,1$.

Frage 10: Wie ist die Erreichbarkeit der Formblätter?

Man ist fast einstimmig der Meinung, dass man die benötigten Formblätter gut erreichen kann.

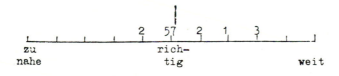

Frage 13: Ändern Sie, bevor Sie anfangen zu arbeiten, die Höhe Ihres Stuhls?

Die Sitzhöhe wird meistens geändert:

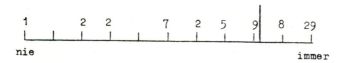

Diejenigen, die die Sitzhöhe ihres Stuhls nie oder fast nie einstellen, bemerken, dass das Einstellen beschwerlich, bezw. manchmal unmöglich sei (der Mechanismus sei oft defekt).

R

Diskussion

Aus dem vorstehenden geht hervor, dass in der untersuchten Situation die Kunden keineswegs Bedenken gegen die sitzende Bedienungsweise hatten. Sie bezeugten alle ihre Zufriedenheit über den Verlauf der Dinge.

Die Angestellten waren zufrieden über die Höhe ihrer Schreibplatte, stellen jedoch dabei die Sitzhöhe ihrer Stühle so ein, dass diese für die kleineren Personen in vielen Fällen deutlich unerwünscht hoch war (Figur 3).

Die gewählten Sitzhöhen ergaben Augenhöhen mit einer relativ geringen interindividuellen Streuung. Der Abstand zwischen den Augen der Angestellten und die der Passagiere betrug im Durchschnitt etwa 100 cm: senkrecht ~ 35 cm und waagerecht ~ 95 cm (Figuren 2 und 3). Die Mehrzahl der Angestellten zeigte sich mit diesem Abstand zufrieden. Vor allem kleinere Personen fanden dieser Abstand jedoch sowohl mit Rücksicht auf das Überreichen von Dokumenten als bei Gesprächen mehr oder weniger hinderlich. Lärm war hier offenbar ein Faktor (Fliessbänder; nahe Gespräche).

Abschliessend kann festgestellt werden:

1. dass die grosse Mehrzahl der Beteiligten bereit war die Sitzhöhe immer wieder einzustellen:

2. dass dabei, wie auch anderswo festgestellt worden ist, eher die erwünschte Augenhöhe (bezw. die Tischhöhe) als die angenehmste Sitzhöhe massgebend war;

3. dass im Hinblick auf die Kontakte der Angestellten mit den Fluggästen der Boden auf dem ihr Stuhl stand, besser höher hätte sein können. Die konkreten erwünschten und möglichen Verbesserungen sollen an dieser Stelle nicht erwogen werden, weil dabei Einflussgrössen, die oben ausser Betracht gelassen wurden (wie die erwünschte Höhe der Transportbänder, die Höhe der Koffer und ein Pedal), mit eine Rolle spielen müssten.

4. dass der Abstand zwischen dem Angestellten und dem Kunden nicht unnötig gross gewählt werden soll. Faktoren können dabei sein überzureichende Gegenstände und die Notwendigkeit ungestört Gespräche zu führen.

5. dass Beobachtungen, Interviews und Fragebogen—und vor allem die kombination dieser Methoden—bei der ergonomischen Bewertung von " sozio-technischen Systemen " nützliche Daten ergeben können.

Die Kombination der sitzenden Bedienungsweise und des stehenden Kunden hat oft deutliche Vorteile. Bedenken können zusammenhängen mit Schwierigkeiten beim Suchen einer Lösung, die eine gute Sitzhaltung ermöglicht und wobei die Differenz zwischen der Augenhöhe des Angestellten und die der Kunden nicht unerwünscht gross ist. Weiter wird manchmal noch behauptet, dass es dem stehenden Kunden gegenüber unhöflich sei, dass der Angestellte seine Arbeit sitzenderweise ausübt.

Mit Rücksicht auf diese Problematik wird eine " Bewertungsstudie " behandelt, wobei die benötigten Daten mittels Beobachtungen, kurze Interviews und Fragebogen gesammelt wurden.

The combination of the sitting employee and the standing customer has in many situations clear advantages. Objections may be related to difficulties that are experienced when a solution must be found that permits a satisfactory sitting posture, as well as a difference between eye height of the employee and that of the customers that is not undesirably large. Moreover it is still sometimes felt that it is impolite to the standing customer for the employee to be seated.

In view of this problem, a validation study is described. The required data were collected by means of direct observation, short interviews and questionnaires.

La situation relationelle de l'employé assis et du client debout présente certains avantages évidents. Cependant des difficultés apparaissent lorsqu'il s'agit de trouver une solution qui permet à la fois une posture assise satisfaisante et une différence d'altitude des yeux qui ne soit pas trop importante entre l'employé et le client. En outre, le fait d'être un client debout en face d'un employé assis est parfois ressenti comme étant impoli.

En vue de trouver une solution à ce problème, on s'est livré à une étude de validation. Les données ont été fournies par l'observation directe, par de brèves entrevues et par des questionnaires.

Literatur

Damon, A., Stoudt, H. W., and McFarland, R. A., 1966: *The Human Body in Equipment Design* (Cambridge, Mass.: Harvard University Press).

Röntgenuntersuchung* über die Stellung von Becken und Wirbelsäule beim Sitzen auf vorgeneigten Flächen

Von U. Burandt

Erlangen–Bruck, Schwedlerstrasse 43, Deutschland

1. Einführung

Vorgeneigte Sitzflächen werden seit langem dort verwendet, wo eine aufrechte Körperhaltung ohne Rückenabstützung erforderlich ist. Sie sind bekannt von Kutscherböcken und Orgelbänken; Drescher (1929) beschreibt sie für Arbeiten an Pressen, für Tätigkeiten, von denen man häufig aufstehen muß, und Arbeiten mit starkem Beineinsatz. Für den gleichen Anwendungsbereich wurden sie vom Ministry of Labour and National Service in London (1951) empfohlen. Auf vorgeneigten Flächen ruhen die Sitzhöcker auf Pendel- und Stehsitzen, auf den abgerundeten Vorderkanten der üblichen Arbeitssitze und auf hinten konkav ansteigenden Sitzmulden.

Die Wirkungen solcher Flächen auf die Mechanik des Beckens und der Wirbelsäule, sowie die davon abhängigen Einflüsse auf die Lage und Funktion der inneren Organe sind von der Orthopädie, der Arbeitsmedizin und der Arbeitsphysiologie bisher kaum ernsthaft untersucht worden. Nur von Schlegel (1956) liegen Versuchsergebnisse vor. Er stellte fest, daß beim Sitzen auf einer um 8 Grad vorgeneigten Fläche 'eine übermäßige Kyphosierung praktisch unmöglich war. Es wurde eine leicht vorgeneigte vordere Sitzlage erzielt, die ein wesentlich besseres Arbeiten ohne Ermüdung am Tisch und Schreibtisch ermöglichte'. Er fand mit Hilfe von Röntgenbildern, daß die Aufrichtung des Beckens geringer ist und die Hüft- und Kniewinkel offener gehalten werden als beim Sitzen auf waagerechten Flächen. Er resümiert: 'die Neigung der Sitzfläche nach vorne schafft eine günstige Arbeitshaltung von großer prophylaktischer und therapeutischer Bedeutung'. Leider ist die Veröffentlichung

1. weder methodisch noch statistisch genügend unterbaut;
2. unvollständig, weil die Beckenstellung beim Sitzen auf einer um 8 Grad zurückgeneigten Fläche, die auch untersucht wurde, nicht diskutiert wird.

Dieser zweite Punkt ist insofern bedauerlich, als in Bezug auf die vorgeneigte Sitzfläche positiv hervorgehoben wird, daß das Becken weniger zurückkippt als beim Sitzen auf der waagerechten Fläche, daß aber auf einer vergleichenden Abbildung das auf einer zurückgeneigten Sitzfläche lagernde Becken noch steiler steht als das Becken auf der vorgeneigten Fläche.

2. Fragestellung

Welche Unterschiede in Bezug auf die Stellung des Beckens und der Wirbelsäule bestehen beim Sitzen auf 6 Grad vor- und 6 Grad zurückgeneigten Sitzflächen beim Arbeiten an einem Steckbrett ohne Anlehnen des Rumpfes?

* Die Röntgenarbeiten wurden in der betriebsärztlichen Dienststelle des Wernerwerks für Medizinische Technik (WW Med), Erlangen, durchgeführt. Wir danken Herrn Dr. med. Bösche für seine Unterstützung.

3. Versuchsanordnung

Vor der Stützwand eines Wirbelsäulenaufnahmegerätes stand die in Bild 1 gezeigte Versuchsvorrichtung mit verstellbarer Sitzhöhe (bzw. Fußstütze) und Sitzflächenneigung. Die Sitzfläche war nicht größer als 60 × 10 cm und stützte nur den Bereich unter den Sitzhöckern. Sie wurde in zwei Varianten eingestellt

$$\text{Variante } vF = 6 \text{ Grad vorgeneigt,}$$
$$\text{Variante } zF = 6 \text{ Grad zurückgeneigt.}$$

Als weiteres Zubehör, zur Ablenkung der Versuchspersonen, wurde ein Steckbrett verwendet.

Bild 1. Versuchsvorrichtung mit verstellbarer Fußstütze und einstellbarer Sitzbrettneigung. Die Tischplatte mit dem Steckbrett liegt 28 cm über der Sitzfläche.

4. Versuchspersonen

Als Versuchspersonen wurden Frauen und Männer aus der Fertigung verwendet. Die Auswertung der Versuche beschränkte sich auf 21 Frauen im Alter von 19 bis 61 Jahren und mit Körperlängen von 142 bis 181 cm.

5. Versuchsverlauf

1. Notieren persönlicher Daten, Messen von Körperlänge und Kniehöhe (beides ohne Schuhe).

2. Erläutern und Einüben der Steckbrettarbeit an einem gewöhnlichen Tisch.

3. Einstellen der Sitzhöhe nach Kniehöhe und Sitzflächenneigung

      Sitzhöhe für $vF = 0,8$ · Kniehöhe $+ 4$ (cm),

      Sitzhöhe für $zF = 0,7$ · Kniehöhe $+ 4,5$ (cm).

4. Umsetzen der Vp auf die Versuchsvorrichtung, wenn sie in die Steck-brettarbeit eingearbeitet ist (nach etwa 3 min).

5. Fortsetzen der Steckbrettarbeit auf der Versuchsvorrichtung.

6. Wiederholte Aufforderungen an die Vp, Becken und Lende vor- und zurückzukippen, in einer zwanglosen, bequemen Haltung zu verweilen, die Arme nicht aufzustützen.

7. Zwei Minuten nach der letzten Aufforderung eine Aufnahme.

8. Absteigen der Vp von der Versuchsvorrichtung.

9. Sitzflächenneigung und Sitzhöhe umstellen.

10. Aufsteigen der Vp und Wiederholung der Positionen 5. bis 8.

11. Abschlußfrage: ' Welche Variante war bequemer, die 1. oder die 2.?'

6. Auswertungsvefahren

Die Röntgenaufnahmen wurden auf Transparentpapier durchgezeichnet, so daß die in Bild 2 eingetragenen Winkel auf der Medianebene gemessen werden konnten.

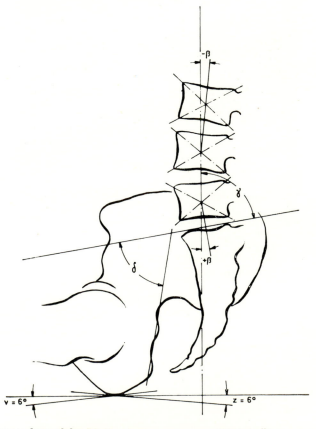

Bild 2. Bezeichnung der auf der Medianebene gemessenen Winkel: Übergangswinkel β, Becken-neigungswinkel γ, Kreuzbeinwinkel δ.

Der *Übergangswinkel* β kennzeichnet die Stellung der Achse zwischen den Mittelpunkten benachbarter Wirbelkörper in Bezug zur Senkrechten. β ist positiv, wenn die Achse ventral und negativ, wenn sie dorsal geneigt steht.

In dieser Weise wurden alle Übergangswinkel von L 5 bis C 6 gemessen. S 1 zu L 5 wurde rechtwinklig zur Kreuzbeindeckplattenebene definiert.

Der *Beckenneigungswinkel* γ kennzeichnet die Lage der Kreuzbeindeckplattenebene zur Senkrechten.

Der *Kreuzbeinwinkel* δ kennzeichnet die Lage der Kreuzbeindeckplattenebene in Bezug zum Becken—definiert durch eine die Sitzbeine tangierende Ebene.

Zur Prüfung der Unterschiede der Stichproben wurden verwendet der Zeichen-Test und der Wilcoxon-Test für 2 Stichproben.

7. Ergebnisse

7.1. *Präferenzen*

Welche Variante war bequemer?

Die Ergebnisse sind in Tabelle 1 zusammengestellt.

Tabelle 1. Häufigkeit der Präferenzen

Entscheidung	Häufigkeit
für vF	14
für zF	5
unentschieden	2

7.2. *Kreuzbeinwinkel δ*

Wie liegt die Kreuzbeindeckplattenebene in Bezug zum Becken?

Die Ergebnisse der Messungen an den 21 Vpn sind in Tabelle 2 angegeben und rangiert. Der Durchschnitt beträgt 66, 2 Grad.

Tabelle 2. Ergebnisse der Befragungen über Präferenzen der vor- (v) und zurückgeneigten (z) Sitzflächen, sowie der Messungen des Kreuzbeinwinkels δ und dei Beckenneigungswinkels γ beim Sitzen auf beiden Varianten

Vp	Prä-ferenz	δ	Rang	γ z	γ v	γ z – v
1	v	81	21	77	86	– 9
2	v	61	6	75	87	–12
3	v	68	13,5	79	85	– 6
4	v	64	9,5	67	80	–13
5	v	71	17	70	74	– 4
6	=	55	1,5	78	76	+ 2
7	v	66	11	66	76	–10
8	z	60	5	71	68	+ 3
9	v	64	9,5	80	82	– 2
10	z	69	15,5	85	93	– 8
11	v	56	3	77	79	– 2
12	v	68	13,5	74	82	– 8
13	v	69	15,5	88	91	– 3
14	v	75	18	67	90	–23
15	z	59	4	77	81	– 4
16	z	67	12	90	85	+ 5
17	v	78	19	96	98	– 2
18	=	63	8	74	86	–12
19	v	79	20	86	96	–10
20	v	62	7	77	81	– 4
21	z	55	1,5	66	54	+12

7.3. Beckenneigungswinkel γ

Wie steht die Kreuzbeindeckplattenebene in Bezug zur Senkrechten?

Die Meßergebnisse sind in Tabelle 2 zusammengestellt. Der Durchschnitt der γ_z-Werte (sitzen auf zF) beträgt 77,1°, der der γ_v-Werte 82,4°. Der größte Unterschied zwischen den Messungen bei einer Person betrug 23 Grad (Vp 14), bei allen anderen Vpn lagen die Unterschiede unter 14 Grad. 17 von 21 Unterschieden sind negativ (Beckenneigungswinkel γ größer bei vF). Die Becken waren also beim Sitzen auf vF stärker zurückgerollt und weniger steil stehend als beim Sitzen auf zF. Die Beobachtung ist mit $2\alpha < 0,01$ signifikant. Positive Unterschiede (γ größer bei zF) kamen bei Vpn, die vF präferierten, nicht vor.

Signifikant mit $2\alpha < 0,01$ ist ferner die Beobachtung, daß die präferierte Variante jener entspricht, bei der das Becken flacher stand, γ also größer war. Die beiden Vpn, für die diese Aussage nicht zutrifft (Vp 10 und Vp 15), zählen zu den wenigen, die zF präferierten.

7.4. Übergangswinkel β

Wie beeinflussen die Varianten den Verlauf der Wirbelsäule?

Die Varianten vF und zF bewirkten bei jeder Vp unterschiedliche Wirbelsäulenverläufe. Prinzipiell können die beiden sitzflächenabhängigen Verläufe die in Tabelle 3 dargestellten Lagebeziehungen haben: Nebeneinander verlaufend, einmal gekreuzt, zweimal gekreuzt.

Tabelle 3. Angaben und Anzahl der Versuchspersonen, bei denen die sechs schematisch dargestellten individuellen Lagebeziehungen der Wirbelsäule beim Sitzen auf den vor (vF) und zurückgeneigten (zF) Flächen beobachtet wurden

In der Tabelle sind die Häufigkeiten der Lagebeziehungen und die Vpn eingetragen, und zwar in der oberen Zeile für γ kleiner bei vF und in der unteren Zeile für γ kleiner bei zF.

Die Durchschnitte von jeweils 21 Übergangswinkeln sind in Bild 3 eingetragen. Die Signifikanz der Unterschiede zwischen den Varianten wurde mit dem Zeichentest nach der Häufigkeit positiver bzw. negativer Differenzen geprüft.

Signifikante Unterschiede zeigten sich nur im Lendenbereich.

Diese Erscheinung korrespondiert mit dem Beckenneigungswinkel γ. Sowohl die Lendenauskickung als auch die Beckenkippung sind stärker beim Sitzen auf vF.

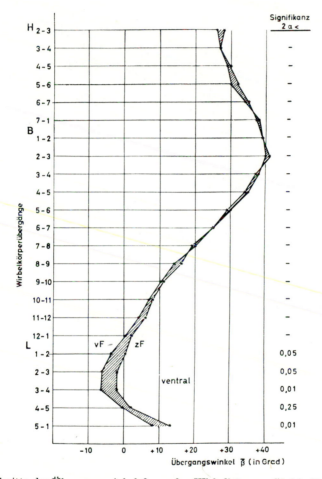

Bild 3. Durchschnitte der Übergangswinkel β von den Wirbelkörpern C2 bis S1 bei 21 Frauen unter dem Einfluß einer 6° vor (vF) und einer 6° zurückgeneigten (zF) Sitzfläche.

Die summierten Durchschnitte der Übergangswinkel in Bild 4 zeigen einen parallelen Verlauf der Wirbelsäulenprofile im Brust- und Halsbereich. Bei fast allen Wirbelkörperübergängen sind die summierten Werte der Variante vF signifikant kleiner.

Die in den Bildern 3 und 4 gezeigten Ergebnisse besagen, daß die Neigung der Sitzfläche den Verlauf der Wirbelsäule direkt nur im Bereich des Beckens und der Lende beeinflußte. Die darüberliegenden Segmente zeigten im Bezug zur Vertikalen keine systematischen Unterschiede. Sie verschoben sich über dem rollenden Becken und der sich biegenden Lende parallel.

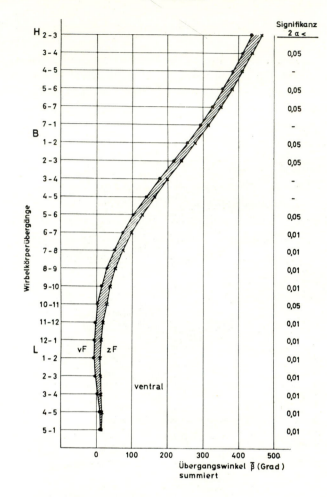

Bild 4. Summierte Durchschnitte der Übergangswinkel β aus Bild 3.

8. Diskussion der Ergebnisse

Die vorliegenden Ergebnisse bestätigen die von Schlegel unkommentierte Beobachtung (*loc. cit.* Abb. 3), wonach das Becken beim Sitzen in der vorderen Sitzlage auf einer zurückgeneigten Sitzfläche steiler steht als auf einer vorgeneigten Sitzfläche. Den dort von Schlegel gezeigten Wirbelsäulenverläufen können wir allerdings nicht zustimmen. Die Sitzflächenneigungen haben weder den Verlauf der oberen Wirbelsäule noch die Kyphose systematisch beeinflußt. Als Grundlage für die Diskussion der Wirkung verschieden geneigter Sitzflächen auf die Sitzhaltung kann somit nur die Abhängigkeit zwischen Sitzfläche und Becken gelten.

Beim Hinsetzen wird die Last des Rumpfes von den Hüftgelenken auf die Sitzhöcker übertragen. Sobald die Sitzhöcker auf die Sitzfläche aufgesetzt werden, knickt der Rumpf ein. Das Becken rollt dabei zurück und richtet den Lendenstiel dorsal, die darüber liegenden Segmente werden vorverlagert, um in der vorderen Sitzlage das Gleichgewicht zu stabilisieren. Der Rücken fällt in eine Totalkyphose. Eine Aufrichtung von Becken und Rumpf aus dieser Stellung mit Muskelkraft ist nur vorübergehend möglich. Der Grad der äußersten Ausknickung ist individuell, bei jungen und männlichen Personen gewöhnlich stärker als bei weiblichen und alten (Åkerblom (1948), Schoberth (1962)).

Mechanisch verhält sich das Becken auf den Sitzflächen folgendermaßen.

Die zurückgeneigte Sitzfläche entspricht einer schiefen Ebene, auf der das Becken in die gleiche Richtung abwärts rollen kann, in die es auch durch die natürliche Rumpfmechanik gedrückt wird. Das Biegemoment in der Lende, das Kippmoment am Becken und die schiefe Ebene unterstützen sich gleichsinnig. Bei der vorderen Sitzlage erzwingen sie die Kyphose. Die hintere Sitzlage ist ohne Rückenstütze unmöglich.

Die beobachtete geringe Beckenneigung γ bei der vorderen Sitzlage beruht wahrscheinlich auf einer aktiven Reaktion auf die beschriebene Tendenz. Der Rumpfschwerpunkt wird vorverlagert und die Oberschenkel stärker, die Sitzhöcker weniger belastet. Das Becken ist dadurch weniger stark zwischen Rumpflast und den Stützwiderstand eingespannt und leichter beweglich. Es kann somit über einen kritischen Punkt aufgerichtet werden und findet dort ein ventral orientiertes indifferentes Gleichgewicht. Je stärker das Becken aus dieser Lage zurückgekippt wird, desto größer werden die Kräfte, die diese Bewegung unterstützen.

Jedes Aufrichten aus der extremen Beckenkippung ist anstrengend, weil diese Kräfte überwunden werden müssen.

Die vorgeneigte Sitzfläche entspricht einer schiefen Ebene, auf der das Becken durch die Kräfte der Rumpflast aufwärts gerollt wird. Die Wirkung der Steigung wirkt dem Biegemoment der Lende und dem Kippmoment des Beckens entgegen. Das Becken kann passiv gegen die Steigung angekippt werden, bis es ein dorsal orientiertes indifferentes Gleichgewicht findet. Aus dieser Position läßt es sich mit geringer Muskelanstrengung vor- und zurückschaukeln.

Die Rücken- und Hüftmuskulatur kann in dynamischer Aktion bleiben.

Die waagerechte Sitzfläche hat weder den Nachteil der zurückgeneigten noch die Vorteile der vorgeneigten Fläche. Der Beckenneigungswinkel γ ist nach Schlegel größer als beim Sitzen auf vor- oder zurückgeneigten Flächen. Wahrscheinlich liegt das daran, daß diese Fläche für die vordere Sitzhaltung gerade noch bequem genug ist und die für die zurückgeneigte Fläche beschriebene Reaktion nicht auslöst.

Diese mechanischen Überlegungen unterstützen die These, daß vorgeneigte Sitzflächen für nach vorne orientierte Arbeiten in der vorderen Sitzhaltung günstiger sind als die bisher allgemein üblichen zurückgeneigten Sitzflächen. Sie wurden in dieser Untersuchung auch häufiger als bequem bevorzugt. Die gleiche Feststellung machten Burandt und Grandjean (1964), als sie Sitzkeile mit waagerechten Flächen verglichen. Sie befragten 52 Personen nach dem bequemsten Neigungswinkel der Sitzfläche unter den Sitzhöckern bei aufrechtem Sitzen in der mittleren Sitzlage und notierten Werte zwischen 5° und 22° Vorneigung. Keine Versuchsperson wählte eine waagerechte oder zurückgeneigte Sitzfläche. Es zeigte sich jedoch, daß starke Neigungswinkel Rückenschmerzen auslösen. Wahrscheinlich ist es notwendig, die Sitzflächenneigung von Arbeitsstühlen verstellbar zu machen, damit die optimale Beckenlage individuell ausgewogen werden kann.

Die Untersuchung wurde an 21 Frauen durchgeführt, die unangelehnt saßen und eine Steckbrettarbeit durchführten. Die beiden geprüften Sitzflächen (6° vor- und 6° zurückgeneigt) beeinflußten die Stellung des Beckens und den Verlauf der Lendenwirbelsäule. Das Becken stand auf der vorgeneigten Fläche signifikant häufiger flach; die Vpn fanden diese Lage, bei der die Kreuzbeindeckplattenebene waagerechter lag, bequemer. Die steile Stellung des Beckens auf der zurückgeneigten Fläche wird als Reaktion auf die Kräfte gedeutet, die das Ausknicken der Lende erzwingen. Die zurückgeneigte Fläche erleichtert das Ausknicken, weil das Becken darauf bergab rollt.

Twenty-one female workers sitting on forward and backward tilted seats without backrests were X-rayed. Both kinds of seat influenced the position of the pelvis and the curvature of the lumbar spine. When the workers sat on the forward tilted surface the position of the pelvis was tilted backwards much more. This position was preferred.

The more upright position of the pelvis on the backward tilted surface has been interpreted as a reaction against the forces buckling the lumbar spine and the tendency of the pelvis to roll down the slope.

Les auteurs ont fait passer des radiographies à vingt-et-une ouvrières travaillant assises sur des sièges sans dossier pouvant être inclinés en avant ou en arrière. Les deux sortes de sièges influent sur la position du bassin et la coubure de la colonne lombaire. Sur les sièges inclinés vers l'avant, le bassin est fortement penché en arrière; c'est cette position qui est préférée.

La position plus droite du bassin sur les sièges inclinés en arrière est interprêtée comme une réaction contre les forces qui tendent à courber la colonne lombaire et la tendance du bassin à glisser sur la surface inclinée.

Literatur

ÅKERBLOM, B., 1948, *Standing and Sitting Posture* (Stockholm: NORDISKA BOKHANDELN).

BURANDT, U., und GRANDJEAN, E., 1964, Die Wirkung verschiedenartig profilierter Sitzflächen von Bürostühlen auf die Sitzhaltung. *Int. Z. angew. Physiol. einschl. Arbeitsphysiol.,* **20,** 551–452.

DRESCHER, E. W., 1929, Der Arbeitssitz. *Reichsarbeitsblatt III, Arbeitsschutz,* **9,** 153–175.

Factory Department Ministry of Labour and National Service, 1951, *Seats for Workers in Factories* (London: H.M.S.O.).

SCHLEGEL, K. F., 1956, Sitzschäden und deren Vermeidung durch eine neuartige Sitzkonstruktion. *Med. Klinik,* **51,** 1940–1942.

SCHOBERTH, H., 1962, *Sitzhaltung—Sitzschaden—Sitzmöbel.* (Berlin: SPRINGER).

Schlussfolgerungen

Von E. Grandjean

ETH, Zürich, Schweiz

1. Historische Gesichtspunkte

Seit der Mensch auf zwei Beinen geht, war er immer bestrebt zu sitzen. Der wichtigste Grund ist physiologischer Natur: In der Sitzhaltung kann er die Bein- und Hüftmuskulatur entlasten; gleichzeitig ist die Beanspruchung von Herz und Kreislauf geringer.

Die Kulturgeschichte des Menschen zeigt, dass die von ihm hergestellten Sitze nicht nur Sitzgelegenheiten, sondern auch Vorwand für Kunstwerke waren. Mit der künstlerischen Ausstattung zusammen schlich sich allmählich eine weitere Funktion ein: Der Sitz wurde vielfach zu einem sozialen Statussymbol. Je schöner und prunkvoller ein Sitz war, um so höher war der soziale Rang; das hat sich bis heute nicht geändert.

Die Fortschritte der Wissenschaften—vor allem der Medizin, der Physiologie und der Psychologie—haben vor der Gestaltung von Möbeln nicht Halt gemacht. Die Auffassung, dass ein Sitz nicht nur ästhetisch befriedigend, solid und billig sein muss, sondern dass er auch dem Menschen angepasst d.h. bequem und gesundheitsfördernd zu sein hat, – diese Auffassung setzt sich in neuerer Zeit mehr und mehr durch.

Die ersten Rufer in der Wüste waren die Orthopäden: Der schwedische Arzt Dr. Åkerblom hat als erster vor rund 20 Jahren auf Grund ärztlicher Ueberlegungen und Untersuchungen gefordert, dass Stauungen in den Beinen durch eine maximale Sitzhöhe von 40 cm vermieden werden sollten und dass durch ein anatomisch richtig gestaltetes Profil der Rückenlehne dem Rücken die natürliche Form bei entspannter Muskulatur gesichert werden müsse. Seither haben sich zahlreiche andere Orthopäden (Keegan in den U.S.A., Schoberth in Deutschland, u.a.m.) auf Grund ihrer Beobachtungen über schlechte Haltungen eingehend mit der Gestaltung medizinisch richtiger Schul- und Arbeitssitze befasst.

Neuerdings sind zur Erforschung der Sitzhaltung und zur Entwicklung physiologisch richtiger Sitze exaktere wissenschaftliche Methoden angewandt worden. Man hat das Sitzverhalten (durch Multimoment- oder Filmaufnahmen) bei Schulkindern, in Hochschulhörsälen, in Büros und in der Eisenbahn analysiert. Man hat mit psychologischen Methoden die Bequemlichkeits- bzw. Unbequemlichkeitsempfindungen von Versuchspersonen auf Sitzmaschinen oder auf verschiedenen Sitztypen untersucht. Man hat mit elektrophysiologischen Methoden die Aktivität der Rückenmuskulatur bei verschiedenen Sitzhaltungen erforscht. Man hat exakte Erhebungen über Länge und Breite bestimmter Körperteile (Unterschenkel, Oberschenkel, Gesässbreite, usw.) in Abhängigkeit von Alter und Geschlecht bei zahlreichen Völkern ausgeführt, und schliesslich hat man die verschiedenen Stellungen und Bewegungsräume von Kopf, Rumpf, Beinen und Armen bei komplizierten Tätigkeiten wie das Autofahren oder Flugzeuglenken genau analysiert, um danach die funktionell richtigen Sitze zu entwickeln.

2. Die wissenschaftlichen Ergebnisse des Symposiums

Man könnte—wie dies Wissenschaftler oft tun—sich mit der Feststellung begnügen, dass die Wissenschaft zur Einsicht gekommen sei, man wisse noch zu wenig, um etwas aussagen zu können, und weitere Forschungen seien notwendig. Eine solche Betrachtungsweise wäre bestimmt ungerecht, denn die vorliegende Monographie zeigt, dass wir bereits über so viele Informationen verfügen, dass konkrete Angaben für den Bau von Sitzmöbeln möglich sind. Gute und weitgehend fundierte Grundlagen können zu folgenden Fragen gegeben werden:

(a) *Erfassung der Bequemlichkeit.* Einigkeit herrscht bei den Arbeitsphysiologen und Ergonomen in der Frage der Bequemlichkeit bzw. Unbequemlichkeit einer Sitzhaltung. Wir wissen, dass es keinen Wundersitz gibt, der allen Personen unter jeder Bedingung bequem ist. Trotzdem hat sich die Einsicht durchgesetzt, dass die Methode, körperliche Unbequemlichkeiten beim Sitzen auf ein Minimum zu reduzieren, ein zweckmässiges Verfahren darstellt zur Entwicklung von Schulsitzen, Fahrzeugsitzen, Arbeitssitzen und Ruhesitzen.

Die Frage, ob bequeme Sitze auch aus der Sicht der Orthopädie als gesund oder gesundheitsfördernd gelten, ist zwar von dieser Seite nicht bestritten worden; trotzdem wünschen die Orthopäden, dass bei der Gestaltung von Sitzen nicht nur die Bequemlichkeit, sondern auch die Haltung der Wirbelsäule und des Beckens berücksichtigt werden.

(b) *Schulmöbel.* Auf dem Gebiet der Schulmöbel gibt es zahlreiche Angaben über die zweckmässigen Dimensionen (Sitzhöhe, Sitztiefe, Tischhöhe) in Abhängigkeit der Körpergrösse. Es ist unbestritten, dass diese anthropometrischen Grundlagen die wichtigste Voraussetzung für sinnvolle Schulmöbel darstellen. Auf dem Gebiet der Hörsaalbestuhlung sind Vorschläge ausgearbeitet worden, die einerseits anthropometrisch und andererseits auf Ergebnissen psychologischer Experimente begründet sind.

(c) *Arbeitssitze.* Arbeitsphysiologen und Ergonomen sind sich einig in der Forderung, dass der Arbeitssitz in seiner Beziehung zum Arbeitsplatz vor allem in Zusammenhang mit der Arbeitshöhe zu betrachten sei. Vorschläge für die Dimensionen, für den Verstellbereich der Arbeitssitze und für die Beziehung zur Arbeitshöhe stimmen weitgehend überein. Man ist sich auch einig darüber, dass Sitz und Tisch eher grossen Personen angepasst werden müssen, während für kleine Personen Fussstützen vorzusehen sind.

(d) *Fahrzeugsitze.* Die ergonomische Analyse der bestehenden Motorfahrzeuge hat ergeben, dass viele Modelle—vor allem die Kleinwagen—den Anforderungen der Ergonomie nicht genügen. Auf Grund anthropometrischer Masse und unter Berücksichtigung der Griff- und Arbeitsbereiche von Armen und Beinen sowie der Sehbedingungen sind sehr detaillierte Vorschläge für die Gestaltung von Motorfahrzeugsitzen unter Einbeziehung der Bedienungselemente ausgearbeitet worden.

(e) *Ruhesitze.* Bei den Ruhesesseln sind die Fachleute in der grundsätzlichen Forderung einig, dass der ganze Sitz so zu gestalten sei, dass eine grösstmögliche Muskelentspannung bei gleichzeitig gesicherter natürlicher Körperhaltung (vor allem der Wirbelsäule) anzustreben ist. Durch systematische psychophys'ologische Experimente an Versuchspersonen sind konkrete Vorschläge für Sitzprofile ausgearbeitet worden, die sich im wesentlichen mit den älteren Forderungen der Orthopäden (Åkerblom) decken.

Ungelöst und unbeachtet blieb die Frage, wie Ruhesessel für alte und invalide Personen zu gestalten seien.

Eine Reihe weiterer Fragen ist weitgehend noch ungelöst und bedarf wissenschaftlicher Abklärung. Dazu gehören vor allem folgende Problemkreise:

1. Das Ausmass der Polsterung und deren Bedeutung für die Sitzhaltung.

2. Die Einflüsse der Bezugsmaterialien (Leder, Stoff, Kunststoff) auf die Bequemlichkeit, das Sitzverhalten, die Kleiderabnützung, den Wärmehaushalt des Körpers und die Feuchtdurchlässigkeit.

3. Die Form der Sitzfläche be Arbeitssitzen; zur Diskussion stehen der flache Sitz, der hinten aufgewinkelte Sitz, der Sitz, der zur aktiven Lordosierung Anlass gibt, und der nach vorn abwärts geneigte Sitz.

4. Die Gestaltung der Rückenlehne bei Arbeits- und Fahrzeugsitzen. Muss der Lendenbausch verstellbar sein? Genügt eine kleine Stütze in der Lendengegend? Kann der Rücken auch bei Arbeitsstellung gestützt werden?

5. Der Einfluss der Sitzdauer auf das Sitzverhalten und die Beurteilung der Bequemlichkeit. Shackel und Ch. Jones beobachteten eine Zunahme der Unbequemlichkeit in Abhängigkeit der Sitzzeit, während die relative Beurteilung verschiedener Sitzmodelle unbeeinflusst blieb.

6. Schaukelsitze, bewegliche Sitze und Liegestühle sind bis heute noch kaum untersucht worden.

Die Organisatoren des Symposiums hoffen, dass die vorliegenden Arbeiten einen Markstein für die Entwicklung der Sitze darstellen. Der Mensch benützt Sitze seit mehr als 4 000 Jahren. Wir hoffen, dass es nicht noch einmal so lange dauert, bis für die oben erwähnten ungelösten Sitzprobleme wissenschaftlich gesicherte Kenntnisse vorliegen!